Supplements to the 2nd Edition of

RODD'S CHEMISTRY OF CARBON COMPOUNDS

VOLUME I

ALIPHATIC COMPOUNDS

★

VOLUME II

ALICYCLIC COMPOUNDS

★

VOLUME III

AROMATIC COMPOUNDS

★

VOLUME IV

HETEROCYCLIC COMPOUNDS

★

VOLUME V

MISCELLANEOUS

GENERAL INDEX

★

Supplements to the 2nd Edition of

RODD'S CHEMISTRY OF CARBON COMPOUNDS

ELSEVIER SCIENCE PUBLISHERS B.V.
Molenwerf 1
P.O. Box 211, 1000 AE Amsterdam, The Netherlands

Distributors for the United States and Canada:

ELSEVIER SCIENCE PUBLISHING COMPANY INC.
52, Vanderbilt Avenue
New York, N.Y. 10017

Library of Congress Card Number: 64-4605

ISBN 0-444-42485-7

Printed in The Netherlands

Supplements to the 2nd Edition (Editor S. Coffey) of

RODD'S CHEMISTRY OF CARBON COMPOUNDS

A modern comprehensive treatise

Edited by
MARTIN F. ANSELL
Ph.D., D.Sc. (London) F.R.S.C. C. Chem.
Reader Emeritus, Department of Chemistry,
Queen Mary College, University of London, Great Britain

Supplement to

VOLUME IV HETEROCYLIC COMPOUNDS

Part B:
Five-membered Heterocylic Compounds with a
Single Hetero-Atom in the Ring: Alkaloids, Dyes
and Pigments

ELSEVIER
Amsterdam – Oxford – New York – Tokyo 1985

CONTRIBUTORS TO THIS VOLUME

DOUGLAS J. FRY, B.Sc.

Formerly of Ilford Ltd.

DAVID J. ROBINS, B.Sc., Ph.D.

Department of Chemistry, The University, Glasgow G1L8QQ

MALCOLM SAINSBURY, Ph.D., D.Sc., C.Chem., F.R.S.C.

Department of Chemistry, The University, Bath, BA2 7AY

KEITH S.J. STAPLEFORD, M.Sc., Ph.D.

Halton College, Widnes, Cheshire WA8 7QQ

JACK G. WOOLLEY, B.Sc., Ph.D.

School of Pharmacy, The Polytechnic,
Leicester, LE1 9BH

RAYMOND E. FAIRBAIRN, B.Sc., Ph.D., F.R.S.C.

Formerly of Research Department, Dyestuff Division,
I.C.I. (Index)

PREFACE TO SUPPLEMENT IVB

The publication of this volume continues the supplementation of the second edition of Rodd's Chemistry of Carbon Compounds thus keeping this major work of reference up to date. This supplement covers the Chapters in volume IVB with the exception of Chapters 12 and 13 for which the manuscripts were not received in time. These chapters will appear in a subsequent supplement. Although each chapter in this book stands on its own, it is intended that it should be read in conjunction with the parent chapter in the second edition.

At a time when there are many specialist reviews, monographs and reports available, there is still in my view an important place for a book such as "Rodd", which gives a broader coverage of organic chemistry. One aspect of the value of this work is that it allows the expert in one field to quickly find out what is happening in other fields of chemistry. On the other hand a chemist looking for the way into a field of study will find in "Rodd" an outline of the important aspects of that area in chemistry together with leading references to other works to provide more detailed information.

As editor I have been fortunate in that each contributor to this supplement has produced a very readable critical assessment of their particular area of chemistry. As an organic chemist I have enjoyed reading the chapters and profited from learning of advances in areas of chemistry that are outside my own special interests.

This volume has been produced by direct reproduction of the manuscripts. I am most grateful to the contributors for all the care and effort both they and their secretaries have put into the production of the manuscripts, including the diagrams. I also wish to thank the staff at Elsevier for all the help they have given me and for seeing the transformation of authors' manuscripts to published work.

February 1985 Martin Ansell

CONTENTS

VOLUME IV B

Five-membered Heterocyclic Compounds with a Single Hetero-Atom in the Ring: Alkaloids, Dyes and Pigments

Preface VII
Official publications; Scientific journals and periodicals XIII
List of common abbreviations and symbols used XIV

Chapter 7. Five-membered Monoheterocyclic Compounds: Alkaloids (continued): Pyrrolidine Alkaloids
by J.D. ROBINS

1. Pyrrolidine Bases 1
2. Pyrrolidones 7
3. *N*-Acylpyrrolidines 9
4. *N*-Acylpyrrolidones 13

Chapter 8. Five-membered Monoheterocyclic Compounds: Alkaloids (continued): Pyrrolizidine Alkaloids
by D.J. ROBINS

Introduction 15
1. Necines 17
   (a) Unhydroxylated derivaties 17
   (b) Monohydroxylated derivatives 18
   (c) Dihydroxylated derivatives 22
   (d) Trihydroxylated derivatives 24
2. Necic acids 26
   (a) C_6-Acids 26
   (b) C_7-Acids 26
   (c) C_8-Acids 26
   (d) C_9-Acids 28
   (e) C_{10}-Acids 29
3. Alkaloids 33
4. Miscellaneous types of pyrrolizidine alkaloids 67
5. Pharmacology of the pyrrolizidine alkaloids 68

Chapter 9. The Indole Alkaloids
by K.S.J. STAPLEFORD

Introduction 71
1. Alkaloids lacking a tryptamine unit 72
   (a) Simple indoles 72
   (b) Carbazole derivatives 76

2. Alkaloids containing a tryptamine unit 79
(a) Compounds without an isoprene moiety 79
(i) Simple tryptamine and tryptophan derivatives, 79 — (ii) Eserine types, 84 —
(b) Compounds containing isoprene (but not terpene derived moiety) 85
(i) The ergot alkaloids, 85 — (ii) Mould metabolites, 87 —
(c) Compounds containing a terpene derived moiety 97
(i) Alkaloids with the Coryanthe-Strychnos unit 97 — (ii) Alkaloids with a seco-type unit, 120 — (iii) Alkaloids with the Aspidosperma unit, 121 — (iv) Alkaloids containing the Iboga unit, 127 — (v) Novel types, 132 —
3. Bis-indole alkaloids 135
(a) Compounds containing two identical "halves" linked symmetrically 135
(b) Compounds not composed of identical halves nor linked symmetrically 138
(i) Sesquimeric compounds, 138 — (ii) The secamines and presecamines, 141 — (iii) Bis-indoles from Vinca rosea, 141 — (iv) Other representative bis-indole alkaloids, 145 —

Chapter 10. Five-membered Monoheterocyclic Compounds: Amaryllidaceae Alkaloids
by M. SAINSBURY

1. Introduction 151
2. Biosynthesis 151
3. New alkaloids and plant sources 154
(a) Lycorine and pretazettine 154
(b) 4,5-Etheno-8,9-methylenedioxy-6-phenanthridone 157
(c) Carinatine and goleptine 158
(d) Alkaloids from plants of the genus *Crinum* 159
(e) Havanine, varadine, zaidine and caribine 160
(f) Ungvedine and ungspiroline 162
(g) Hippadine, hippafine and hippagine 163
(h) Galanthamine and its *N*- and *O*-demethyl derivatives 163
(i) Clivatine and clivacetine 164
4. Crystal structure determinations 166
(a) Maritidine 166
(b) Norgalanthamine 166
(c) Cocculine and cocculidine 166
(d) Lycorine chlorohydrin 166
(e) Lycorine (9). 167
5. Spectroscopy 167
6. Photochemical reactions 168
7. Synthesis 168
(a) The synthesis of lycorine and related structures 168
(b) Clividine and clivonine 177
(c) Clivimine 178
(d) Maritidine, (+)- and (−)-galanthamine 179
(e) (±)-Lycoramine (dihydrogalanthamine) 182
(f) Apogalanthamine analogues 184
(g) Tetrahydrometinoxocrinine and crinine 184

(h) (±)-Crinamine, (±)-6-hydroxycrinamine, (±)-criwelline and (±)-macronine . . . 186
(i) Lycoricidine and related compounds . . . 188
(j) Elswesine . . . 191
(k) (±)-Tazettine . . . 192

Chapter 11. Five-membered Monoheterocyclic Compounds: Alkaloids (continued): Tropane Alkaloids
by J.G. WOOLLEY

Introduction . . . 199
1. Synthesis . . . 206
2. New alkaloids . . . 219
(a) Proteaceae . . . 219
(b) 2,3-Disubstituted tropanes . . . 223
(c) 2,3,6-Trisubstituted tropanes . . . 225
(d) 2,3,7-Trisubstituted tropanes . . . 226
(e) 3,6-Disubstituted tropanes . . . 227
(f) Pyranotropanes . . . 227
(g) Erythroxylaceae . . . 228
(h) Solanaceae . . . 234
(i) Littorine and related topics . . . 234
(j) Heterodiesters and related topics . . . 239
(k) *N*-oxides . . . 241
(l) Secotropanes . . . 246
3. Spectroscopy . . . 246

Chapter 12. Five-membered Monoheterocyclic Compounds (continued) — The Pyrrole Pigments

Key references . . . 251

Chapter 13. Five-membered Monoheterocyclic Compounds (continued): Azaporphyrins; Benzoporphyrins; Benzoazoporphyrins; Phthalocyanines and Related Structures

Key reviews . . . 252

Chapter 14. Five-membered Monoheterocyclic Compounds: The Indigo Group
by M. SAINSBURY

Introduction . . . 253
1. Indigo and its derivatives . . . 253
(a) Methods of synthesis . . . 253
(b) Spectroscopy and *E—Z* isomerism of indigo and thioindigo derivatives . . . 260
(c) Tyrianpurple and its biological precursors . . . 262
(d) Deoxyindigo (structural reassignment) . . . 264
(e) Reaction between indigo and hydrazine . . . 266
(f) Indigo diimine . . . 266

Chapter 15. Cyanine Dyes and related Compounds
by D.J. FRY

Introduction 267
1. Cyanines 269
1.1 Fluoro-substituted dyes 269
1.2 Long-chain dyes 270
1.3 Physical properties 272
1.4 Reaction of cyanines 278
1.5 Miscellaneous syntheses and uses 282
2. Acetylenic dyes 283
3. Aza-, phospha- and arsa-cyanines 288
3.1 Azacyanines 288
3.2 Phospha- and arsa-cyanines 290
4. Merocyanines 290
5. Oxonols 291
6. Styryl dyes 292
7. Pyrilium dyes 293
8. Reaction mechanisms 294
Index 297

OFFICIAL PUBLICATIONS

B.P.	British (United Kingdom) Patent
F.P.	French Patent
G.P.	German Patent
Sw.P.	Swiss Patent
U.S.P.	United States Patent
U.S.S.R.P.	Russian Patent
B.I.O.S.	British Intelligence Objectives Sub-Committee Reports
F.I.A.T.	Field Information Agency, Technical Reports of U.S. Group Control Council for Germany
B.S.	British Standards Specification
A.S.T.M.	American Society for Testing and Materials
A.P.I.	American Petroleum Institute Projects
C.I.	Colour Index Number of Dyestuffs and Pigments

SCIENTIFIC JOURNALS AND PERIODICALS

With few obvious and self-explanatory modifications the abbreviations used in references to journals and periodicals comprising the extensive literature on organic chemistry, are those used in the World List of Scientific Periodicals.

LIST OF COMMON ABBREVIATIONS AND SYMBOLS USED

A	acid
Å	Ångström units
Ac	acetyl
a	axial; antarafacial
as, *asymm.*	asymmetrical
at	atmosphere
B	base
Bu	butyl
b.p.	boiling point
C, mC and μC	curie, millicurie and microcurie
c, *C*	concentration
C.D.	circular dichroism
conc.	concentrated
crit.	critical
D	Debye unit, 1×10^{-18} e.s.u.
D	dissociation energy
D	dextro-rotatory; dextro configuration
DL	optically inactive (externally compensated)
d	density
dec. or decomp.	with decomposition
deriv.	derivative
E	energy; extinction; electromeric effect; Entgegen (opposite) configuration
E1, E2	uni- and bi-molecular elimination mechanisms
E1cB	unimolecular elimination in conjugate base
e.s.r.	electron spin resonance
Et	ethyl
e	nuclear charge; equatorial
f	oscillator strength
f.p.	freezing point
G	free energy
g.l.c.	gas liquid chromatography
g	spectroscopic splitting factor, 2.0023
H	applied magnetic field; heat content
h	Planck's constant
Hz	hertz
I	spin quantum number; intensity; inductive effect
i.r.	infrared
J	coupling constant in n.m.r. spectra; joule
K	dissociation constant
kJ	kilojoule

LIST OF COMMON ABBREVIATIONS

k	Boltzmann constant; velocity constant
kcal	kilocalories
L	laevorotatory; laevo configuration
M	molecular weight; molar; mesomeric effect
Me	methyl
m	mass; mole; molecule; *meta*-
ml	millilitre
m.p.	melting point
Ms	mesyl (methanesulphonyl)
[M]	molecular rotation
N	Avogadro number; normal
nm	nanometre (10^{-9} metre)
n.m.r.	nuclear magnetic resonance
n	normal; refractive index; principal quantum number
o	*ortho*-
o.r.d.	optical rotatory dispersion
P	polarisation, probability; orbital state
Pr	propyl
Ph	phenyl
p	*para*-; orbital
p.m.r.	proton magnetic resonance
R	clockwise configuration
S	counterclockwise config.; entropy; net spin of incompleted electronic shells; orbital state
S_N1, S_N2	uni- and bi-molecular nucleophilic substitution mechanisms
S_Ni	internal nucleophilic substitution mechanisms
s	symmetrical; orbital; suprafacial
sec	secondary
soln.	solution
symm.	symmetrical
T	absolute temperature
Tosyl	*p*-toluenesulphonyl
Trityl	triphenylmethyl
t	time
temp.	temperature (in degrees centigrade)
tert.	tertiary
U	potential energy
u.v.	ultraviolet
v	velocity
Z	zusammen (together) configuration

LIST OF COMMON ABBREVIATIONS

α	optical rotation (in water unless otherwise stated)
$[\alpha]$	specific optical rotation
α_A	atomic susceptibility
α_E	electronic susceptibility
ε	dielectric constant; extinction coefficient
μ	microns (10^{-4} cm); dipole moment; magnetic moment
μ_B	Bohr magneton
μg	microgram (10^{-6}g)
λ	wavelength
ν	frequency; wave number
χ, χ_d, χ_μ	magnetic, diamagnetic and paramagnetic susceptibilities
~	about
(+)	dextrorotatory
(-)	laevorotatory
(±)	racemic
$\ominus$	negative charge
$\oplus$	positive charge

Chapter 7

FIVE-MEMBERED MONOHETEROCYCLIC COMPOUNDS: ALKALOIDS (CONTINUED):

PYRROLIDINE ALKALOIDS

DAVID J. ROBINS

Annual reviews in this area are available in "The Alkaloids", Vols 1-10, The Chemical Society, London, 1971-1980. Pyrrolidine alkaloids are included in a review of nitrogen-containing compounds in tobacco (I. Schmeltz and D. Hoffmann, Chem. Rev., 1977, 77, 295).

1. Pyrrolidine Bases

Hygrine (1) and cuscohygrine (2) are present in roots of *Salpichroa origanifolia* (W.C. Evans, A. Ghani, and V.A. Woolley, Phytochem., 1972, 11, 469). Cuscohygrine has also been obtained from roots of *Solanum carolinense* (Evans and A. Somanabandhu, *ibid*., 1977, 16, 1859); *Convolvulus erinacius* (S.F. Aripova, V.M. Malikov, and S. Yu. Yunusov, *Chem. Abs*., 1972, 77, 162 010); *Scopolia tangutica* (S.A. Minina and I. Barene, *ibid*., 1972, 77, 111 461); and *Cyphomandra betacea* (Evans, Ghani, and Woolley, J. Chem. Soc. Perkin 1, 1972, 2017). The roots of nine *Datura* species (*idem*, Phytochem., 1972, 11, 2527) and five of the twelve known species of *Solandra* contain cuscohygrine (*idem*, *ibid*., 1972, 11, 470). The pharmacological activity of cuscohygrine has been studied (Minina and Barene, *loc. cit*.).

Stachydrine (3) occurs in *Asphodelus microcarpus* (F.M. Hammouda, A.M. Rizk, and M.M. Abdel-Gawad, Curr. Sci., 1971, 40, 631); *Capparis spinosa* (S. Mukhamedova, S.T. Akramov, and Yunusov, Khim. Prir. Soedin., 1965, 67); *Courbonia glauca* (D.A. Taylor and A.J. Henry, Phytochem., 1973, 12, 1178); *Desmodium triflorum* (S. Ghosal *et al*; Planta Med., 1973, 23,

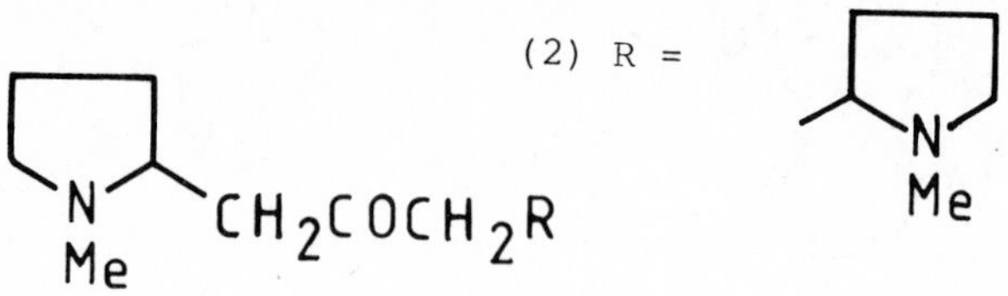

(1) R = H

(2) R =

CH_2COCH_2R

321) and (named cadabine) in Cadaba fruticosa (V.U. Ahaman, A. Basha, and A. Rahman, Phytochem., 1975, 14, 292). The alkaloid content of 36 species of the Capparidaceae family has been studied. Stachydrine (3) is present in 27 species, and 3-hydroxystachydrine (4) has been found in 18 species (P. Dealaveau, B. Koudogbo, and J.L. Pousset, ibid., 1973, 12, 2893).

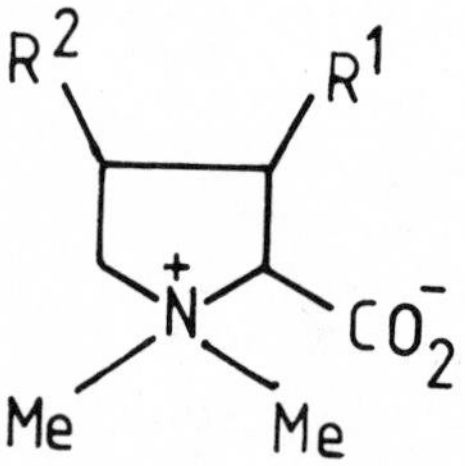

(3) $R^1 = R^2 = H$

(4) $R^1 = OH, R^2 = H$

(5) $R^1 = H, R^2 = OH$

Combretine, $C_7H_{13}NO_3$, discovered in Combretum micranthum, appears to be a stereoisomer of betonicine (5) (A.U. Ogan, Planta Med., 1972, 21, 210).

Shihunine, $C_{12}H_{13}NO_2$, has been isolated from Dendrobium lohohense (Y. Inubishi et al., Chem. pharm. Bull. Japan, 1964, 12, 749) and D. pierardii (M. Elander, L. Gawell, and K. Lander, Acta Chem. Scand., 1971, 25, 721). The ^1H-nmr spectrum of shihunine in non-polar solvents suggests a phthalide structure (6), while the spectrum taken in methanol or water indicates a betaine structure (7), which is presumably the form present in the plant. The synthesis of shihunine has been achieved utilising the rearrangement of a cyclopropylimine to generate the pyrrolidine ring in the key step (E. Breuer and S. Zbaida, Tetrahedron, 1975, 31, 499). Shihunine has m.p. 78-79°, and the picrate m.p. 154-155.5°.

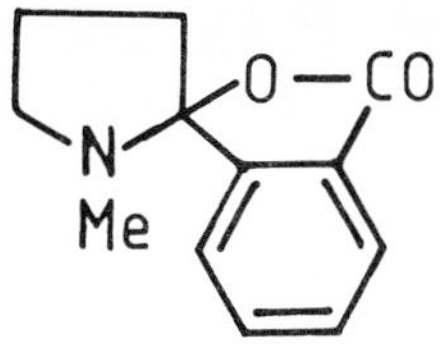

(6)

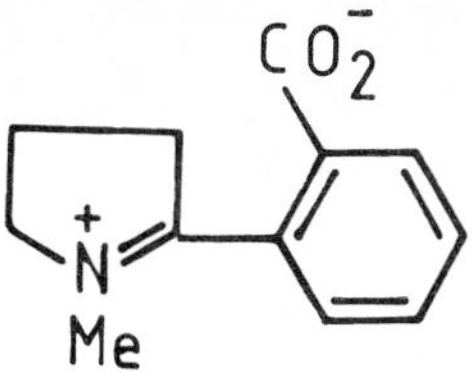

(7)

Gerrardine, $C_{11}H_{19}NO_2S_4$, is present in Cassipourea gerrardii (Rhizophoraceae) (W.G. Wright and F.L. Warren, J. Chem. Soc. (C), 1967, 283; 284). Its structure (8) has been established by X-ray crystallography (G. Gafner and L.J. Admiraal, Acta Crystallogr., Sect. B, 1971, 27, 565). Gerrardine has m.p. 180°. The hydrochloride, $[\alpha]_D^{23}$ -172° (H_2O), has m.p. 207°.

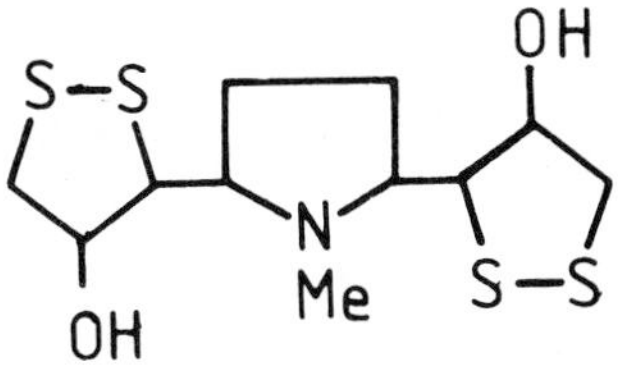

(8)

A tetraol, $C_6H_{13}NO_4$, isolated from the leaves of Derris elliptica (Leguminosae) has been shown to be 3,4-dihydroxy-2,5-dihydroxymethylpyrrolidine (9) by nmr and mass spectrometry (A. Welter et al., Phytochem., 1976, 15, 747). The relative configuration of the tetraol, $[\alpha]_D^{20}$ +56.4° (H_2O), follows from the ^{1}H-nmr spectral coupling constants, and a half-chair conformation for the molecule has been suggested.

Codonopsine, $C_{14}H_{21}NO_4$, and codonopsinine, $C_{13}H_{19}NO_3$, are present in Codonopsis clematidea. The structures (10) and (11), respectively, are assigned to these two compounds on the basis of spectroscopic studies, Hofmann degradations, and permanganate oxidations (S.F. Matkhalikova et al., Chem. Abs.,

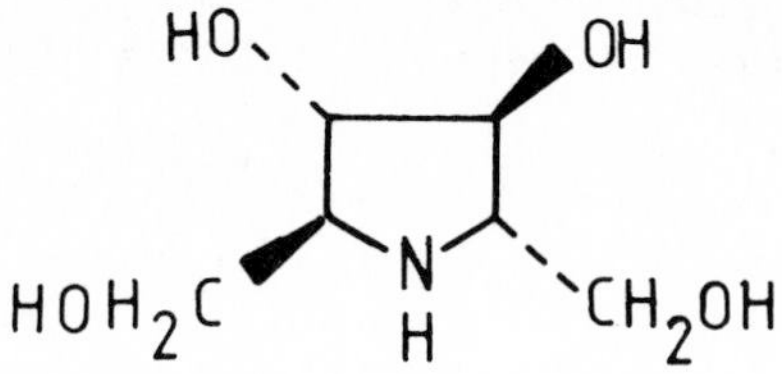

(9)

1970, 73, 15 050; 1971, 75, 36 409). The relative stereochemistry for each base has been deduced from double-resonance ^{1}H-nmr studies, and again a half-chair conformation is indicated for each compound (M.R. Yagudaev *et al.*, *ibid.*, 1972, 77, 164 902). The pharmacology of codonopsine has been investigated (M.T. Khanov, M.B. Sultanov, and T.A. Egorova, *ibid.*, 1972, 77, 135 091).

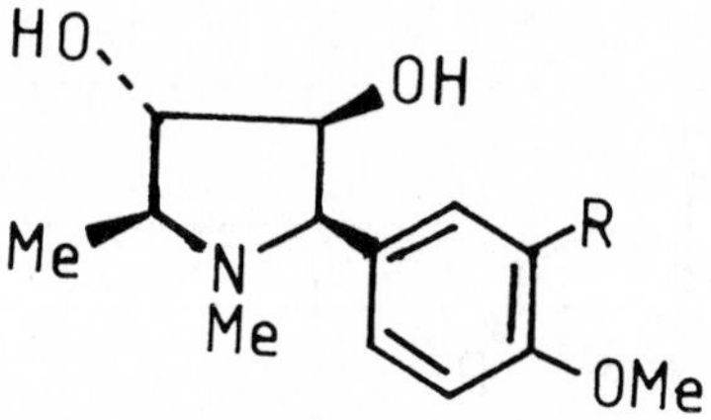

(10) R = OMe

(11) R = H

Eight related alkaloids have been isolated from the stems and leaves of *Darlingia darlingiana*. The structure of darlingianine (12) has been established by X-ray diffraction analysis of the base, and by synthesis of the racemate from (±)-hygrine (1) and cinnamaldehyde (B.F. Anderson *et al.*, Chem. Ind. (London), 1977, 764). The structures of the remaining alkaloids follow from their spectra, aided by comparison with the spectra of model compounds (I.R.C. Bick, J.W. Gillard, and H.-M. Leow, Austral. J. Chem., 1979, 32, 2523). The physical properties of these eight alkaloids are listed in Table 1.

TABLE 1

Darlingia Alkaloids: Physical Properties

Alkaloid	m.p. (°C)	$[\alpha]_D^{19}$ (°) ($CHCl_3$)
Darlingianine (12)	93.5	+62
Dehydrodarlingianine (13)	54-55	+53
Isodarlingianine (14)	50-52	+47
Dihydrodarlingianine (15)	72-74	+34
Tetrahydrodarlingianine (16)	oil	+22
Dehydrodarlinine (17)	42-44	+64
Darlinine (18)	59-61	+75
Epidarlinine (19)	66-68	+59

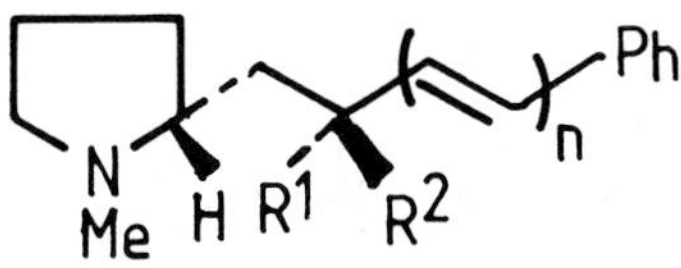

(12) R^1 = H, R^2 = OH, n = 2(E,E)

(13) R^1, R^2 = O, n = 2(E,E)

(14) R^1 = H, R^2 = OH, n = 2(E,Z)

(15) R^1 = H, R^2 = OH, n = 2(-,E)

(16) R^1 = H, R^2 = OH, n = 2(-,-)

(17) R^1, R^2 = O, n = 1(E)

(18) R^1 = H, R^2 = OH, n = 1(E)

(19) R^1 = OH, R^2 = H, n = 1(E)

Three related bases are present in the roots of Ruspolia hypercrateriformis (Acanthaceae). Treatment of nor-ruspolinone (20) with diazomethane gives ruspolinone (21). Catalytic hydrogenation of the N,O-diacetyl derivatives of ruspolinone and nor-ruspoline (22) gives the same product. Spectral studies on these bases, their degradation products, and synthetic model compounds established the structures (20) - (22) (F. Roessler *et al.*, Helv., 1978, 61, 1200). Nor-ruspolinone is the only crystalline compound, m.p. 175°.

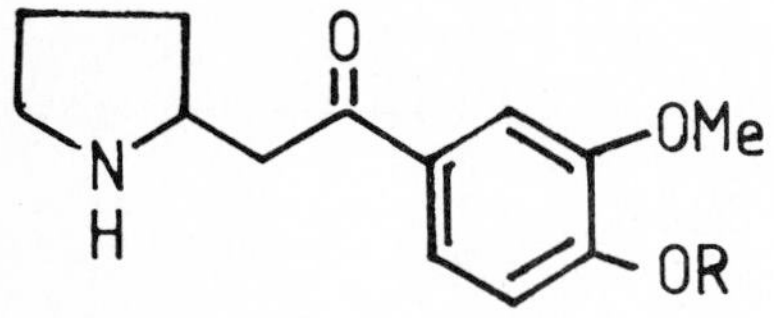

OMe
OR
N
H
O

OMe
OH
N
H

(20) R = H

(21) R = Me

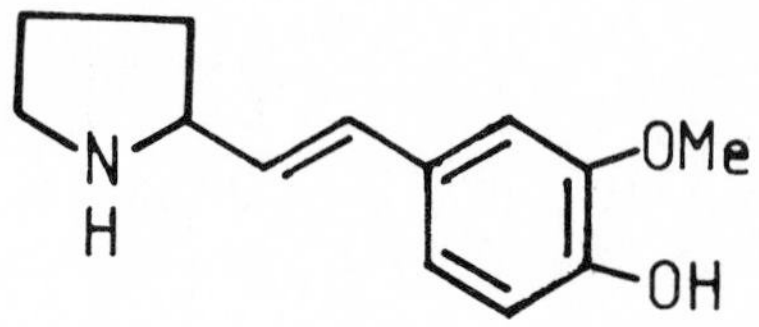

(22)

Desdanine, pyracrimycin A, and cyclamidomycin, obtained from various Streptomyces species are identical (A.D. Argoudelis, H. Hoeksema, and H.A. Whaley, J. Antibiot., 1972, 25, 432). The synthesis of desdanine (24) has been achieved (W. Martin and W. Dürckheimer, Tetrahedron Letters, 1973, 1459). Condensation of 2-methyl-1-pyrroline-1-oxide with the hemiacetal of ethyl glyoxylate gave the ester (23), which afforded desdanine on ammonolysis, electrolytic reduction, and dehydration. Desdanine has m.p. 198-199°.

Another pyrroline derivative from a Streptomyces species polymerises rapidly at room temperature. Its structure (25) has been deduced from the ultraviolet chromophore, the ^{1}H- and ^{13}C-nmr spectra, and Hofmann degradations (M. Onda *et al.*, Chem. Pharm. Bull. Japan, 1974, 22, 2916). The picrate of (25) has m.p. 168-169°, and the hydrochloride, $[\alpha]_D^{20}$ +21.5° (H_2O), has m.p. 150°.

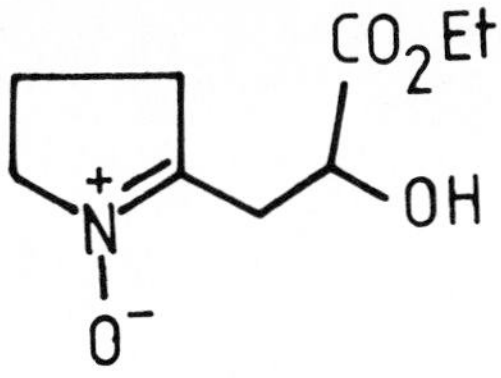

(23)

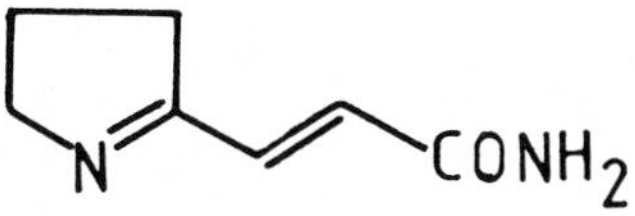

(24)

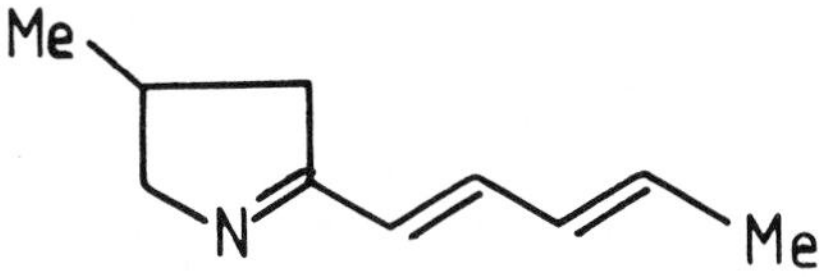

(25)

2. Pyrrolidones

Pterolactam, $C_5H_9NO_2$, (26), present in bracken (*Pteridium aquilinium*) has been identified by spectroscopic methods (K. Takatori *et al.*, *ibid.*, 1972, *20*, 1087). Pterolactam has m.p. 56-57° and $[\alpha]_D^{25}$ +2.0° ($CHCl_3$).

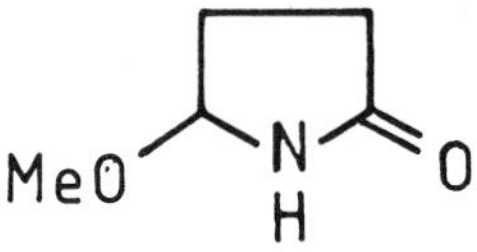

(26)

Corydolactam (=alkaloid P), C_6H_9NO, occurs in Corydallis pallida var. tenuis. The structure (27) follows from its spectral data and from hydrogenation of the alkaloid to a dihydroderivative which has been synthesised (H. Kaneko and S. Naruto, J. pharm. Soc., 1971, 91, 101; T. Kametami, M. Ihara, and T. Honda, J. Chem. Soc. (C), 1970, 1060). Other syntheses of alkaloid P from the methyl ketone (28) (prepared by a number of different routes) have been achieved by reduction and dehydration steps (Kametami and Ihara, ibid., 1971, 999; Chem. Pharm. Bull. Japan, 1971, 19, 2256). Alkaloid P has m.p. 172-174°.

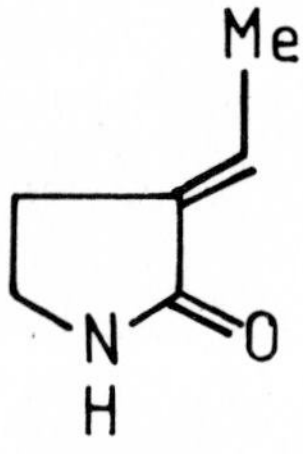

Me N H O

(27)

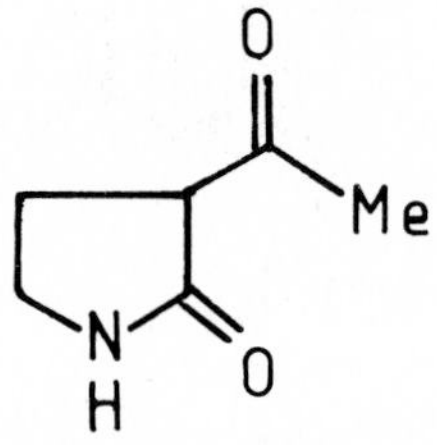

O Me N H O

(28)

Mycosporine-2, $C_{13}H_{19}NO_6$, (29) has been obtained from the fungus Botrytis cinerea and characterized by spectroscopic means (N. Arpin, J. Favre-Bonvin, and S. Thivend, Tetrahedron Letters, 1977, 819).

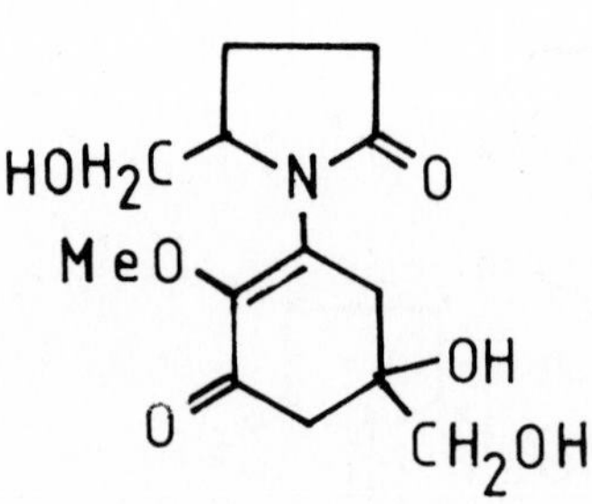

HOH_2C N O MeO OH O CH_2OH

(29)

3. N-Acylpyrrolidines

The E- and Z-isomers of N-cinnamoylnorcuscohygrine, $C_{21}H_{28}N_2O_2$, have been isolated from Dendrobium chrysanthum and the structures (30) and (31) established by spectroscopy and by synthesis of the racemate of the E-isomer (30) (U. Ekevag et al., Acta Chem. Scand., 1973, 27, 1982). The absolute configuration at C-2 has been assigned by comparison of the c.d. curves of the geometrical isomers with those of L-prolinol derivatives. Both isomers are oils; the E-isomer has $[\alpha]_D^{22}$ -11° ($CHCl_3$), and the Z-isomer has $[\alpha]_D^{22}$ -19° ($CHCl_3$).

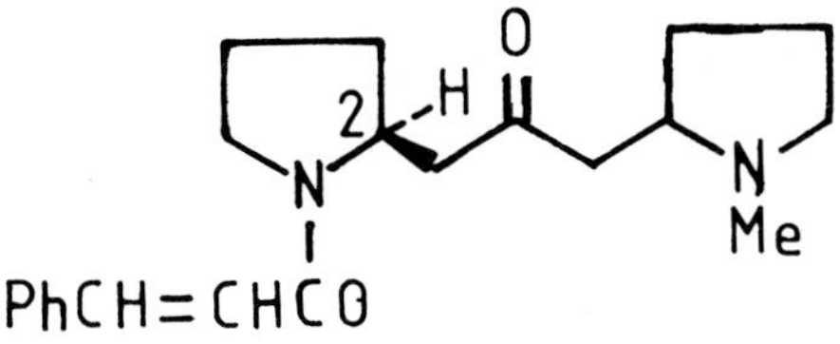

(30) E-isomer

(31) Z-isomer

Trichonine, $C_{24}H_{43}NO$, present in Piper trichostachyon, has been characterized as (32) by spectral means and by acid hydrolysis of trichonine to eicosanoic acid and pyrrolidine hydrochloride. The tetrahydroderivative has been synthesized (J. Singh, K.L. Dhar, and C.K. Atal, Tetrahedron Letters, 1971, 2119). Trichonine has been synthesized utilising a Wittig reaction in the key step (B. Vig and A.C. Mahajan, Indian J. Chem., 1972, 10, 564). Trichonine has m.p. 65-67°.

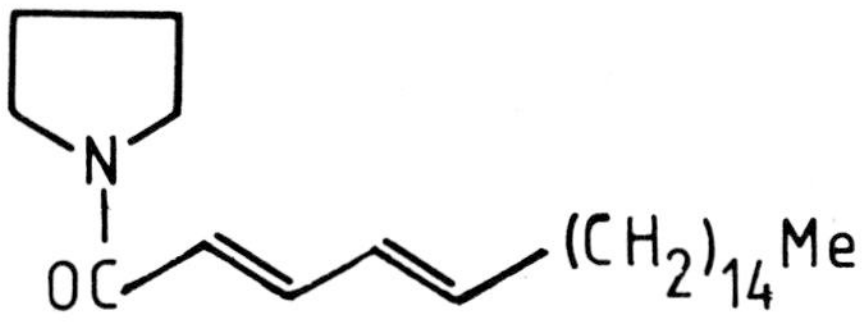

(32)

Three related amides have been isolated from *Piper* species. Peepuloidine, $C_{16}H_{19}NO$, (33) is present in *P. peepuloides* (Atal, P.N. Moza and A. Pelter, Tetrahedron Letters, 1968, 1397), and the other two (34) and (35) occur in kava roots (*P. methysticum*) (H. Achenbach and W. Karl, Ber., 1970, *103*, 2535). Syntheses of all three compounds are readily achieved from pyrrolidine and the appropriate acid chloride (F. Dallacker and J. Schubert, *ibid*., 1971, *104*, 1706; Achenbach and Karl, *loc. cit*.). Peepuloidine has m.p. 149-150°.

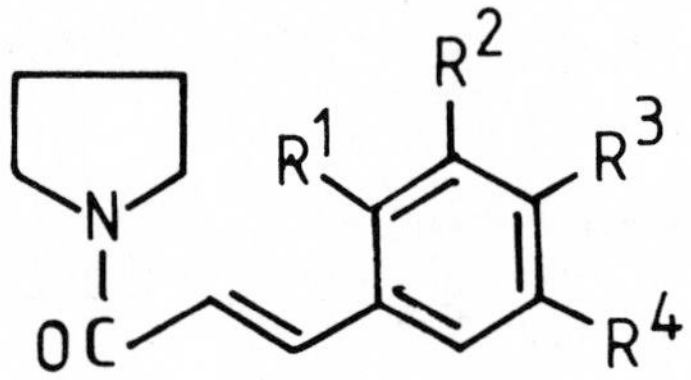

(33) $R^1 = R^2 = OMe$, $R^3, R^4 = OCH_2O$

(34) $R^1 = R^2 = R^3 = R^4 = H$

(35) $R^1 = R^3 = R^4 = H$, $R^2 = OMe$

A further group of related amides are also found in *Piper* species. Trichostachine, $C_{16}H_{17}NO_3$, occurs in *P. trichostachyon* (Sing, Dhar, and Atal, Tetrahedron Letters, 1969, 4975); *P. nigrum* (R. Grewe *et al*., Ber., 1970, *103*, 3752); and *P. guineense* (D. Dwuma-Badu *et al*., Lloydia, 1976, *39*, 60). The structure (36) for trichostachine has been established spectroscopically; by hydrolysis to pyrrolidine and piperic acid; and by synthesis from piperoyl chloride and pyrrolidine (Singh, Dhar and Atal, *loc. cit*.). Trichostachine has m.p. 142-143°.

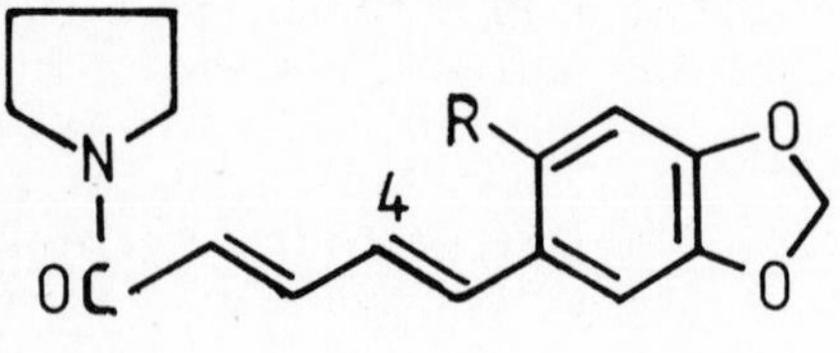

(36) R = H

(37) R = OMe

(38) R = OMe, no Δ^4

Wisanidine (= Okolasine), $C_{17}H_{19}NO_4$, is present in the seeds of *P. guineense*. Permanganate oxidation of wisanidine to 3,4-methylenedioxy-6-methoxybenzaldehyde establishes the position of the methoxyl-group and leads to the formulation of wisanidine as 6'-methoxytrichostachine (37) (B.L. Sondengam *et al.*, Tetrahedron Letters, 1977, 367). The synthesis of wisanidine by standard methods has been reported (H.-D. Scharf *et al.*, Ann., 1978, 573). Wisanidine has m.p. 172-174°.

4,5-Dihydrowisanidine (38) also occurs in *P. guineense*. It yields wisanidine (37) on dehydrogenation with 2,3-dichloro-5,6-dicyanobenzoquinone (Sondengam, S.F. Kimbu, and J.D. Connolly, Phytochem., 1977, 16, 1121). 4,5-Dihydrowisanidine has m.p. 82-84°.

1-Piperethylpyrrolidine, $C_{18}H_{19}NO_3$, and tricholeine, $C_{20}H_{27}NO_3$, have been isolated from *P. trichostachyon*. Their structures (39) and (40) were established by hydrolysis to the constituent acid and basic moieties (Singh, D.D. Santani, and Dhar, Phytochem., 1976, 15, 2018). 1-Piperethylpyrrolidine has m.p. 147-149°.

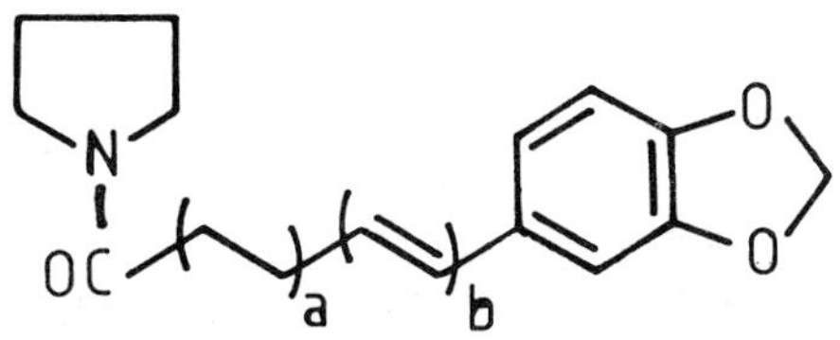

(39) a = 0, b = 3

(40) a = 6, b = 1

Cyclostachine A (41) is also present in the stems of *P. trichostachyon*. Its structure has been established by spectral and degradative studies. Confirmation of the structure and stereochemistry of cyclostachine A has been obtained by X-ray diffraction analysis of the sulphate of the amine, formed by reduction of cyclostachine A (B.S. Joshi *et al.*, Experientia, 1975, 31, 880). Cyclostachine A has m.p. 136-138°.

Cyclostachine B, obtained from the same source, is a stereoisomer of cyclostachine A with a *trans*-ring junction. Cyclostachine B has m.p. 135-136°.

Cyclostachine A and B have both been synthesized by internal Diels-Alder reaction of the ester (42), followed by separation of the diastereoisomeric products, conversion into the respective acid chlorides and treatment with pyrrolidine (Joshi *et al.*, Helv., 1975, *58*, 2295).

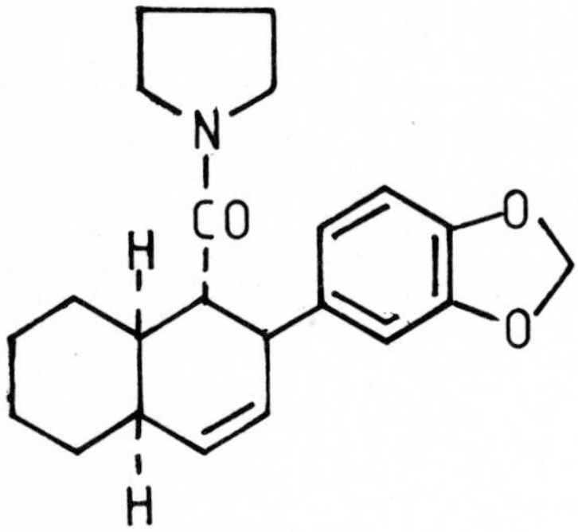

(41) 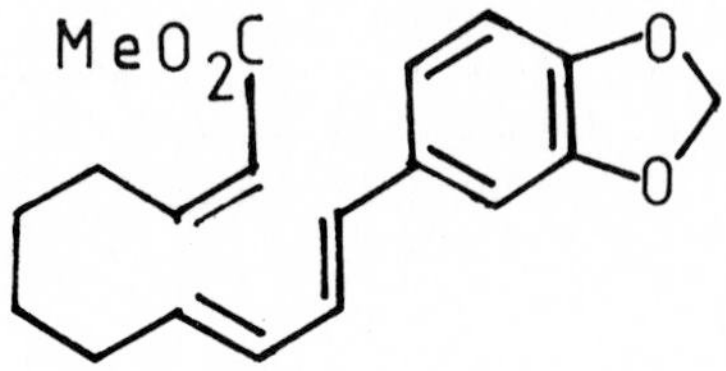 (42)

Roxburghilin, $C_{18}H_{24}N_2O_2$, (43) has been isolated from *Aglaia roxburghiana* (Meliaceae). The isolation of (+)-2-methylbutanoic acid on alkaline hydrolysis of roxburghilin indicates that the configuration at C-2' is *S*. The configuration at C-2 is not known. The structure (43) has been confirmed by the synthesis of the racemic dihydroderivative from L-proline. A minor component also present in this species is probably 2'-hydroxyroxburghilin (K.K. Purushothaman *et al.*, J. Chem. Soc. Perkin 1, 1979, 3171). Roxburghilin, $[\alpha]_D$ +34°, has m.p. 205°.

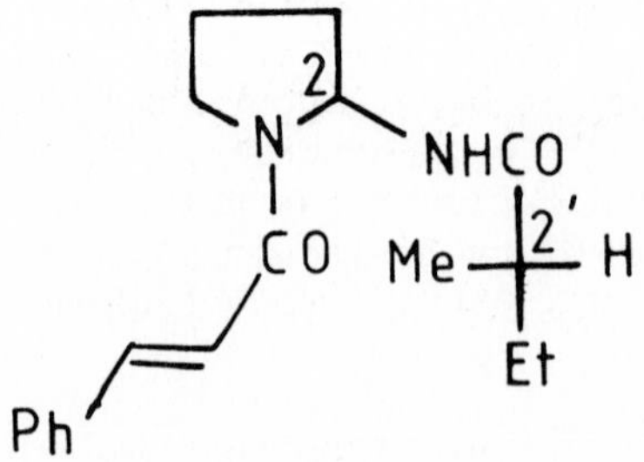

(43)

N-Ethoxycarbonyl-L-prolinamide, $C_8H_{14}N_2O_3$, occurs in the leaves of *Arnica montana*. Its structure, including absolute configuration, (44), has been established by synthesis from L-proline (M. Holub *et al.*, Coll. Czech. chem. Comm., 1977, **42**, 151). *N*-Ethoxycarbonyl-L-prolinamide, $[\alpha]_D^{20}$ -55.8°, has m.p. 103-103.5°.

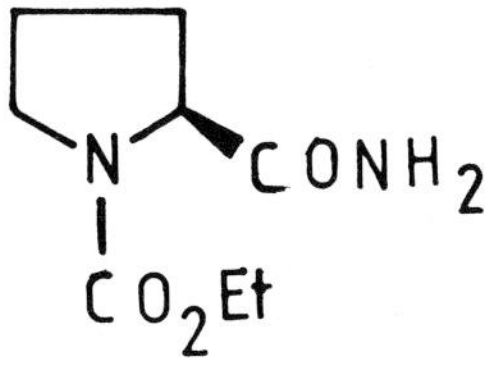

(44)

4. N-Acylpyrrolidones

Variotin, $C_{17}H_{25}NO_3$, (45) is an antifungal antibiotic obtained from *Paecilomyces varioti*. The synthesis of variotin depends upon *N*-acylation of 2-pyrrolidone under mild conditions using an *N*-trimethylsilylated derivative (M. Sakakibara and M. Matsui, Agric. biol. Chem., 1973, **37**, 911). Three analogues have also been prepared (*idem*, *ibid*., 1973, **37**, 1131).

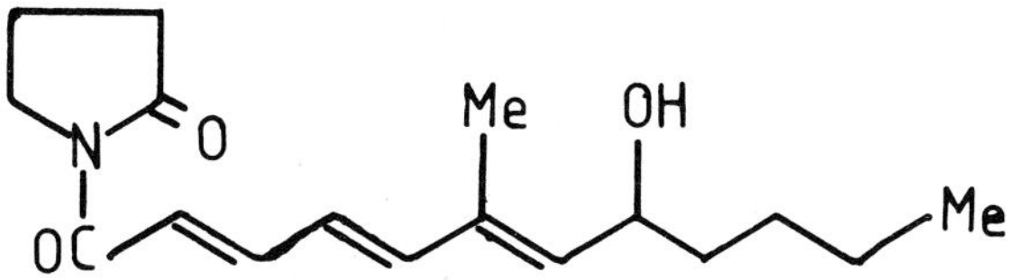

(45)

The marine sponge Dysidea herbacea contains a novel chlorine-containing compound. The structure (46) has been deduced from spectroscopic and degradative studies and confirmed by X-ray diffraction analysis (W. Hofheinz and W.E. Oberhänsli, Helv., 1977, 60, 660). The metabolite (46), $[\alpha]_D^{25}$ +141° ($CHCl_3$), has m.p. 127-129°.

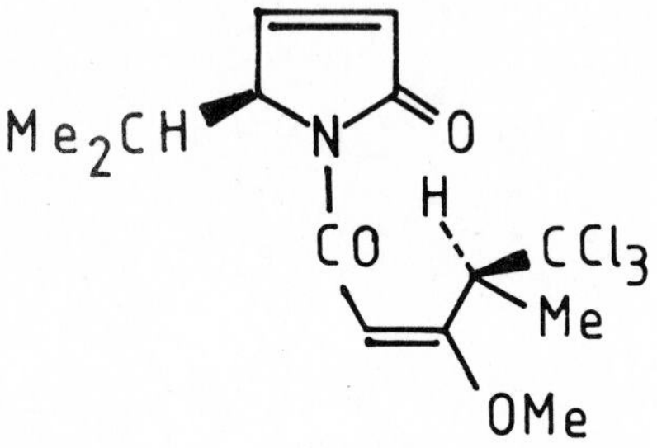

(46)

Chapter 8

FIVE-MEMBERED MONOHETEROCYCLIC COMPOUNDS: ALKALOIDS (CONTINUED):

PYRROLIZIDINE ALKALOIDS

DAVID J. ROBINS

This chapter follows the pattern set out in the previous article on this topic (C.C.C. 2nd Edn., Vol IVB, p.7). The past years have seen steady growth in the number of known pyrrolizidine alkaloids - over 200 have now been characterised. Additional plant families which are now known to contain pyrrolizidine alkaloids are the Apocynaceae, Celastraceae, Euphorbiaceae, Ranunculaceae, and Scrophulariaceae.

Progress in this area is reviewed annually in "The Alkaloids", Vols. 1-10, The Chemical Society, London, 1971-80. Other reviews are available on the chemistry of pyrrolizidines (D.J. Robins, Adv. heterocyclic Chem., 1979, 24, 247) and the chemotaxonomic importance of pyrrolizidine alkaloids (C.C.J. Culvenor, Bot. Notiser, 1978, 131, 473).

Data are available on spectroscopic properties. Ultraviolet spectra of 20 pyrrolizidine derivatives (V.P. Gupta, S.K. Handoo, and R.S. Sawhney, Indian J. Pure Appl. Phys., 1975, 13, 776), and infrared spectra of 25 pyrrolizidine esters (Gupta, Handoo, and Sawhney, Curr. Sci., 1975, 44, 451) have been recorded. ^{13}C Nuclear magnetic resonance spectrometric assignments have been made for a number of pyrrolizidine alkaloids (N.V. Mody *et al.*, J. natural Products, 1979, 42, 417; E.J. Barreiro *et al.*, J. chem. Res. (S), 1980, 330). The mass spectra of pyrrolizidine alkaloids have been discussed (Ya. V. Rashkes, U.A. Abdullaev, and S. Yu. Yunusov, Chem. Abs., 1978, 89, 163 803); alkaloids containing otonecine as base have characteristic fragment ions at m/e 168, 151, 150, 122, 110, and 94 (M.P. Cava *et al.*, J. org. Chem., 1968,

33, 3570; Abdullaev, Rashkes and Yunusov, Chem. Abs., 1976, 85, 177 726). More circular dichroism data have been compiled (J. Hrbek *et al.*, Coll. Czech. Chem. Comm., 1972, 37, 3918).

Analytical methods have been developed. Ion exchange resins have been employed for the extraction (J.T. Deagen and M.L. Deinzer, Lloydia, 1977, 40, 395) and separation of alkaloids (H.J. Huizing and T.M. Malingre, J. Chromatog., 1979, 176, 274). The conversion of *N*-oxides into tertiary bases using a redox polymer on an anion exchanger gave better yields than reduction with zinc in acidic solution (Huizing and Malingre, J. Chromatog., 1979, 173, 187). Separation by high performance liquid chromatography (hplc) has been achieved on alkaloid mixtures (H.J. Segall, Toxicol. Letters, 1978, 1, 279; Segall and R.J. Molyneux, Res. Comm. Chem. Pathol. Pharmacol., 1978, 19, 545; C.W. Qualls and Segall, J. Chromatog., 1978, 150, 202) and the *N*-oxides (G. Tittel, H. Hinz, and H. Wagner, Planta Med., 1979, 37, 1). Improved separations of alkaloid mixtures are obtained on reverse-phase hplc (Segall, J. liquid Chromatog., 1979, 2, 429; 1319; G.P. Dimenna, T.P. Krick, and Segall, J. Chromatog., 1980, 192, 474). Sensitive methods for detection of pyrrolizidine alkaloids include glc-mass spectrometry of silylated derivatives (J.V. Evans, A. Peng, and C.J. Nielsen, Biomed. Mass Spectrom., 1979, 6, 38), electron capture glc of fluorinated retronecine derivatives (Deinzer *et al.*, Biomed. Mass Spectrom., 1978, 5, 175) and ^{1}H nmr spectrometric determination of the vinyl proton at C-2 in retronecine (1) derivatives (Molyneux *et al.*, J. agric food Chem., 1979, 27, 494).

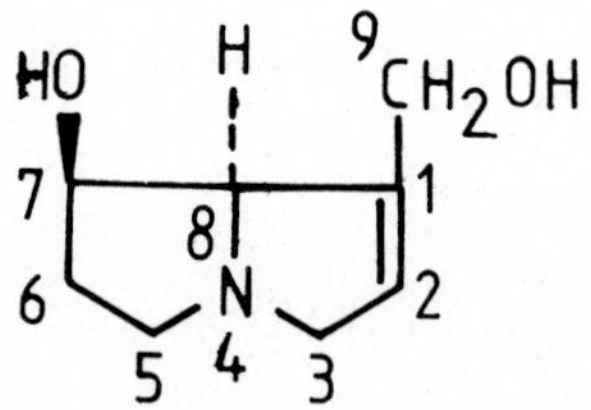

(1)

1. Necines

The numbering scheme used for the pyrrolizidine nucleus is as shown for retronecine (1).

(a) Unhydroxylated derivatives

A range of *endo*-1-substituted pyrrolizidines (3) has been prepared by transannular electrophilic attack on the amine (2) (S.R. Wilson and R.A. Sawicki, J. org. Chem., 1979, 44, 287). The ring system of loline (5) has been constructed by an extension of this strategy from the bicyclic amine (4). Loline (5) could not be prepared from (6) (Wilson and Sawicki, Tetrahedron Letters, 1978, 2969; see also R.G. Glass, D.R. Deardoff and L.H. Gains, *ibid*, 1978, 2965).

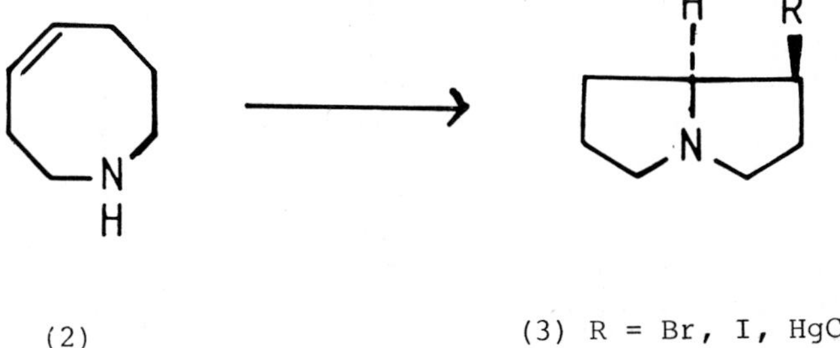

(2)

(3) R = Br, I, HgCl, PhS or PhSe

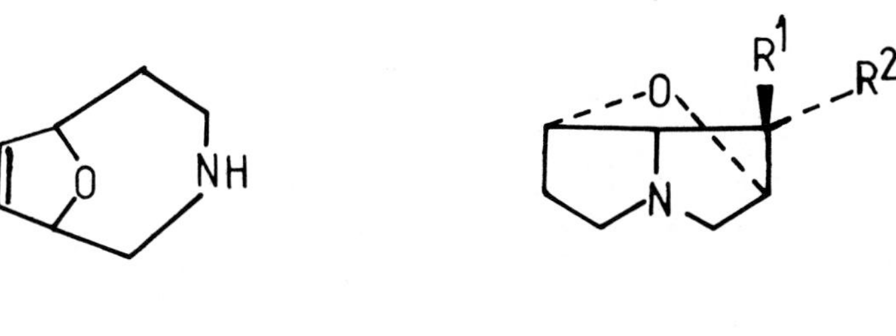

(4)

(5) R^1 = NHMe, R^2 = H

(6) R^1 = H, R^2 = Br

(b) Monohydroxylated derivatives

The first synthesis of the optically forms of 1-hydroxymethylpyrrolizidine has been reported (Robins and S. Sakdarat, Chem. Comm., 1979, 1181). Regiospecific 1,3-dipolar cycloaddition of ethyl propiolate to the N,O-diformyl derivative of natural (-)-4-hydroxy-L-proline gives a (+)-ester (7). Deformylation and hydrogenation yielded a single (+)-ester (8). After removal of the hydroxy group the (+)-ester (9) was converted into (+)-isoretronecanol (10) by reduction, and (+)-laburnine (11) by epimerisation of the ester group followed by reduction. Epimerisation of the chiral centre in (7)

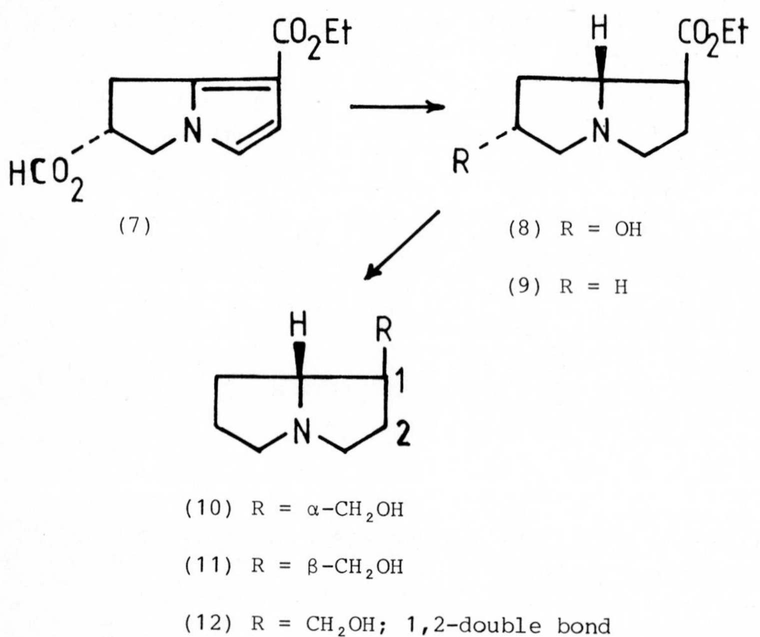

established a route to the corresponding (-)-alcohols. Thermal elimination of a phenylseleno-group has been used for converting saturated esters such as (9) into their 1,2-didehydro-analogues (Robins and Sakdarat, J. chem. Soc., Perkin I, 1979, 1734). Thus, (+)-supinidine (12) has been synthesised.

Intramolecular opening of activated cyclopropanes has been used to synthesise necine bases (S. Danishefsky, Acc. chem. Res., 1979, 12, 66). Intramolecular alkylation of an activated cyclopropane (13) followed by lactamisation yielded the

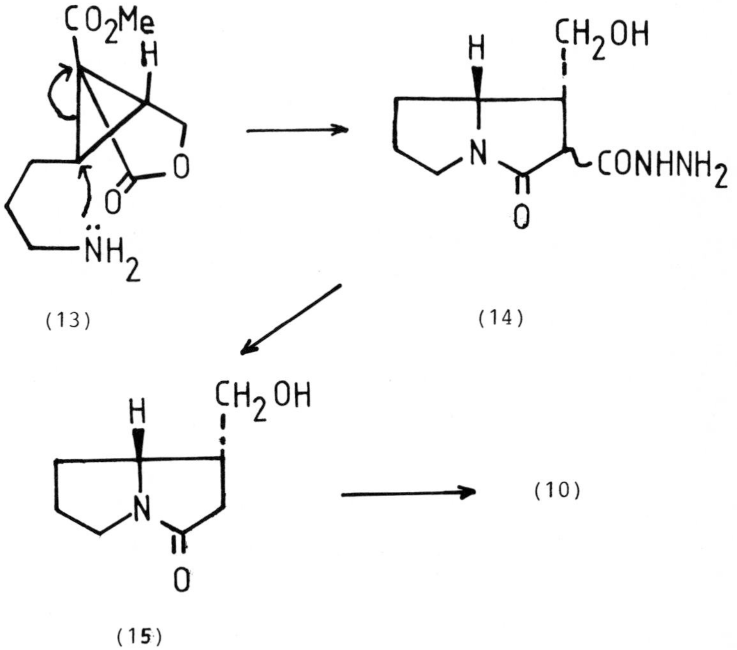

(13)

(14)

(15)

(10)

pyrrolizidine (14). Deacylation afforded the lactam (15) and reduction gave (±)-isoretronecanol (10) (Danishefsky, R. McKee and R.K. Singh, J. Am. Chem. Soc., 1977, 99, 4783).

Acid-catalysed rearrangement of cyclopropylimines has also been used to construct necine bases (R.V. Stevens, in 'The Total Synthesis of Natural Products', ed. J. ApSimon, Wiley-Interscience, New York, 1977, Vol. 3, p. 515). The stabilised endocyclic enamine system (17) is obtained by acid-catalysed rearrangement of the cyclopropylimine (16), and is transformed into (±)-isoretronecanol (10) by cyclisation in acid, followed by reduction and desulphurisation steps (Stevens, Y. Luh, and J.-T. Sheu, Tetrahedron Letters, 1976, 3799). A related route has been described from the cyclopropylimine (18) (H.W. Pinnick

PhS–C(cyclopropyl)–CH=N(CH$_2$)$_3$CH(OEt)$_2$ → PhS–(pyrroline)–N(CH$_2$)$_3$CH(OEt)$_2$ → (10)

(16) (17)

(10)

and Y.-H. Chang, Tetrahedron Letters, 1979, 837). Acid-catalysed rearrangement of (18) afforded an unsaturated pyrrolizidine ester, from which (±)-isoretronecanol (10) was obtained by hydrogenation and hydride reduction. The same authors have developed another route to (±)-isoretronecanol (Pinnick and Chang, J. org. Chem., 1978, 43, 4662). Hydrogenation and hydride reduction of the lactam (20) obtained by cyclisation of the diester (19) gives (±)-isoretronecanol (10).

CO_2Et CO_2Et

N → N → (10)

(18)

CO_2Et CO_2Et

CH_2CO_2Et →

NH N → (10)

O

(19) (20)

A reductive cyclisation technique has been employed (R.F. Borch and B.C. Ho, J. org. Chem., 1977, 42, 1225). Oxidation of the diastereomeric cycloheptenone aminoesters (21) with osmium tetroxide and periodate, followed by reductive cyclisation with sodium cyanoborohydride gave the diastereomeric esters (22). Separation and reduction afford (±)-isoretronecanol (10) and (±)-trachelanthamidine (11).

NH_2 CO_2Me → H CO_2Me N → (10) and (11)

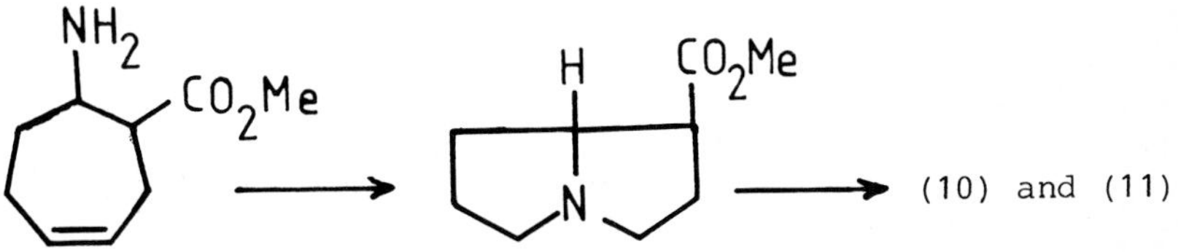

(21) (22)

The regiospecific 1,3-dipolar cycloaddition of 1-pyrroline-1-oxide to dihydrofuran has been utilised in a synthesis of (±)-isoretronecanol (T. Iwashita, T. Kusumi, and H. Kakisawa, Chem. Letters, 1980, 1337). Cyclisation of the acetylene (23) under acidic conditions gives the diastereomeric pyrrolizidines (24) which can be separated and reduced to give (±)-isoretronecanol (10) and (±)-trachelanthamidine (11) (P.M.M. Nossin and W.N. Speckamp, Tetrahedron Letters, 1979, 4411).

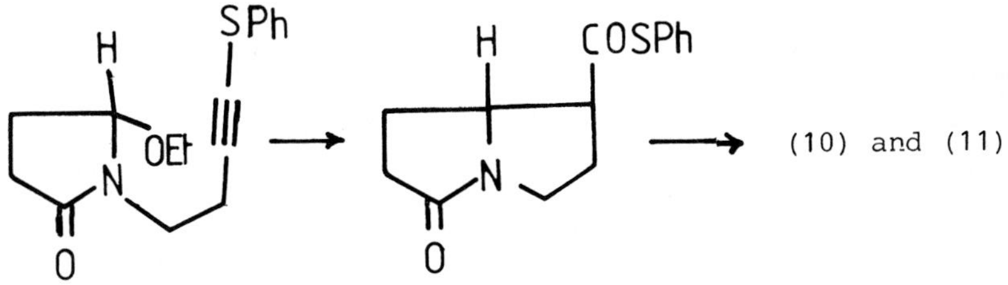

H SPh OEt N O → H COSPh N O → (10) and (11)

(23) (24)

(c) Dihydroxylated derivatives

(-)-Petasinecine, $C_8H_{15}NO_2$, is the base constituent of petasinine and petasinoside (K. Yamada *et al.*, Tetrahedron Letters, 1978, 4543). The 2-hydroxy-1-hydroxymethylpyrrolizidine gross structure follows from its mass spectrum. The four possible racemates of 2-hydroxy-1-hydroxymethylpyrrolizidine had been synthesised previously (A.J. Aasen and C.C.J. Culvenor, J. org. Chem., 1969, **34**, 4143). The diacetate of petasinecine is identical with the diacetate of (±)-2β-hydroxy-1β-hydroxymethyl-8α-pyrrolizidine (25). The absolute configuration is not known.

(25)

(±)-Hastanecine (26) and (±)-dihydroxyheliotridane (27) have been synthesised by an extension of the method involving intramolecular opening of activated cyclopropanes (Danishefsky, *loc. cit.*) using suitably hydroxylated precursors (Danishefsky, McKee, and Singh, J. Amer. chem. Soc., 1977, **99**, 7711).

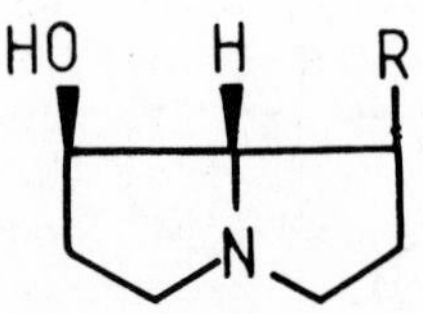

(26) R = β-CH_2OH

(27) R = α-CH_2OH

In like manner, the strategy of 1,3-dipolar cycloaddition of nitrones to alkenes used in the synthesis of (±)-supinidine (12) (J.J. Tufariello and J.P. Tette, J. org. Chem., 1975, 40, 3866) has been extended into a synthesis of (±)-retronecine (1) by utilising substituted nitrones (Tufariello and G.E. Lee, J. Amer. chem. Soc., 1980, 102, 373).

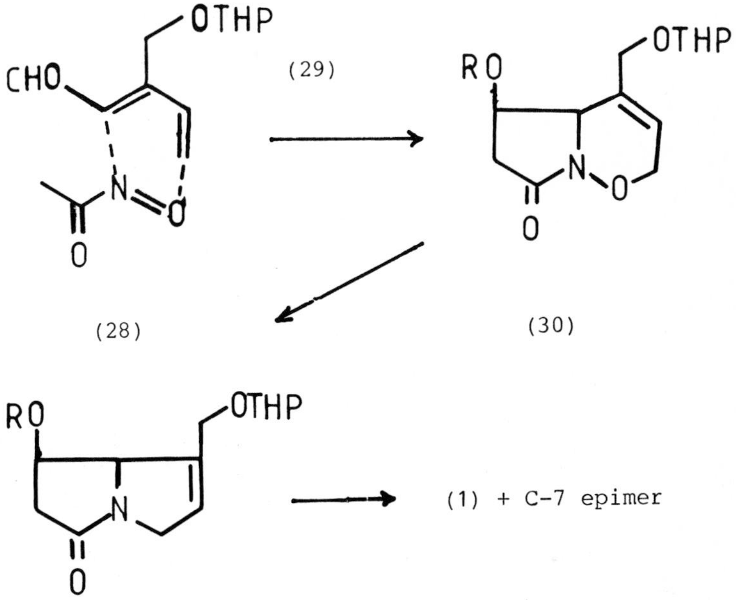

(29)

(28)

(30)

(1) + C-7 epimer

(31)

Another synthesis of (±)-retronecine (1) and its C-7 epimer (±)-heliotridine has been reported (G.E. Keck and D.G. Nickell, J. Amer. chem. Soc., 1980, 102, 3632). Cycloaddition of an acylnitroso moiety (28) generated from the dimethylanthracene Diels-Alder adduct, to the diene (29) produces the diastereomeric oxazine derivatives (30). Reductive cleavage of the N-O bond followed by elimination of water gives the diastereomeric pyrrolizidine (31) which can be separated and converted into (±)-retronecine (1) and (±)-heliotridine by reduction and deprotection steps.

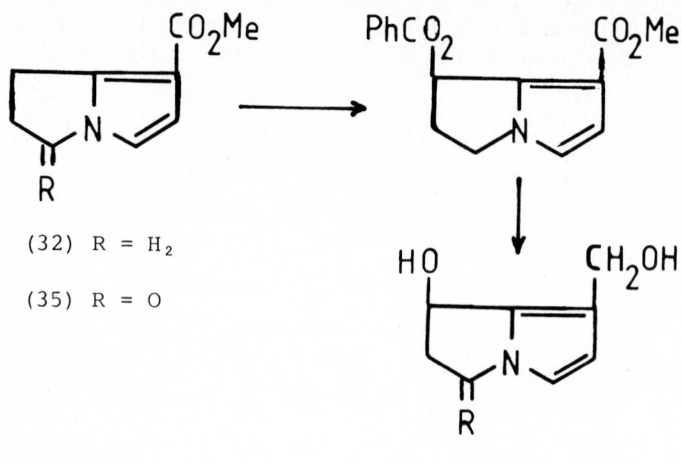

(33) R = H_2

(34) R = O

The unsaturated ester (32) is easily prepared by 1,3-dipolar cycloaddition of N-formyl-L-proline to methyl propiolate. The direct introduction of an oxygen function into the allylic position has been achieved using peroxyesters. (±)-Dehydroheliotridine (33) is obtained on reduction (F. Bohlmann, W. Klose and K. Nickisch, Tetrahedron Letters, 1979, 4411). Both (+)- and (−)-forms of the dihydropyrrolizinone (34) occur in *Senecio* species. The compound (±)-(34) has been prepared from the corresponding unsaturated ester (35).

(d) Trihydroxylated derivatives

Croalbinecine, $C_8H_{15}NO_3$, (36) is the necine component of croalbidine. The stereochemistry at C-1, C-7, and C-8 has been determined by correlation with turneforcidine (37).

The location of the third hydroxyl group at C-2 has been deduced from ^1H nmr spectroscopic data and confirmed by decoupling experiments. From the magnitude of the coupling constants $J_{1\beta,2} = J_{2,3\beta} = 8.0$ Hz, due to the almost diaxial arrangement of H-1 and H-2, the 2-hydroxy group is assigned the β configuration (Sawhney *et al*., Austral. J. Chem., 1974, *27*, 1805).

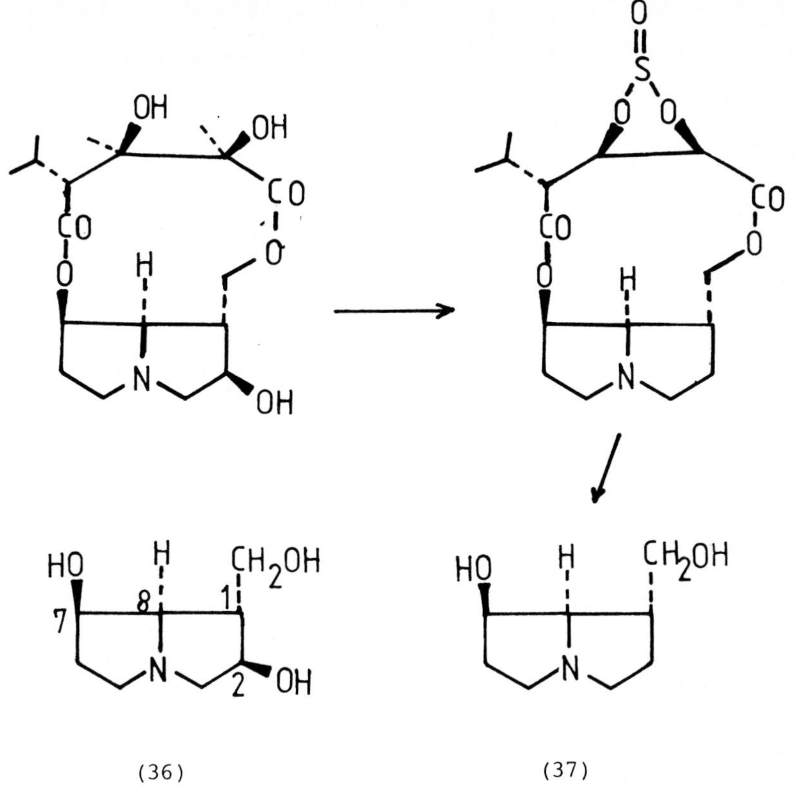

(36) (37)

TABLE 1

NECINES: PHYSICAL PROPERTIES

Necine	m.p. (°C)	$[\alpha]_D^{(°)}$
Petasinecine (25)	132-4	-20 (EtOH)
Croalbinecine (36)	165-6 (HCl salt)	-

2. Necic Acids

(a) C_6-Acids

(–)-2,3-Dihydroxy-3-methylpentanoic acid (38), $C_6H_{12}O_4$, obtained from strigosine by alkaline hydrolysis, has been shown to have the (2R,3R) absolute configuration. The (±)-*erythro*-isomer is synthesised by *trans*-hydroxylation of (E)-3-methylpent-2-enoic acid with pertungstic acid. Resolution via the quinine salts yields the (–)-isomer which is identical with the naturally occurring (–)-acid (38). The absolute configuration has been established by degradation of the (–)-isomer to a compound of known absolute configuration (D.H.G. Crout and D. Whitehouse, J. chem. Soc., Perkin I, 1977, 544).

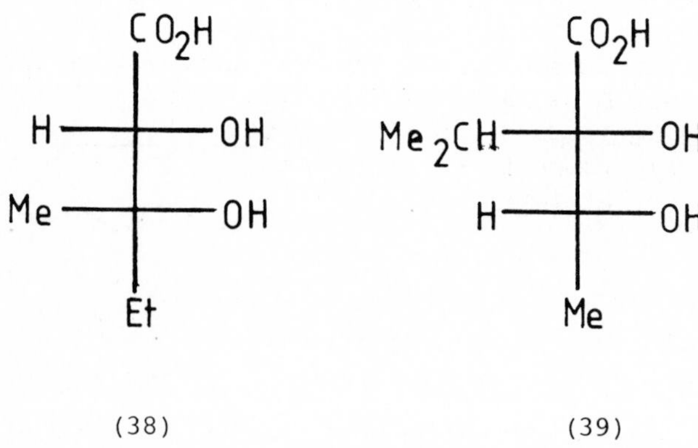

(38) (39)

(b) C_7-Acids

(+)-Viridifloric acid (39), $C_7H_{14}O_4$, has been obtained from the alkaloid coromandalin (S. Mohanraj *et al*., Chem. Comm., 1978, 423).

(c) C_8-Acids

(+)-Homoviridifloric acid (40), $C_8H_{16}O_4$, the first C_8-monocarboxylic necic acid, has been obtained by hydrolysis of curassavine (Mohanraj *et al*., *loc. cit*.) and identified spectroscopically. The ^{1}H nmr spectrum of (+)-homoviridifloric acid shows the presence of $MeCH_2$-, MeCH- and MeCHOH groups, and oxidation of the methyl ester of (40) with periodate gives $MeCH_2CHMeCOCO_2Me$ (mass spectral data). The absolute configuration of this acid is not known but the configuration of the glycol group is assumed to be *erythro*, since the acid showed similar electrophoretic mobility to *erythro*-viridifloric acid.

$$MeCH(OH)COH(CO_2H)CHMeCH_2Me$$

(40)

Crispatic acid, $C_8H_{14}O_5$, (41), has been assigned the *R*-meso configuration, and fulvinic acid (42) the *S*-meso configuration on the basis of asymmetric syntheses (T. Matsumoto, K. Fukui, and J.D. Edwards, Chem. Letters, 1973, 283). The two remaining stereoisomeric forms (43) of 3-hydroxy-2,3,4-trimethylglutaric acid, cromaduric and isocromaduric acids, are present in cromadurine (P.G. Rao, R.S. Sawhney, and C.K. Atal, Indian J. Chem., 1975, *13*, 870) and isocromadurine (*idem*, Experientia, 1975, *31*, 875), respectively, although the absolute configuration of each enantiomer is not known.

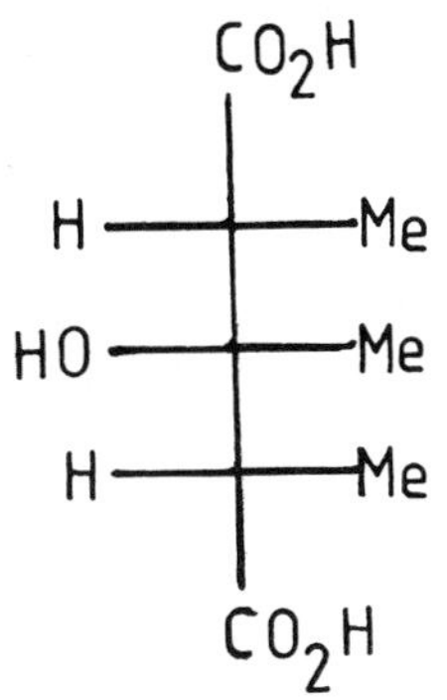

(41) 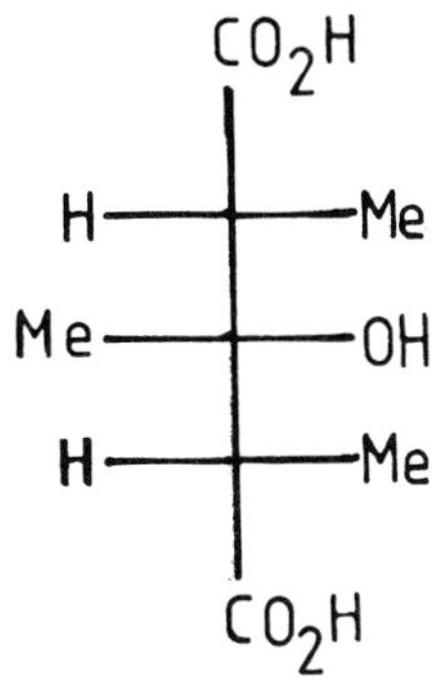 (42) 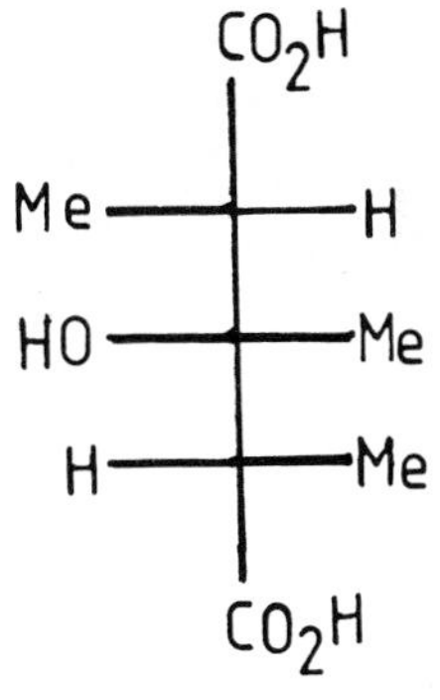 (43)

All eight stereoisomeric forms of monocrotalic acid have been synthesized and the (*2R*,*3R*,*4R*)- absolute configuration of natural (-)-monocrotalic acid (44) has been confirmed (Matsumoto, M. Takahashi, and Y. Kashihara, Bull. chem. Soc. Japan, 1979, 52, 3329).

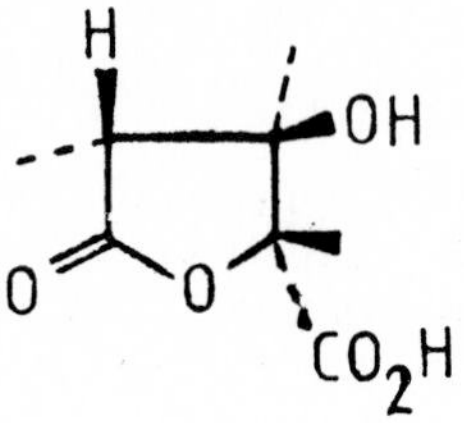

(44)

(d) C_9-Acids

Crotananic acid, (45), $C_9H_{16}O_5$, is obtained on hydrolysis of crotananine with barium hydroxide. The presence of an α-hydroxy acid is indicated by a positive $FeCl_3$ test, and in the ^1H nmr spectrum all three methyl signals are doublets. The stereochemistry of this acid is not known (M.A. Siddiqi *et al.*, Phytochemistry, 1978, 17, 2143).

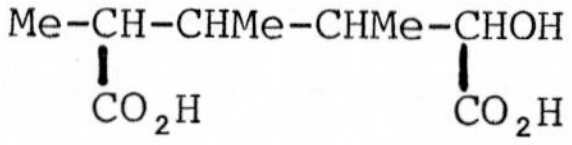

```
Me-CH-CHMe-CHMe-CHOH
   |            |
   CO2H         CO2H
```

(45)

Another C_9 acid, cronaburmic acid, (46), $C_9H_{14}O_4$, is obtained as a δ-lactone by hydrolysis of cronaburmine, and identified by its ^1H nmr spectrum. Again the stereochemistry of this acid has not been established. (Siddiqi *et al.*, Indian J. Chem., 1978, 16B, 1132).

Other C_9-acids are the acid portion (47) of crotafoline (Crout, J. chem. Soc. Perkin I, 1972, 1602), its geometrical

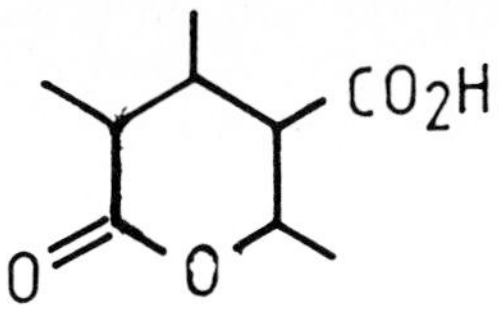

(46)

isomer, striatic acid (48), (Atal *et al.*, Tetrahedron Letters, 1968, 5605; Sawhney and Atal, Planta Med., 1972, 21, 435), and the O-acetyl derivative (49) of striatic acid (V. Batra, R.N. Gandhi, and T. Rajagopalan, Indian J. Chem., 1975, 13, 989).

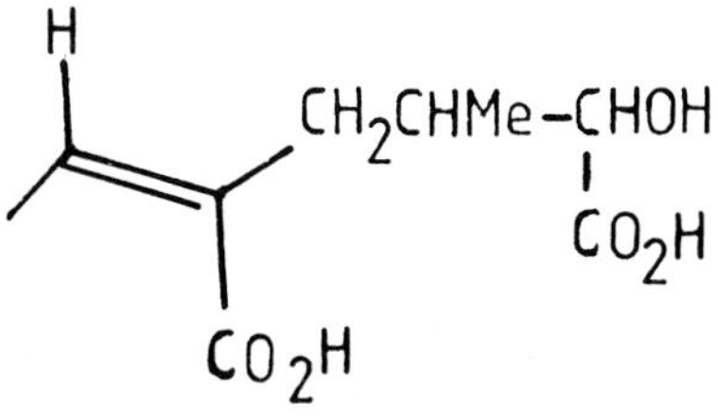

(47)

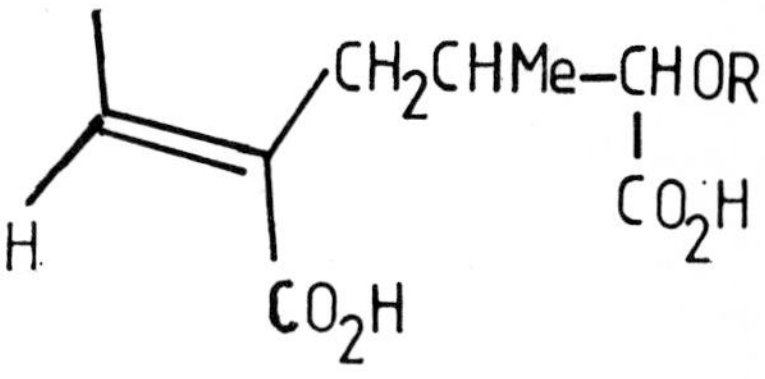

(48) R = H

(49) R = Ac

(e) C_{10}-Acids

cis-Nemorensic acid, (50), $C_{10}H_{16}O_5$, is obtained by alkaline hydrolysis of retroisosenine and doronenine, whereas *trans*-nemorensic acid (51) is formed on base hydrolysis of bulgarsenine. Both of these acids are also present in the free form in the plant *Senecio nemorensis* L. var. *subdecurrens* Griseb. (A. Klasek *et al.*, Coll. Czech. chem. Comm., 1980, 45, 548). The structure and absolute configuration of *cis*-nemorensic acid have been established by X-ray crystallography (A. Kirfel *et al.*, Cryst. Struct. Comm., 1980, 9, 363).

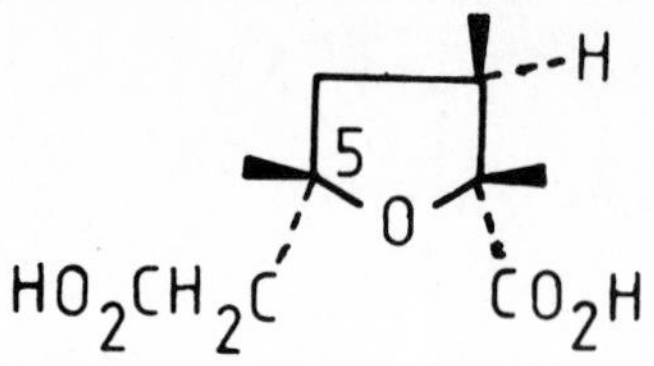

(50)

(51) opposite stereochemistry at C-5

Petasinenic acid, (52), $C_{10}H_{16}O_5$, obtained by hydrolysis of petasitenine is diastereomeric at C-5 with jaconecic acid. Oxidation of petasinenic acid gives a γ-lactone (53) which is identical to that obtained by similar oxidation of jaconecic acid. Thus the stereochemistry at C-2, C-3, and C-5 of petasinenic acid is established.

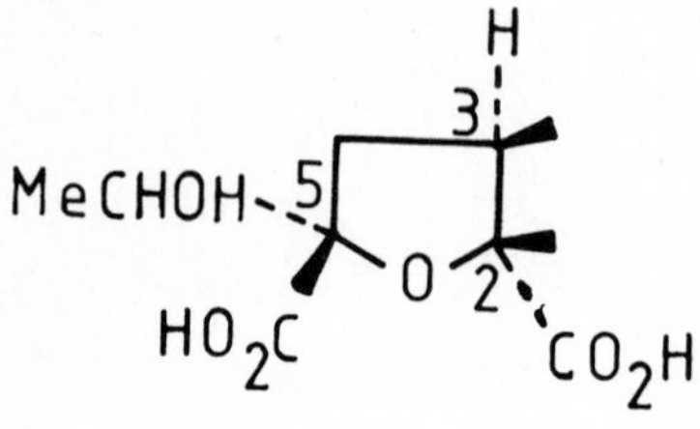

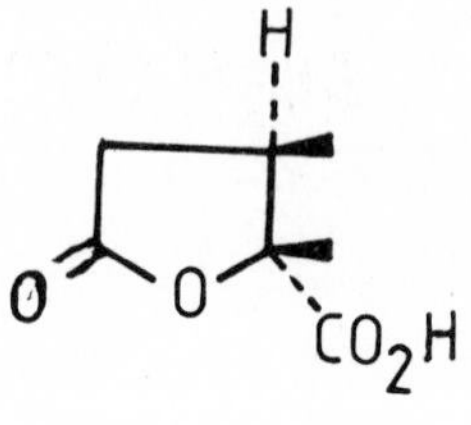

(52) (53)

Syneilesinolides A-C (54-56) are three lactones formed by alkaline hydrolysis of syneilesine. Circular dichroism measurements on these lactones indicates that the configuration at C-2 is R. The formation of the dilactone (56) suggests that the stereochemistry at C-4 is also R. The dihedral angle between the protons on C-3 and C-4 of (56) is estimated to be 36° from the ^{1}H nmr coupling constant of 5.4 Hz, suggesting that the configuration at C-3 is R. The absolute configuration at C-5 is not known (M. Hikichi and T. Furuya,

Chem. pharm. Bull. Japan, 1976, 24, 3178).

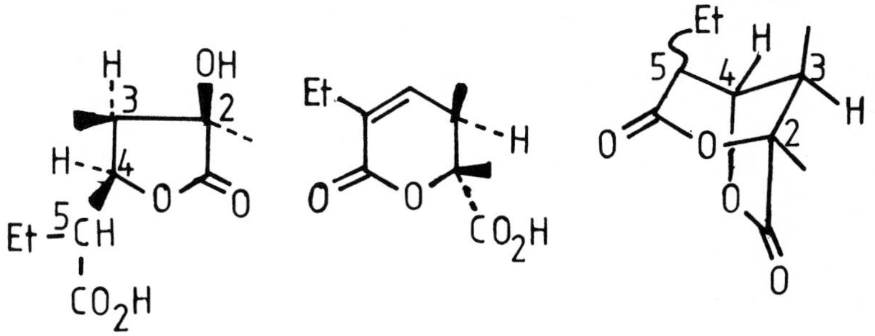

(54) (55) (56)

Ligularidenecic acid (57), $C_{10}H_{14}O_4$, is obtained by alkaline hydrolysis of ligularidine (Hikichi, Y. Asada, and Furuya, Tetrahedron Letters, 1979, 1233). The lactone had physical data which are in agreement with those for the lactone synthesised earlier by Edwards and Matsumoto (J. org. Chem., 1967, 32, 1837).

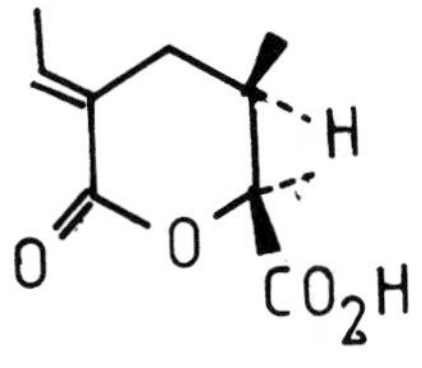

(57)

Senecivernic acid (58), $C_{10}H_{16}O_5$, is obtained by alkaline hydrolysis of senecivernine (E. Röder, H. Wiedenfeld, and U. Pastewka, Planta Med., 1979, 37, 131), while pterophorenecic acid (59) is the esterifying acid of pterophorine. The structure of pterophorenecic acid follows from its 270 MHz ^{1}H and ^{13}C nmr spectra (F. Bohlmann, C. Zdero and M. Grenz, Ber., 1977, 110, 474). The stereochemistry of these acids is unknown; they both belong to an unusual structural group of C_{10} acids.

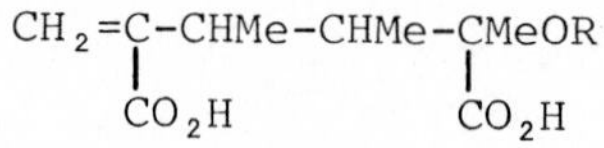

$$\begin{array}{l} CH_2{=}C{-}CHMe{-}CHMe{-}CMeOR \\ \quad\;\; | \qquad\qquad\qquad\quad | \\ \quad CO_2H \qquad\qquad\;\; CO_2H \end{array}$$

(58) R = H

(59) R = Ac

Crotalaric acid (60), $C_{10}H_{16}O_5$, is a γ-lactone of unidentified stereochemistry formed on alkaline hydrolysis of crotalarine (M.A. Ali and G.A. Adil, Pakistan J. Sci. Ind. Res., 1973, 16, 227).

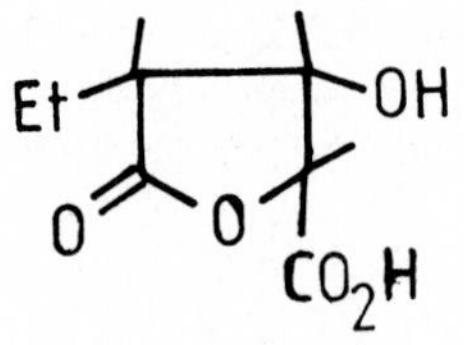

Et, OH, O, O, CO_2H

(60)

Crotaverric acid, $C_{10}H_{16}O_5$, obtained on alkaline hydrolysis of crotaverrine, forms a δ-lactone. The ^{1}H nmr spectrum indicates that it is diastereomeric with integerrinecic acid (61). Lead tetra-acetate oxidation of both acids gives the same ketoacids (identified as their 2,4-dinitrophenylhydrazones). Thus crotaverric acid has the opposite stereochemistry at either C-2 or C-3 to (61) (O.P. Suri *et al.*, Phytochem., 1976, 15, 1061).

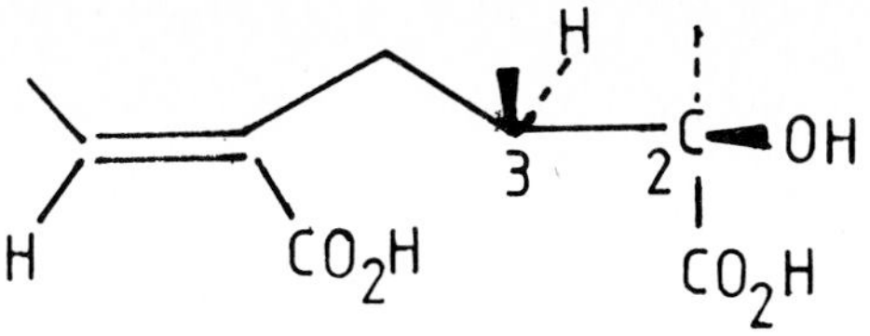

(61)

3. Alkaloids

New alkaloids, together with revised structures, and additional stereochemical details, are subdivided into five groups: monoesters, acyclic diesters, cyclic diesters, necine bases esterified with an aromatic ring, and miscellaneous types.

Structure elucidation is usually carried out spectroscopically, together with hydrolysis or hydrogenolysis of the alkaloid to the component acidic and basic moieties, which are then identified. Increasing use is being made of X-ray crystallography for establishing the absolute configuration. X-Ray studies on 11-membered macrocyclic diesters of retronecine (1) such as monocrotaline (H. Stoeckli-Evans, Acta Crystallogr., 1979, 35B, 231), fulvine (J.L. Sussman and S.J. Wodak, *ibid*., 1973, 29B, 2918), axillarine (Stoeckli-Evans and Crout, Helv., 1976, 59, 2168), and incanine (B. Tashkhodzhaev, M.V. Telezhenetskaya, and S. Yu. Yunusov, Chem. Abs., 1980, 92, 111 199) have shown that the ester carbonyl groups are *syn*-parallel and directed below the plane of the 11-membered ring. However, in trichodesmine, (Tashkhodzhaev, M.R. Yagudaev, and Yunusov, *ibid*., 1980, 92, 111 194) the conformation of the acid portion is quite different. The ester carbonyl groups are nearly *anti*-parallel, as is the case with macrocyclic diesters of retronecine with 12-membered rings such as retrorsine (Stoeckli-Evans, Acta Crystallogr., 1979, 35B, 2798).

The first synthesis of a non-natural 11-membered macrocyclic pyrrolizidine diester has been achieved from (+)-retronecine and 3,3-dimethylglutaric anhydride (Robins and S. Sakdarat, Chem. Comm., 1980, 282).

TABLE 2

NECIC ACIDS: PHYSICAL PROPERTIES

Necic Acid	Formula	m.p. (°C)	$[\alpha]_D^{(°)}$	
(+)-Viridifloric	(39)	122-124	+ 3.1	(EtOH)
Homoviridifloric	(40)	106-110	+ 7.4	(EtOH)
Cromaduric	(43)	138-139	-14.5	(MeOH)
Isocromaduric	(43)	129-130	+14.9	(MeOH)
Crotananic	(45)	142-143	-	
Cronaburmic lactone	(46)	106 (brucine salt)	-	
Acetylstriatic	(49)	189	-	
cis-Nemorensic	(50)	oil	+49	(EtOH)
trans-Nemorensic	(51)	173-177	-	
Petasinecic	(52)	178-180	-	
Syneilesinolide A	(54)	133-134	-	
Syneilesinolide B	(55)	120-121	-	
Syneilesinolide C	(56)	85-86	-	
Ligularidenecic	(57)	131	+10.5	(EtOH)
Senecivernic	(58)	136-138	-	
Crotalaric	(60)	231-232	-	
Crotaverric	isomer of (61)	132-133	-22.3	(EtOH)

TABLE 3

MONOESTER ALKALOIDS

Alkaloid	Structure	Source	Ref.
Heliovicine	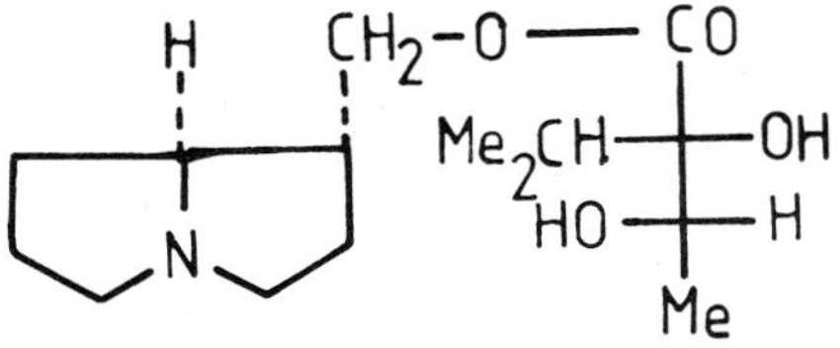 H; CH_2–O — CO; Me_2CH–C–OH; HO–C–H; Me	<u>Heliotropicum curassavicum</u>	1
Coromandalin	As above, but with opposite stereochemistry at C+	<u>H. curassavicum</u>	1
Curassavine	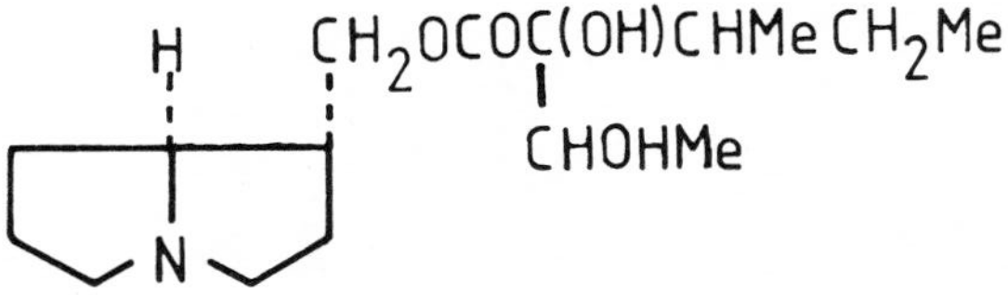 H; $CH_2OCOC(OH)CHMeCH_2Me$ with CHOHMe	<u>H. curassavicum</u>	1

TABLE 3 (CONTINUED)

Alkaloid	Structure	Source	Ref.
Petasinine	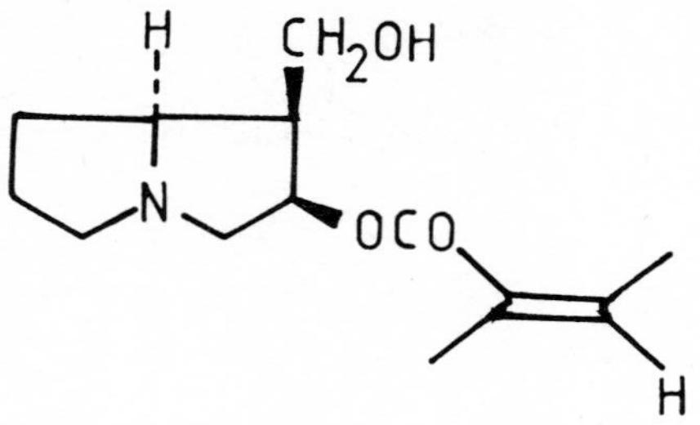 H, CH_2OH, N, OCO, H	*Petasites japonicus*	2
Fuchsisenecionine	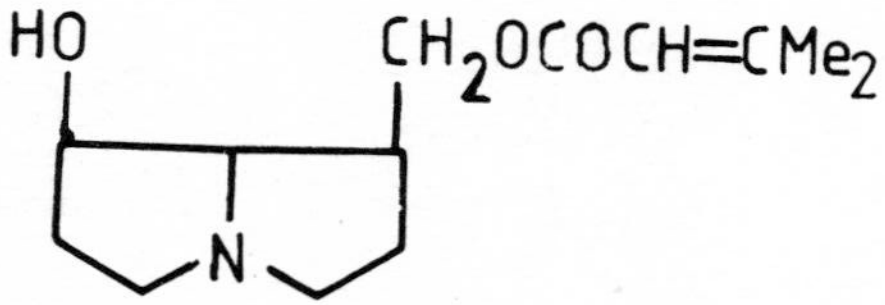 HO, $CH_2OCOCH{=}CMe_2$, N	*Senecio fuchsii*	3
Procerine (?)	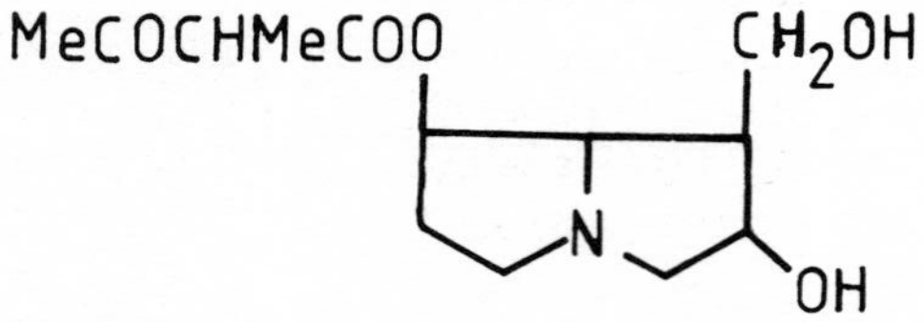 MeCOCHMeCOO, CH_2OH, N, OH	*S. procerus*	4

References to Table 3

1. S. Mohanraj *et al*., Chem. Comm., 1978, 423
2. K. Yamada *et al*., Tetrahedron Letters, 1978, 4543
3. E. Röder and H. Wiedenfeld, Phytochem., 1977, *16*, 1462
4. R.J. Jovceva *et al*., Coll. Czech. chem. Comm., 1978, *43*, 2312

TABLE 4

MONOESTER ALKALOIDS (CONTINUED)

Alkaloid	Molecular Formula	m.p. (°C)	$[\alpha]_D^{(0)}$	Necine	Necic Acid
Heliovicine	$C_{15}H_{25}NO_4$	gum	-2.7 (EtOH)	(-)-Trachelan-thamidine	(-)-Trachelanthic
Coromandalin	$C_{15}H_{25}NO_4$	gum	-6.9 (EtOH)	(-)-Trachelan-thamidine	(+)-Viridifloric
Curassavine	$C_{16}H_{29}NO_4$	-	+0.9 (EtOH)	(-)-Trachelan-thamidine	(+)-Homoviridi-floric
Curassavine N-oxide	$C_{16}H_{29}NO_5$	186-188	-6.6 (EtOH)	(-)-Trachelan-thamidine	(+)-Homoviridi-floric
Petasinine	$C_{13}H_{21}NO_3$	powder	+16 (EtOH)	Petasinecine	Angelic
Fuchsisene-cionine	$C_{13}H_{21}NO_3$	-	-120 (-)	Platynecine or isomer	Senecioic
Procerine	$C_{13}H_{18}NO_5$	238-290	-	-	-
Indicine N-oxide	$C_{15}H_{25}NO_6$	119-120	+34 (EtOH)	(+)-Retronecine	(-)-Trachelanthic

TABLE 5

ACYCLIC DIESTERS

Alkaloid	Structure	Source	Ref.
	R^1O, H, CH_2OR^2 (N)		
7-Acetylechinatine	R^1 = Ac, R^2 = –CO; HO–C–$CHMe_2$; HO–C–H; Me	*Lindelofia spectabilis*	1
Acetyllasiocarpine	R^1 = Angelyl, R^2 = –COC(OH)CHOMeMe with CMe_2OAc on the C(OH)	*Heliotropium europaeum*	2

TABLE 5 (CONTINUED)

Alkaloid	Structure	Source	Ref.
7-Angelylheliotrine	R^1 = Angelyl, R^2 = —CO; HO—C—CHMe$_2$; H—C—OMe; Me	<u>H. eichwaldii</u>	1
	R^{1}O, H, CH$_2$OR^2, N		
7-Acetylintermedine	R^1 = Ac, R^2 = —CO; HO—C—CHMe$_2$; H—C—OH; Me	<u>Symphytum x uplandicum</u>	3

TABLE 5 (CONTINUED)

Alkaloid	Structure	Source	Ref.
7-Acetyllycopsamine	R^1 = Ac, R^2 = –CO; HO–C–CHMe$_2$; HO–C–H; Me	<u>S. x uplandicum</u>	3
Symlandine	R^1 = Angelyl, R^2 as above	<u>S. x uplandicum</u>	3
Uplandicine	R^1 = Ac, R^2 = COC(OH)CHMe$_2$CHOHMe	<u>S. x uplandicum</u>	3
Uluganine	$Me_2C(OH)CH_2CO_2$ … $CH_2OCOC(OH)CHMe_2CHOHMe$	<u>Ulugbekia tschimganica</u>	4

TABLE 5 (CONTINUED)

Alkaloid	Structure	Source	Ref.
	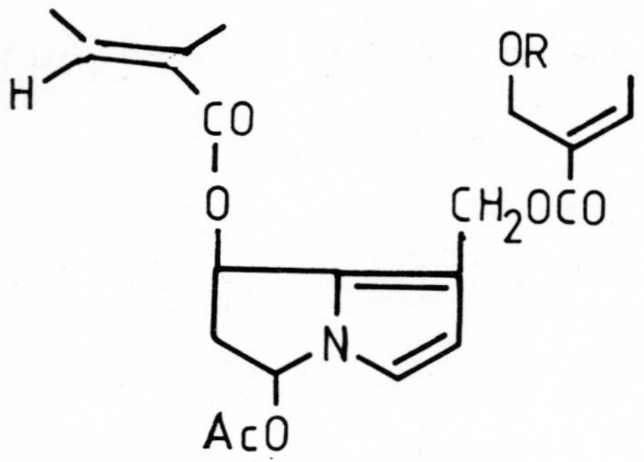		
Senampeline A	R = Senecioyl } mixture	<u>Senecio cissampelinus</u>	5
Senampeline B	R = Tiglyl		

TABLE 5 (CONTINUED)

Alkaloid	Structure	Source	Ref.
Senampeline C	R^1 = Tiglyl, R^2 = Senecioyl } mixture	*S. cissampelinus*	5
Senampeline D	R^1 = R^2 = Tiglyl } mixture	*S. mikanoides*	6
Senampeline E	R^1 = Angelyl, R^2 = Tiglyl } mixture	*S. mikanoides*	6
Senampeline F	R^1 = Angelyl, R^2 = Senecioyl } mixture	*S. mikanoides*	6
Senampeline G	R^1 = R^2 = Angelyl } mixture	*S. mikanoides*	6

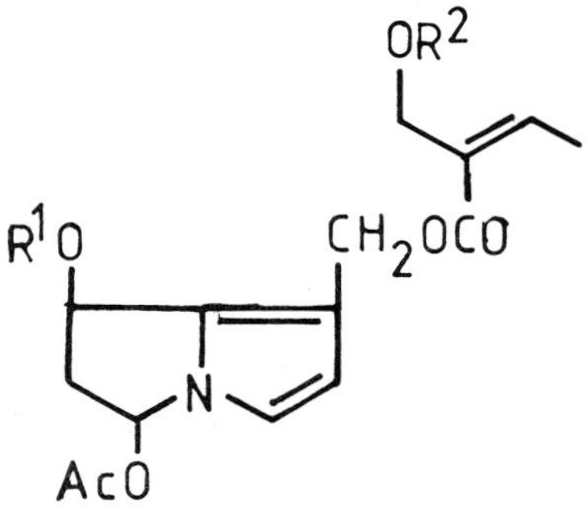

References to Table 5

1. O.P. Suri, R.S. Sawhney, and C.K. Atal, Indian J. Chem., 1975, 13, 505
2. C.C.J. Culvenor, S.R. Johns, and L.W. Smith, Austral. J. Chem., 1975, 28, 2319
3. Culvenor *et al*., *ibid*., 1980, 33, 1105
4. M.A. Khasanova, *et al*., Khim. Prirod. Soedinenii, 1974, 10, 809
5. F. Bohlmann, C. Zdero, and M. Grenz, Ber., 1977, 110, 474
6. Bohlmann *et al*., Phytochem., 1979, 18, 79

TABLE 6

ACYCLIC DIESTERS (CONTINUED)

Alkaloid	Molecular Formula	m.p. (°C)	$[\alpha]_D^{(0)}$	Necine	Necic Acid
7-Acetyl-echinatine	$C_{17}H_{27}NO_6$	-	-	Heliotridine	Acetic and viridifloric
Acetyl-lasiocarpine	$C_{23}H_{35}NO_8$	gum	-0.9 (EtOH)	Heliotridine	Acetic, angelic, and lasiocarpic
7-Angelyl-heliotrine	$C_{21}H_{33}NO_6$	191-193 picrate	-13.8 (EtOH)	Heliotridine	Angelic and heliotric
7-Acetyl-intermedine	$C_{17}H_{27}NO_6$	-	-	Retronecine	Acetic and (+)-trachelanthic
7-Acetyl-lycopsamine	$C_{17}H_{27}NO_6$	-	-	Retronecine	Acetic and (-)-viridifloric
Symlandine	$C_{20}H_{31}NO_6$	-	-	Retronecine	Angelic and (-)-viridifloric
Uplandicine	$C_{17}H_{27}NO_6$	oil	+0.1 (EtOH)	Retronecine	Acetic and echimidinic
Uluganine	$C_{20}H_{33}NO_7$	106-107	-32 (Me_2CO)	Heliotridine or Retronecine	2-Hydroxyisovaleric and trachelanthic

TABLE 7

CYCLIC DIESTERS

Alkaloid	Structure	Source	Ref.
(a) 11-membered rings			
Crocandine and isocrocandine	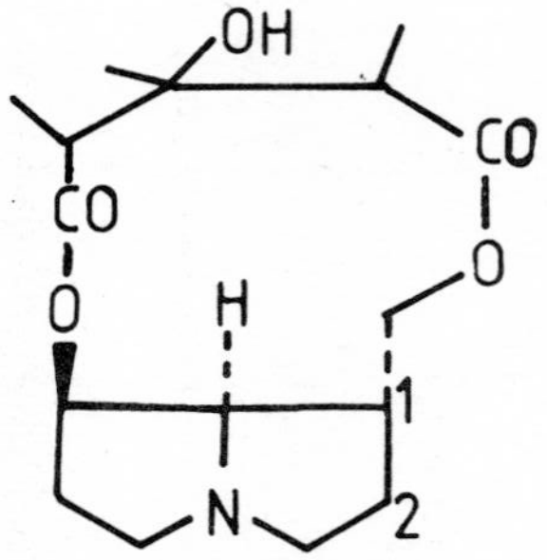	*Crotalaria candicans*	1
Cromadurine and isocromadurine	as above with 1,2-double bond	*C. madurensis*	2 3

TABLE 7 (CONTINUED)

Alkaloid	Structure	Source	Ref.
Croalbidine	HO, OH, CO, O, H, N, OH	*Crotalaria albida*	4
Crotalarine (= croburhine)	OH, OH, CO, O, H, N	*C. burhia*	5

TABLE 7 (CONTINUED)

Alkaloid	Structure	Source	Ref.
Cronaburmine	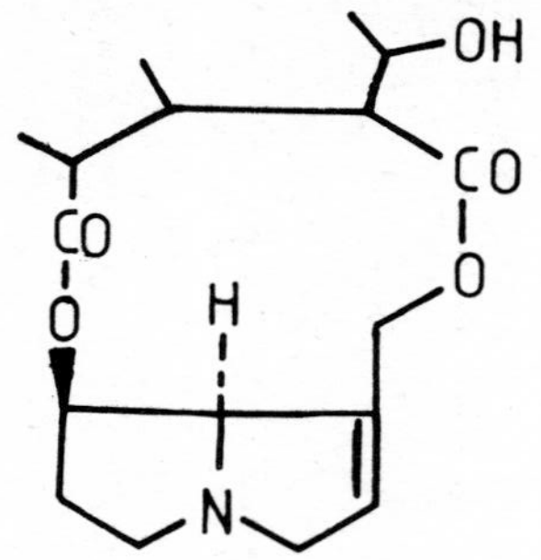	<u>Crotalaria nana</u>	6
Monocrotalinine (Artefact?)	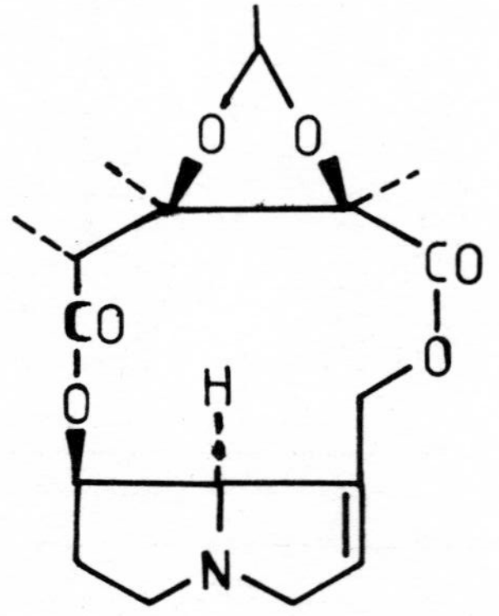	<u>C. grahamiana</u>	7

TABLE 7 (CONTINUED)

Alkaloid	Structure	Source	Ref.
Axillarine	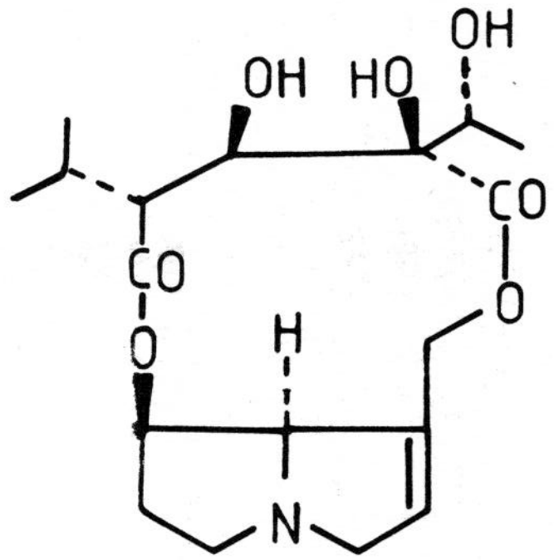 OH, HO, OH, CO, CO, O, O, H, N	*Crotalaria axillaris*	8
Incanine	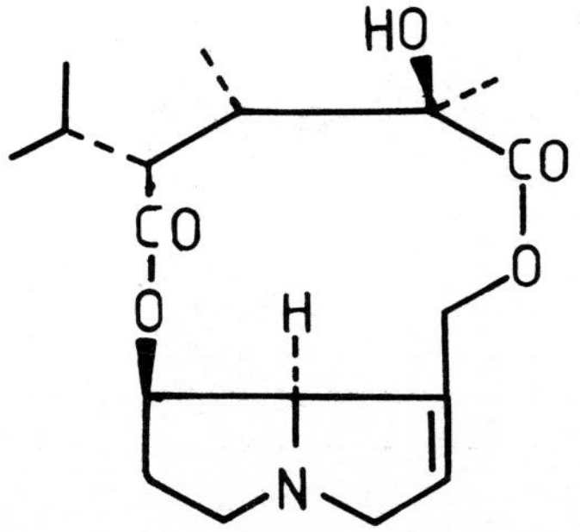 HO, CO, CO, O, O, H, N	*Trichodesma incanum*	9

TABLE 7 (CONTINUED)

Alkaloid	Structure	Source	Ref.
(b) 12-membered rings			
Crotastriatine	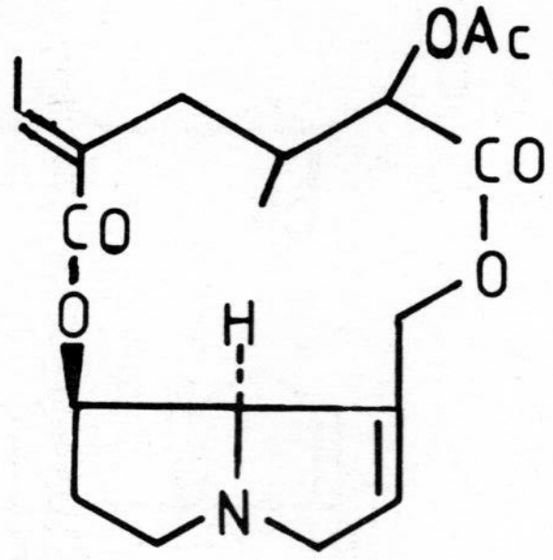	*Crotalaria striata*	10
Crotananine	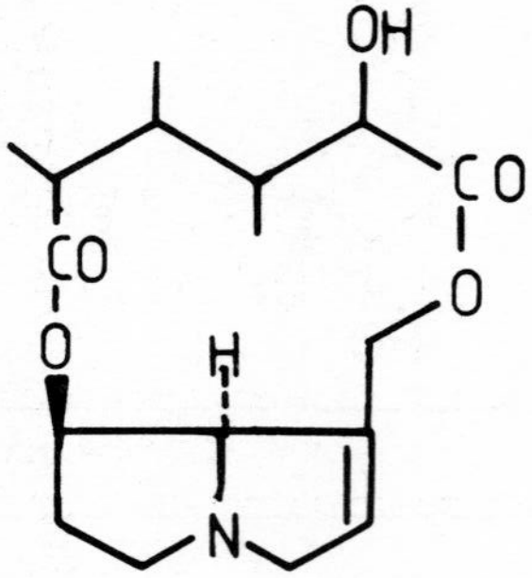	*C. nana*	11

TABLE 7 (CONTINUED)

Alkaloid	Structure	Source	Ref.
Senecivernine	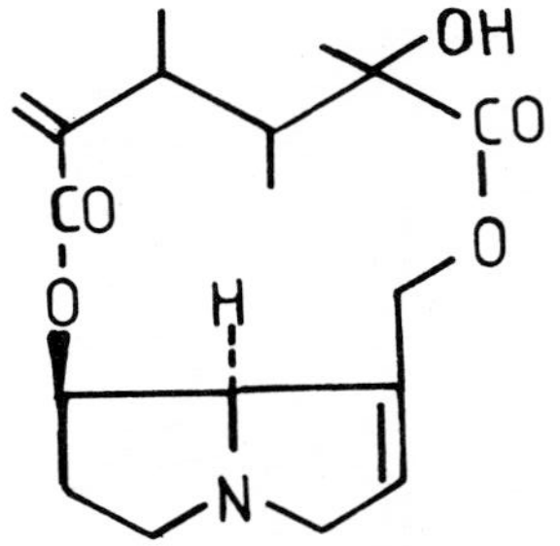	*Senecio vernalis*	12
Yamataimine	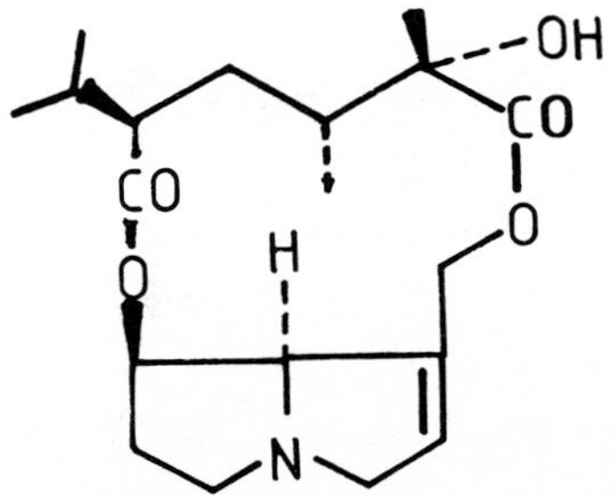	*Cacalia yatabei*	13

TABLE 7 (CONTINUED)

Alkaloid	Structure	Source	Ref.
Senkirkine $R^1 = R^2 = H$	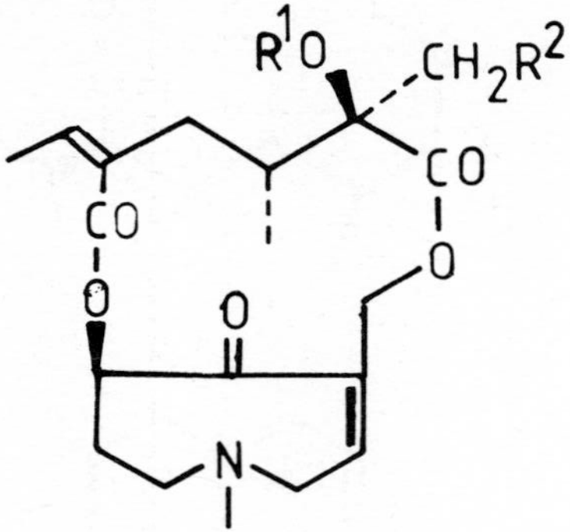 R^1O, CH_2R^2	*Senecio kirkii*	14
O-Acetylsenkirkine $R^1 = Ac, R^2 = H$		*Crotalaria walkeri*	15
Hydroxysenkirkine $R^1 = H, R^2 = OH$		*C. laburnifolia*	16
Neosenkirkine geometrical isomer of senkirkine		*S. auricola*	17
Ligularidine	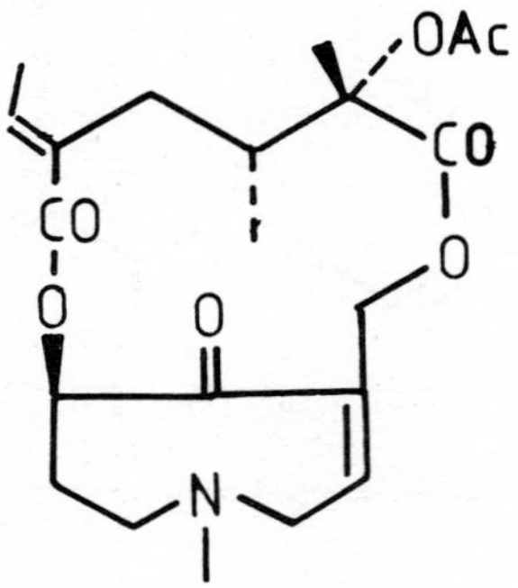 OAc	*Ligularia dentata*	18

TABLE 7 (CONTINUED)

Alkaloid	Structure	Source	Ref.
Crotaverrine R = H		*Crotalaria verrucosa*	19
O-Acetylcrotaverrine R = Ac		and *C. walkeri*	20
Syneilesine R = H		*Syneilesis palmata*	21, 22
Acetylsyneilesine R = Ac		*Syneilesis palmata*	22

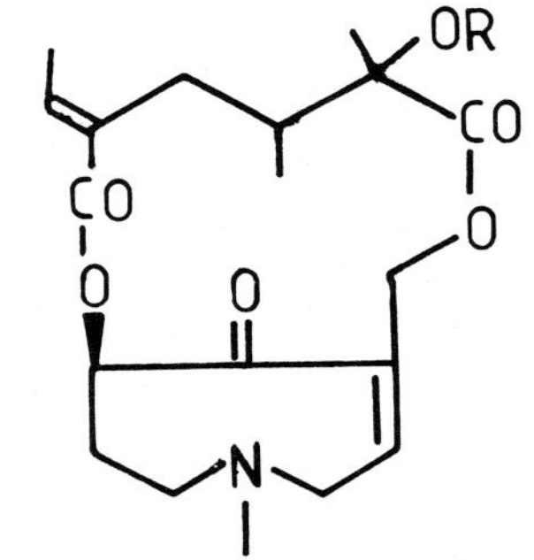

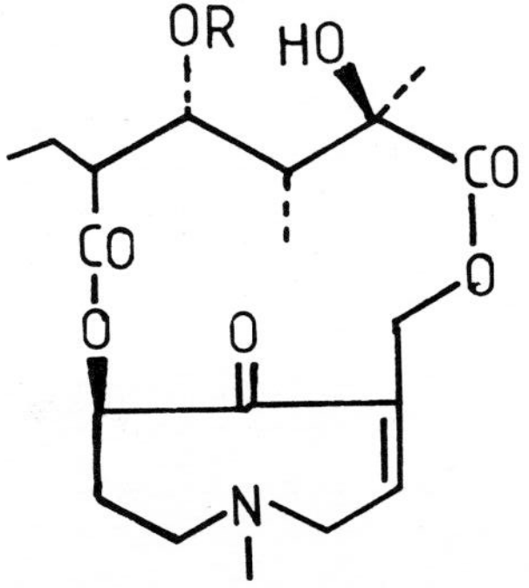

TABLE 7 (CONTINUED)

Alkaloid	Structure	Source	Ref.
Otosenine	R^1, R^2 = α-epoxide, R^3 = H	*Doronicum macrophyllum*	23
Doronine	R^1 = Cl, R^2 = OH, R^3 = Ac	*D. macrophyllum*	23
Petasitenine	R^1, R^2 = β-epoxide, R^3 = H	*Petasites japonicus*	24
(= Fukinotoxin)		*P. japonicus*	25
Neopetasitenine	R^1, R^2 = β-epoxide, R^3 = Ac	*P. japonicus*	24

TABLE 7 (CONTINUED)

Alkaloid	Structure	Source	Ref.
Senaetnine R^1 = H, R^2 = Me	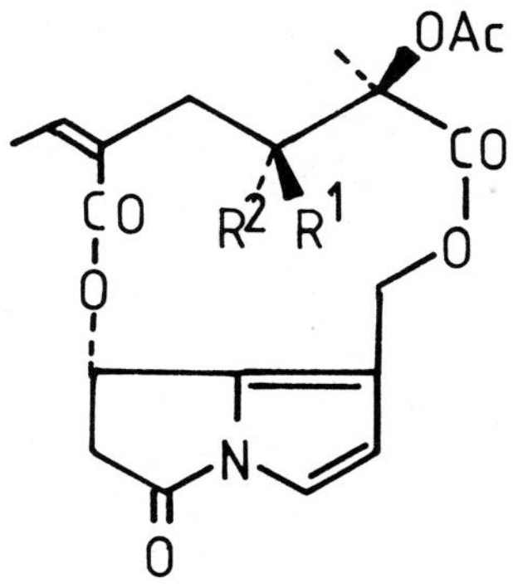	_Senecio aetnensis_	26
Geometrical isomer		_S. aucheri_	27
Dehydrosenaetnine R^1, R^2 = CH_2		_S. barbertonicus_	28
Isosenaetnine R^1 = H, R^2 = Me	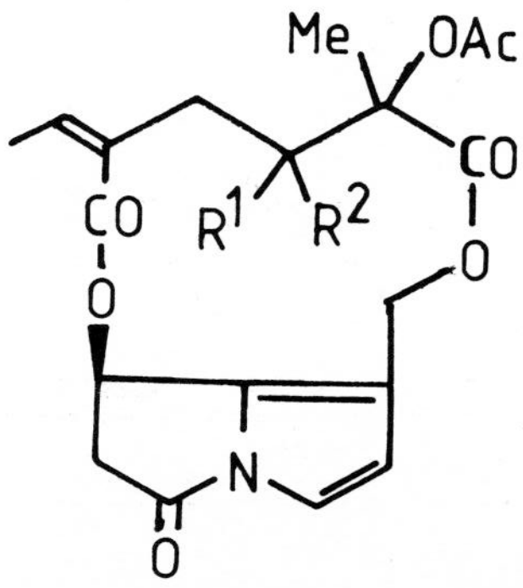	_Kleinia kleinioides_	29
Dehydroisosenaetnine R^1, R^2 = CH_2		_K. kleinioides_	29

TABLE 7 (CONTINUED)

Alkaloid	Structure	Source	Ref.
Pterophorine	OAc, CO, O, CO, O, 7, N, O	*Senecio pterophorus*	30
		S. inaequidens	26
Isopterophorine (C-7 epimer)		*S. pulviniformis*	31
Inaequidenine	CO, O, CO, O, CO, O, N, O	*S. inaequidens*	26

TABLE 7 (CONTINUED)

Alkaloid	Structure	Source	Ref.
(c) 13-membered rings			
Bulgarsenine	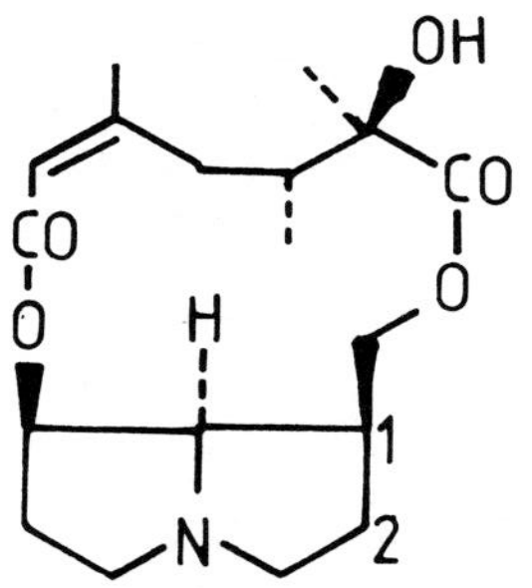	_Senecio_ _nemorensis_ _S. doronicum_	30 31
Doronenine (1,2-double-bond)		_S. doronicum_	33,34
Nemorensine	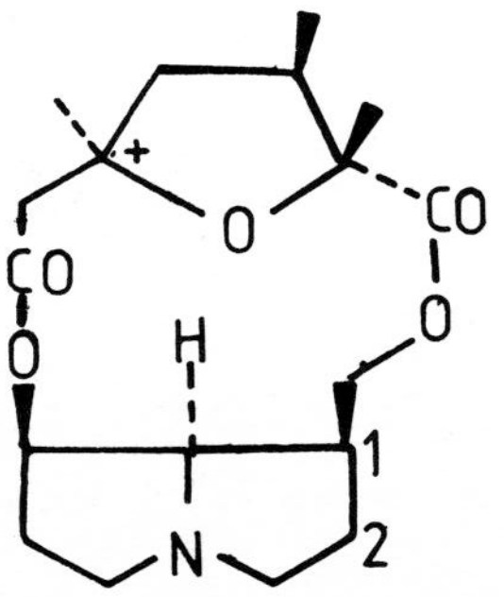	_S. nemorensis_	32,35
Retroisosenine (1,2-double-bond and opposite stereochemistry at C^+)		_S. nemorensis_	32

TABLE 7 (CONTINUED)

Alkaloid	Structure	Source	Ref.
(d) 14-membered rings			
Parsonsine	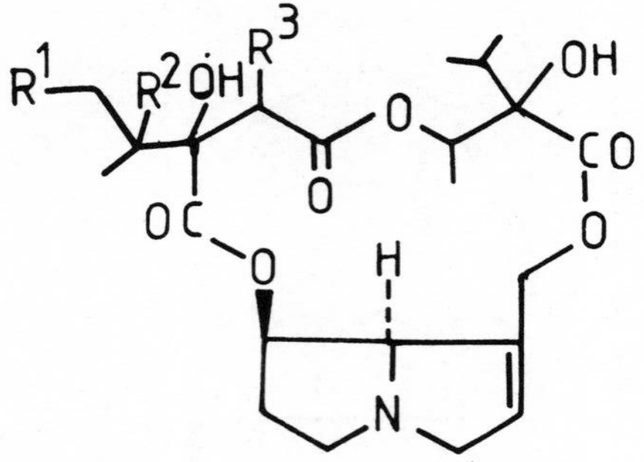	*Parsonsia heterophylla*	36,37
$R^1 = R^2 = R^3 = H$		*P. spiralis*	36,37
Heterophylline		*P. heterophylla*	
$R^1 = Me$, $R^2 = R^3 = H$		and *P. spiralis*	36
Spiraline		*P. spiralis*	36
$R^1 = H$, $R^2 = R^3 = OH$			
Spiranine		*P. spiralis*	36
$R^1 = Me$, $R^2 = H$, $R^3 = OH$			
Spiracine		*P. spiralis*	36
$R^1 = Me$, $R^2 = R^3 = OH$			

References to Table 7

1. M.A. Siddiqi et al., Phytochem., 1979, 18, 1413
2. P.G. Rao, R.S. Sawhney, and C.K. Atal, Indian J. Chem., 1975, 13, 870
3. Rao, Sawhney, and Atal, Experientia, 1975, 31, 878
4. Sawhney et al., Austral. J. Chem., 1974, 27, 1805
5. M.A. Ali and G.A. Adil, Pakistan J. Sci. Ind. Res., 1973, 16, 227; Rao, Sawhney and Atal, Indian J. Chem., 1975, 13, 835
6. Siddiqi et al., Indian J. Chem., 1978, 16B, 1132
7. T.R. Rajagopalan and V. Batra, Indian J. Chem., 1977, 15B, 455
8. H. Stoeckli-Evans and D.H.G. Crout, Helv., 1976, 59, 2168
9. B. Tashkhodzhaev, M.V. Telezhenetskaya, and S. Yu. Yunusov, Chem. Abs., 1980, 92, 111 199
10. Batra, R.N. Gandhi, and Rajagopalan, Indian J. Chem., 1975, 13, 989
11. Siddiqi et al., Phytochemistry, 1978, 17, 2143
12. E. Roder, H. Wiedenfeld, and U. Pastewka, Planta Med., 1979, 37, 131
13. M. Hikichi, T. Furuya, and Y. Iitaka, Tetrahedron Letters, 1978, 767
14. G.I. Birnbaum, J. Amer. chem. Soc., 1974, 96, 6165
15. Atal and Sawhney, Indian J. Pharm., 1973, 35, 1
16. D.H.G. Crout, J. Chem. Soc. Perkin I, 1972, 1602
17. F.M. Panizo and B. Rodriguez, Anales de Quim., 1974, 70, 1043
18. M. Hikichi, Y. Asada, and T. Furuya, Tetrahedron Letters, 1979, 1233
19. O.P. Suri et al., Phytochem., 1976, 15, 1061
20. K.A. Suri, R.S. Sawhney, and C.K. Atal, Indian J. Chem., 1976, 14B, 471
21. Hikichi and Furuya, Tetrahedron Letters, 1974, 3657
22. Hikichi and Furuya, Chem. pharm. Bull. Japan, 1976, 24, 3178
23. Sh. A. Alieva et al., Chem. Abs., 1976, 85, 108 841
24. K. Yamada et al., Chem. Letters, 1976, 461
25. Furuya, Hikichi, and Iitaka, Chem. pharm. Bull. Japan, 1976, 24, 1120
26. F. Bohlmann et al., Phytochem., 1977, 16, 965

References to Table 7 (continued)

27. Bohlmann et al., ibid, 1979, 18, 79
28. Bohlmann, C. Zdero and G. Snatzke, Ber., 1978, 111, 3009
29. Bohlmann and K.-H. Knoll, Phytochem., 1978, 17, 599
30. Bohlmann, Zdero and M. Grenz, Ber., 1977, 110, 474
31. Bohlmann and Zdero, Phytochem., 1979, 18, 125
32. A. Klasek et al., Coll. Czech. chem. Comm., 1980, 45, 548; N.T. Nghia et al., ibid, 1976, 41, 2952
33. Roder, Wiedenfeld, and M. Frisse, Phytochem., 1980, 19, 1275
34. A. Kirfel et al., Cryst. Struct. Comm., 1980, 9, 353
35. Klasek et al., Coll. Czech. chem. Comm., 1973, 38, 2504
36. J.A. Edgar et al., Tetrahedron Letters, 1980, 2657
37. N.J. Eggers and G.J. Gainsford, Cryst. Struct. Comm., 1979, 8, 597; Gainsford, ibid, 1980, 9, 173

TABLE 8

CYCLIC DIESTERS (CONTINUED)

Alkaloid	Molecular Formula	m.p. (°C)	$[\alpha]_D^{(0)}$	Necine	Necic Acid
Croalbidine	$C_{18}H_{29}NO_7$	208-209	-	Croalbinecine	Trichodesmic
Crocandine	$C_{16}H_{25}NO_5$	244-246	+130 (MeOH)	Turneforcidine	Fulvinic
Isocrocandine	$C_{16}H_{25}NO_5$	172-174	+36 (MeOH)	Turneforcidine	Cromaduric
Cromadurine	$C_{16}H_{23}NO_5$	242-243	-	Retronecine	Cromaduric
Isocromadurine	$C_{16}H_{23}NO_5$	135-136	+43 (EtOH)	Retronecine	Isocromaduric
Crotalarine	$C_{18}H_{27}NO_6$	167-168	-80 (EtOH)	Retronecine	Crotalaric
Cronaburmine	$C_{17}H_{25}NO_5$	133-134	-	Retronecine	Cronaburmic lactone
Monocrotalinine	$C_{18}H_{25}NO_6$	160-161	+125 (EtOH)	Retronecine	-
Crotastriatine	$C_{19}H_{25}NO_6$	-	-	Retronecine	Striatic
Crotananine	$C_{17}H_{25}NO_5$	174-175	-80 (MeOH)	Retronecine	Crotananic
Senecivernine	$C_{18}H_{25}NO_5$	105-107	-35 (EtOH)	Retronecine	Senecivernic

TABLE 8 (CONTINUED)

Alkaloid	Molecular Formula	m.p.(°C)	$[\alpha]_D^{(0)}$	Necine	Necic Acid
Yamataimine	$C_{18}H_{27}NO_5$	181-182	+64 (EtOH)	Retronecine	-
Senkirkine	$C_{19}H_{27}NO_6$	196-198	-	Otonecine	Senecic
O-Acetyl-senkirkine	$C_{21}H_{29}NO_7$	195-196	-34 (MeOH)	Otonecine	-
Hydroxy-senkirkine	$C_{21}H_{29}NO_8$	124-125	+5 (EtOH)	Otonecine	Isatinecic
Neosenkirkine	$C_{19}H_{27}NO_6$	223-225	+27 ($CHCl_3$)	Otonecine	Integerrinecic
Ligularidine	$C_{21}H_{29}NO_7$	196	-50 (EtOH)	Otonecine	Ligularidenecic
Crotaverrine	$C_{19}H_{27}NO_6$	142-144	+35 (EtOH)	Otonecine	Integerrinecic isomer
O-Acetyl-crotaverrine	$C_{21}H_{29}NO_7$	oil	+45 (MeOH)	Otonecine	O-Acetyl derivative
Syneilesine	$C_{19}H_{29}NO_7$	194-195	-	Otonecine	Syneilesinolides A-C
Acetyl-syneilesine	$C_{21}H_{31}NO_8$	oil	-	Otonecine	-

TABLE 8 (CONTINUED)

Alkaloid	Molecular Formula	m.p.(°C)	$[\alpha]_D^{(0)}$	Necine	Necic Acid
Otosenine	$C_{19}H_{27}NO_7$	-	-	Otonecine	-
Doronine	$C_{21}H_{30}NO_8Cl$	113-114	+45 ($CHCl_3$)	Otonecine	
Petasitenine	$C_{19}H_{27}NO_7$	129-130	+44 (EtOH) +64 ($CHCl_3$)	Otonecine	Petasinecic
Neopetasitenine	$C_{21}H_{29}NO_8$	-	+49 (EtOH)	Otonecine	O-Acetyl-petasinecic
Senaetnine	$C_{20}H_{23}NO_7$	183-185	+11 (-)	1α-hydroxy-dihydro-pyrrolizinone	O-Acetylsenecic
Dehydro-senaetnine	$C_{20}H_{21}NO_7$	oil	+147 (-)		
Isosenaetnine	$C_{20}H_{23}NO_7$	198.5	-35 (-)	1β-hydroxy-dihydro-pyrrolizinone	
Dehydroiso-senaetnine	$C_{20}H_{21}NO_7$	oil	-46 (-)		
Pterophorine	$C_{20}H_{23}NO_7$	oil	+27 (-)		
Isopterophorine	$C_{20}H_{23}NO_7$	oil	-25 ($CHCl_3$)		
Inaequidenine	$C_{21}H_{25}NO_7$	oil	-		

TABLE 8 (CONTINUED)

Alkaloid	Molecular Formula	m.p.(°C)	$[\alpha]_D^{(0)}$	Necine	Necic Acid
Bulgarsenine	$C_{18}H_{27}NO_5$	114-115	-74 (EtOH) -56 ($CHCl_3$)	Platynecine	
Doronenine	$C_{18}H_{25}NO_5$	124-127	+123 (EtOH)	Retronecine	
Nemorensine	$C_{18}H_{27}NO_5$	132-133	-57 ($CHCl_3$)	Platynecine	Nemorensic
Nemorensine *N*-oxide	$C_{18}H_{28}NO_6$	160-163	-35 ($CHCl_3$)		
Retroisosenine	$C_{18}H_{26}NO_5$	126-127	+117 ($CHCl_3$)	Retronecine	*cis*-Nemorensic
Parsonsine	$C_{22}H_{33}NO_8$	158-198	+20 (MeOH)	Retronecine	Trachelanthic and 2-isopropylmalic
Heterophylline	$C_{23}H_{35}NO_8$	190	-	Retronecine	Trachelanthic and 2-S-butylmalic
Spiraline	$C_{22}H_{33}NO_9$	-	-	Retronecine	
Spiranine	$C_{23}H_{35}NO_9$	-	-	Retronecine	
Spiracine	$C_{23}H_{35}NO_{10}$	-	-	Retronecine	

TABLE 9

AROMATIC ESTERS

Alkaloid	Structure	Source	Ref.
Retronecine-7,9-dibenzoate		*Caccinia glauca*	1
Petasinoside		*Petasites japonicus*	2
Ehretinine		*Ehretia aspera*	3

References to Table 9

1. Siddiqi *et al.*, Phytochem., 1978, *17*, 2049
2. Yamada *et al.*, Tetrahedron Letters, 1978, 4543
3. Suri *et al.*, Phytochem., 1980, *19*, 1273

TABLE 10

AROMATIC ESTERS (CONTINUED)

Alkaloid	Molecular Formula	m.p. (°C)	$[\alpha]_D^{(0)}$	Necine
Retronecine-7,9-dibenzoate	$C_{22}H_{21}NO_4$	oil	-	Retronecine
Petasinoside	$C_{28}H_{37}NO_9$	-	-38 (EtOH)	Petasinecine
Ehretinine	$C_{16}H_{21}NO_3$	187-188	-108 (MeOH)	Retronecanol

4. Miscellaneous types of pyrrolizidine alkaloids

Four more alkaloids have been isolated from the seeds of *Lolium cuneatum*; *N*-formylnorloline (62) (E.Kh. Batirov, V.M. Malikov, and S.Yu. Yunusov, Chem. Abs., 1977, 86, 13 792), *N*-methyl-loline (63), *N*-acetylnorloline (64), and *N*-formylloline (65) (Batirov *et al.*, *ibid*, 1976, 85, 59 556). A dimeric pyrrolizidine alkaloid containing chlorine has also been isolated in trace amounts. The provisional structure (66) was assigned to lolidine (Batirov, Malikov, and Yunusov, *ibid*, 1976, 85, 74 883).

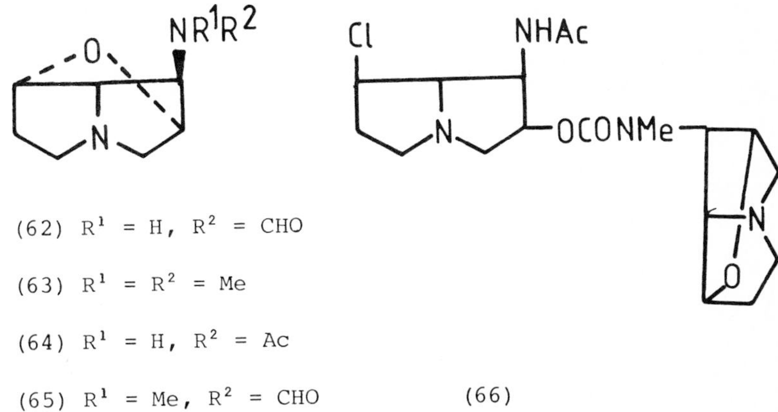

Nitropolyzonamine, $C_{13}H_{22}N_2O_2$, from the defensive secretions of the millipede *Polyzonium rosalbum* has been shown by X-ray analysis to have the opposite absolute configuration to that previously suggested (R.W. Miller and A.T. McPhail, J. Chem. Res. (S), 1978, 76).

A revised structure for peduncularine, $C_{20}H_{24}N_2$, found in *Aristotelia peduncularis*, shows that it is a monoterpene indole alkaloid, and does not contain a pyrrolizidine nucleus (H.-P. Ros *et al.*, Helv., 1979, 62, 481).

Clazamycins A (67) and B (68) have been isolated from *Streptomyces* species (Y. Oriuchi *et al.*, J. Antibiotics, 1979, 32, 762; L.A. Dolak and C. Deboer, *ibid*, 1980, 33, 83). The structure of clazamycin A has been established by X-ray crystallography (H. Nakamura, Y. Iitaka, and H. Umezawa, *ibid*,

1979, 32, 765).

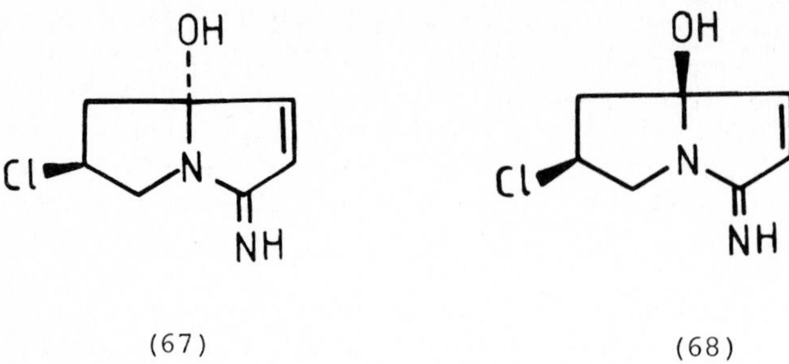

(67) (68)

TABLE 11

MISCELLANEOUS ALKALOIDS: PHYSICAL PROPERTIES

Alkaloid	m.p. (°C)	$[\alpha]_D^{(0)}$
N-formylnorloline	-	+31.3 (Me_2CO)
N-methyl-loline	-	+13.7 (Me_2CO)
N-Acetylnorloline	-	+49.8 ($CHCl_3$)
N-Formyl-loline	93-94	+47.9 ($CHCl_3$)
Lolidine	225-226	+146 ($CHCl_3$)

5. Pharmacology of the pyrrolizidine alkaloids

This area has been recently reviewed (R.J. Huxtable, Gen. Pharmacol., 1979, 10, 159). Ingestion of pyrrolizidine alkaloids by humans is now a considerable health problem, due to the hepatotoxicity and carcinogenicity shown by many of these alkaloids. Accidental poisoning has occurred by consumption of foodstuffs contaminated with seeds of various plants that produce pyrrolizidine alkaloids (O. Mahabbat *et al.*, Lancet, 1976, *2*, 269; Tandon *et al.*, *ibid.*, 1976, *2*, 271). The continued deliberate use of plants containing pyrrolizidine alkaloids as herbal remedies and teas (e.g. coltsfoot, comfrey, etc.) should be strongly discouraged (I. Hirono, H. Mori, and M. Haga, J. Nat. cancer Inst., 1978, *61*, 865). Losses of livestock from pyrrolizidine alkaloid poisoning are increasing ('Effects of Poisonous Plants on Livestock', ed. R.F. Keeler, K.R. Van Kampen, and L.F. James, Academic Press, New York, 1978).

Pyrrolizidine alkaloids are converted into pyrrole derivatives in the liver. These derivatives are highly toxic, probably acting by alkylating sulphydryl groups (R. Schoental, F.E.B.S. Letters, 1976, *61*, 111). Hence the most obvious toxic action is on the liver.

Chapter 9

THE INDOLE ALKALOIDS

K.S.J. STAPLEFORD

Introduction

Since the publication in 1977 of Vol. IV B, there have been advances in the knowledge of this class of natural products. An area of particular interest concerns fungal metabolites, with a variety of complex systems derived from tryptophan having been isolated. A number of halogenated indoles have been isolated, mainly from marine sources.

In the elucidation of structures the increased availability of powerful spectroscopic methods has provided an increased impetus. A number of structures have been assigned from 400 HZ ^{1}H nmr studies and with the increase of information on ^{13}C nmr chemical shifts this technique finds still greater application. In many cases, however, the final structural proof relies upon X-ray crystallographic methods.

The publication of the Specialist Periodical Reports ("The Alkaloids", The Royal Society of Chemistry, London) continues to provide valuable information on the more recent developments and the established specialist publication "The Alkaloids", Academic Press, New York, now edited by R.G.A. Rodgrigo, gives comprehensive information on specific topics. Valuable reviews in this area have appeared in this particular series. Ergot Alkaloids, P.A. Stadler and P. Statz, Vol. XV, p.1; The *Aspidosperma* Alkaloids, G.A. Cordell, Vol. XV11, p.200; The Monoterpenoid Alkaloid Glycosides, R.S. Kapil and R.T. Brown, Vol. XV11, p.546; The Bisindole Alkaloids, Cordell and J.E. Saxton, Vol. XX, p.1; The Eburnamine-Vincamine Alkaloids, W. Döpke, Vol. XX, p.297.

The format of this chapter follows the original work (Vol. IV B) and considers structures and their determination with synthesis largely omitted.

1. *Alkaloids lacking a tryptamine unit*

(a) *Simple indoles*

Simple indoles substituted with units of isoprenoid origin (and hence not strictly alkaloids since they do not contain a basic nitrogen atom) continue to be isolated in a variety of forms. 3,6-Bis ($\gamma\gamma$- dimethylallyl) - indole has been isolated from the stem bark of *Uvaria elliatiana* (H. Achenbach and B. Raffelsberger, Tetrahedron Letters 1979, 2571). An indole sesquiterpene, *polyalthenol*, has been isolated from *Polyalthia oliveri*, a tropical African plant (M. Leboeuf *et al*, *ibid*, 1976, 3559). Presumably this substitution pattern is obtained by condensation of a sesquiterpene pyrophosphate with tryptophan, followed by extrusion of serine (or its biochemical equivalent) and acid catalysed cyclisation of the terpenoid moiety. Migration of a methyl group must also occur.

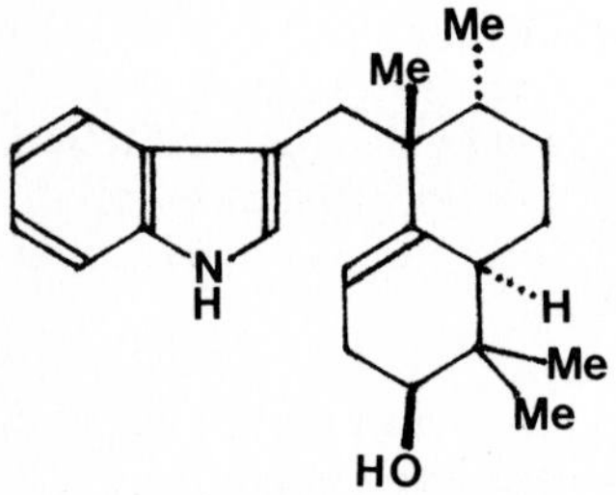

Polyalthenol

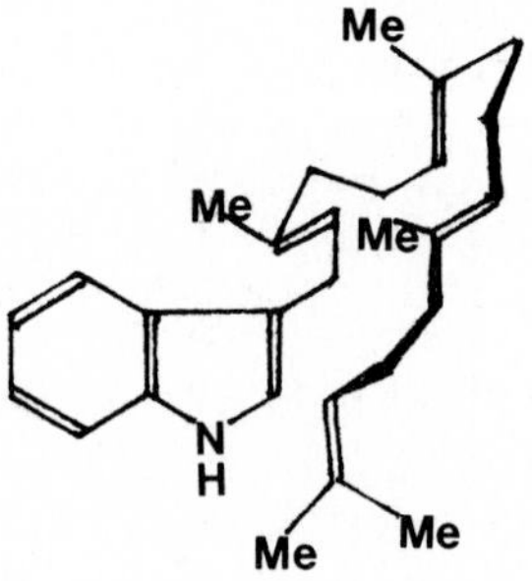

1

Introduction of geranyl-geranyl pyrophosphate by the same process would lead initially to the compound (1). Although this compound has not yet been isolated a number of cyclised analogues have been obtained from the fungus *Claviceps paspali*. The most obvious relative to these is aflavinine, where cyclisation has been accompanied by migration of a methyl group (R.T. Gallagher *et al. ibid.* 1980, 243).

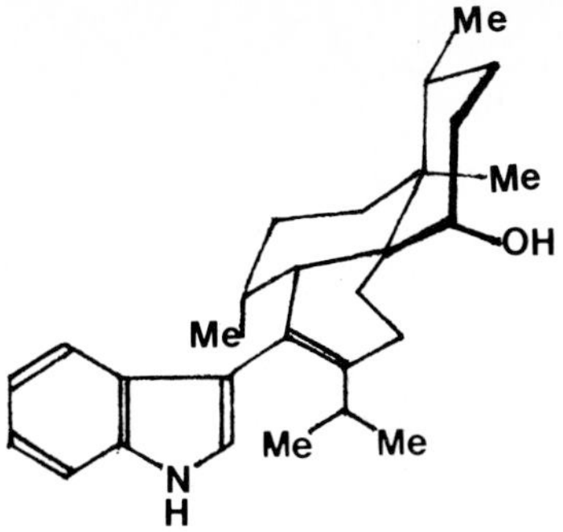

Aflavinine

Other compounds from the same source are paspaline, paspalicine, paspalinine and aflatrem, the latter containing an additional isopentenyl group attached to the aromatic ring.

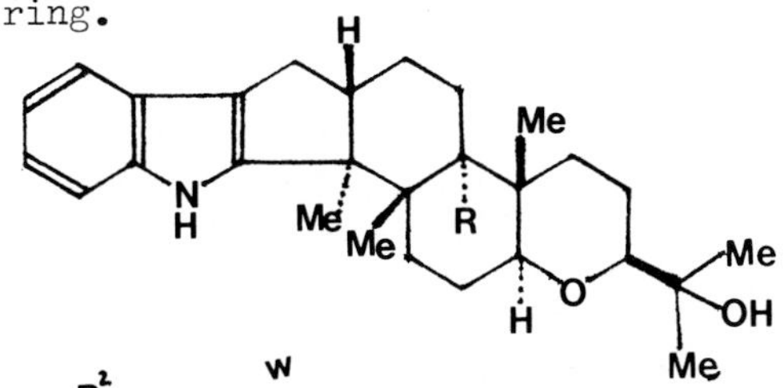

Paspaline R = H

Paxilline R = OH

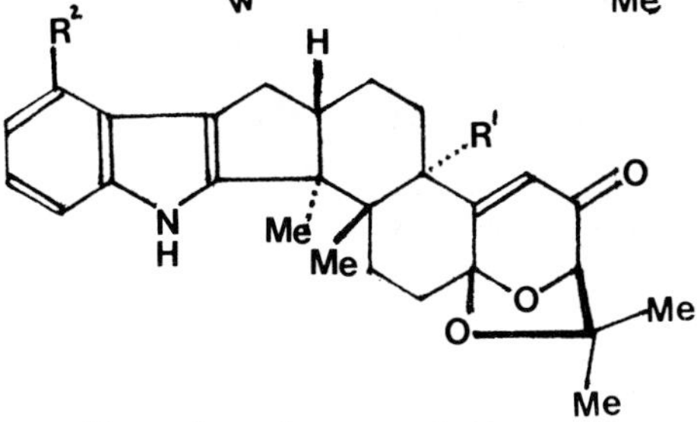

Paspalicine $R^1=R^2=H$

Paspalinine $R^1=OH$ $R^2=H$

Aflatrem $R^1=OH$
$R^2 = -CMe_2-CH = CH_2$

The structures of these compounds have been deduced from X-ray crystallagraphic studies (J.P. Springer and J. Clardy, Tetrahedron Letters, 1980, 231; R.T. Gallagher *et al., ibid.,* 1980, 235,239,243). Paxilline is a metabolite of *Penicillium paxilli* (R.J. Cole, J.W. Kirksey and J.M. Wells, Canad., J. Microbiol., 1974,20 1159). It is notable that those metabolites having an - OH group at the angular position (paspalinine, aflatrem and paxilline) have pronounced tremorgenic properties whereas paspaline is inactive.

Non-isopentenyl units are also evident. For example the phenyl-pentyl unit is obvious in melosatins A and B, isolated from the tumorigenic plant *Melochia tomentosa*(G.J. Kapadia *et al.*, Chem.Comm., 1977, 535).

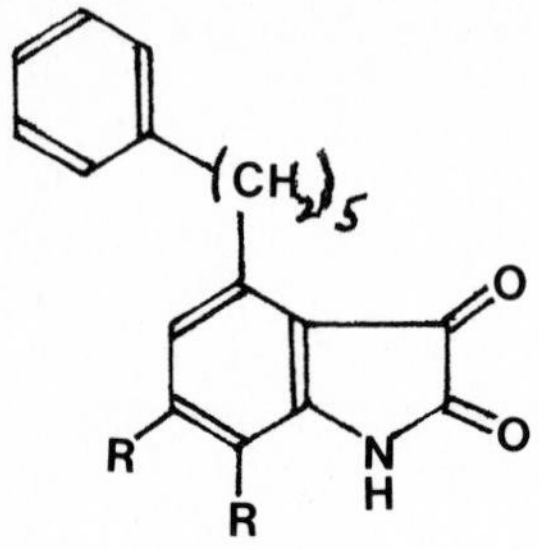

$(CH_2)_5$ O O N H R R

Melosatin A R = OMe

Melosatin B R = H

Melosatin A has been synthesised by the route shown below.

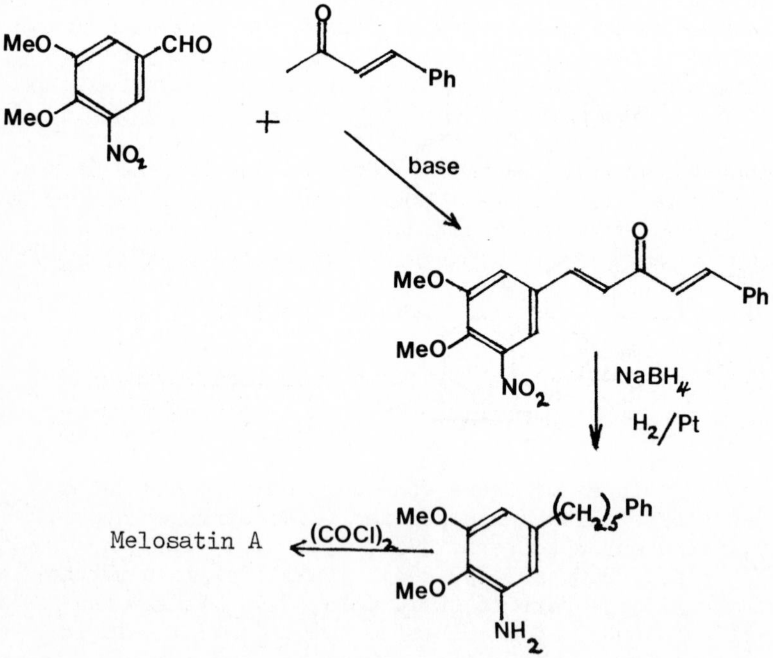

MeO CHO MeO NO_2 + O Ph base MeO MeO NO_2 O Ph $NaBH_4$ H_2/Pt MeO MeO NH_2 $(CH_2)_5Ph$ $(COCl)_2$ Melosatin A

A remarkable antibiotic obtained from *Streptomyces hygroscopicus* is neosidomycin. It is the first reported example of a naturally occurring indole *N*-glycoside and is also the first natural product to contain the 4 - deoxy - *ribo* - hexapyranuronate (R. Furuta, S. Naruto, A. Tamura and K. Yokogawa, Tetrahedron Letters, 1979, 1701).

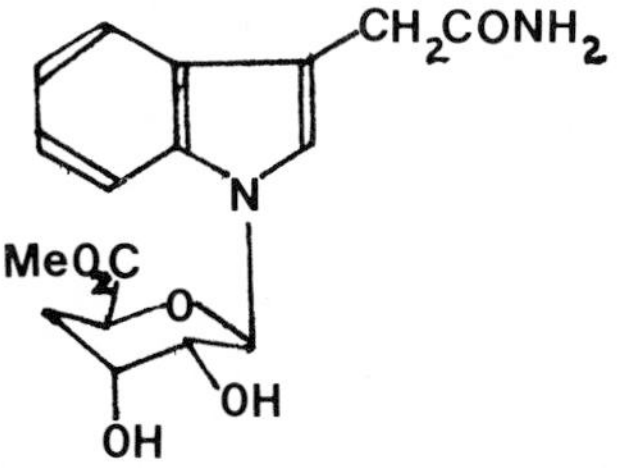

Neosidomycin

Pathogenic micro-organisms have provided a number of indole metabolites. *Balansia epichloe* is a clavicipitaceous fungus on pasture grass and produces ergot - like syndromes in cattle. It has been shown to contain a number of indole polyols, e.g. (2) to (4). The last two of these may be obtained by condensing glyceraldehyde with indole (J.K. Porter *et al.*, J. agric. food Chem., 1977, 25,88). A compound containing a second nitrogen but lacking the tryptamine unit is the keto-lactam (5) which has been isolated from the Caribbean sponge *Halichondria melanodocia* (Y. Gopichand and F.J. Schmitz, J. org. Chem., 1979, 44,4995).

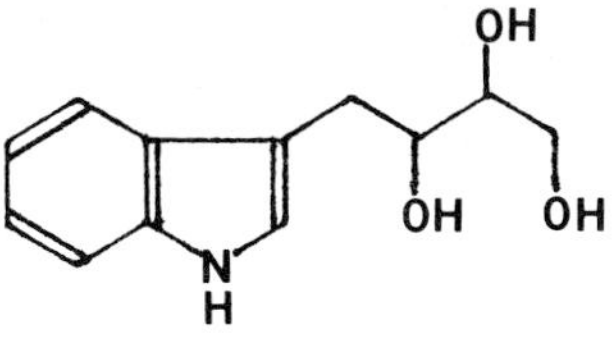

2

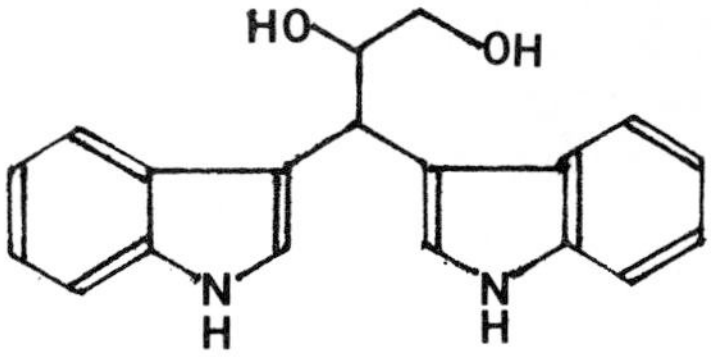

3

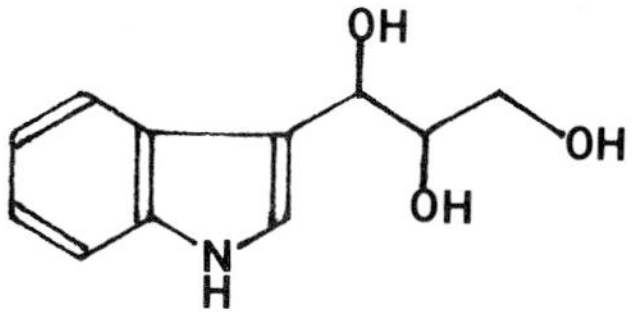

4

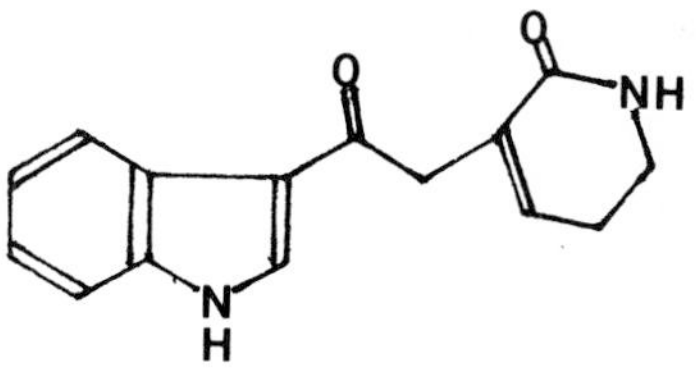

5

(b) *Carbazole derivatives*

3 - Methylcarbazole, believed to be a key intermediate in the biosynthesis of the simple carbazoles from *Rutaceae*, has been isolated from *Clausena heptaphylla* (P. Bhattacharyya and D.P. Chakraborty Phytochem., 1973, 12, 1831; S. Roy, Bhattacharyya and Chakraborty, *ibid.*, 1974, 13, 1017). A simple compound closely related to this is mukonine, isolated from that abundant source of carbazoles, *Murraya Koenigii*, the Indian curry-leaf plant. Mukonine is 1 - methoxy - 3 - carbomethoxy - carbazole and the acid has also been isolated from the same source (Chakraborty *et al*, *ibid.*, 1978, 17, 834). A simple isomer of this system is provided by mukonidine; in this case the aromatic oxygen, present as a phenolic group, is substituted at C - 2 (Chakraborty, J. Ind.chem.Soc., 1978, 55 1114). Yet another isomer is glycozoline which contains the methyl and methoxyl groups in different rings.

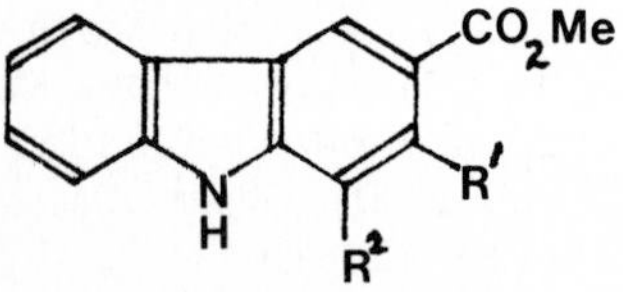

Mukonine R^1=H, R^2= OMe

Mukonidine R^1=OH, R^2=H

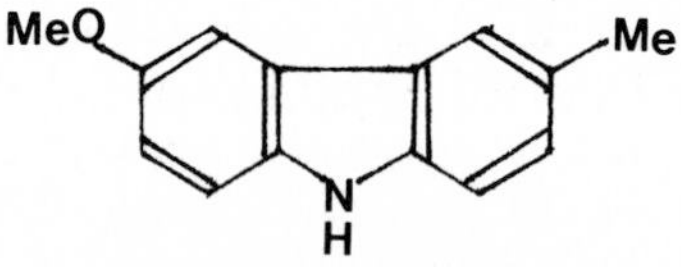

Glycolozine

A new alkaloid of this type has been isolated from the roots of *Clausena indica* and called indizoline (B.S. Joshi and D.H. Gawad, Indian J. Chem., 1974, 12, 437). It has a uv spectrum reminiscent of 3 - formylcarbazole and the product of Huang-Minlon reduction has a uv spectrum resembling that of 1 - methoxycarbazole. The presence of an isopentenyl group is indicated by the ^{1}H nmr spectrum and this must be substituted at C - 2, with the C - 4 proton being observed as a singlet at δ 8.4. The Huang - Minlon reduction product has been synthesised.

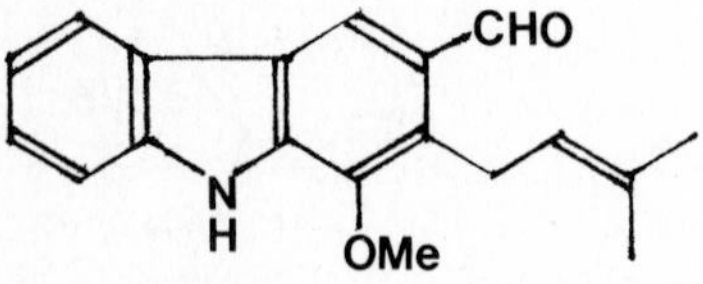

Indizoline

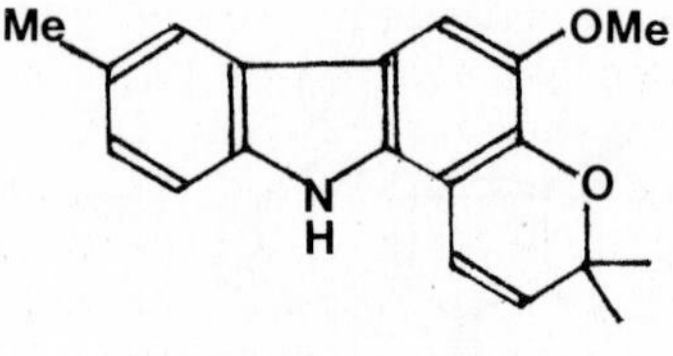

Heptazolidine

Heptazolidine, from *C.heptaphylla,* differs in having the methyl in the other aromatic ring from the oxygen substituents (D.P. Chakraborty, Chem. and Ind.,1974,303). Another isomer of this compound is mupamine which has been shown to be 9-methoxygirinimbine; it was isolated from *C.anisata*(I. Mesher and J. Reisch, Ann.,1977,1725). The structure has been supported by total synthesis (R.B. Sharma *et al.,* Ind. J. Chem., B.1981,701).

Alkaloids of the C_{18} group having the isopentenyl unit substituted differently again are atanisatin and clausanitin from *C. anisata* (D.O. Okorie, Phytochem.,1975,14,2720).

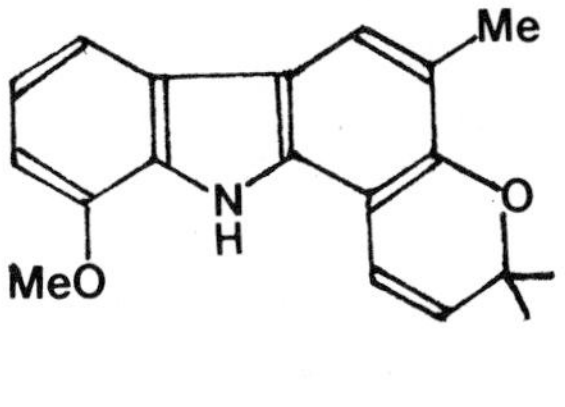

Me
N H O
MeO

Mupamine

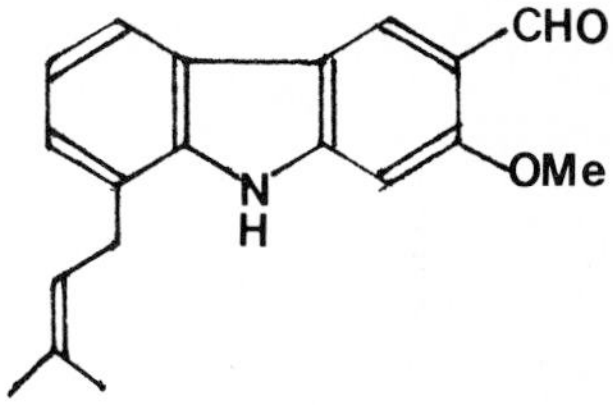

CHO
OMe
N H

Atanisatin

CHO
OH
N H

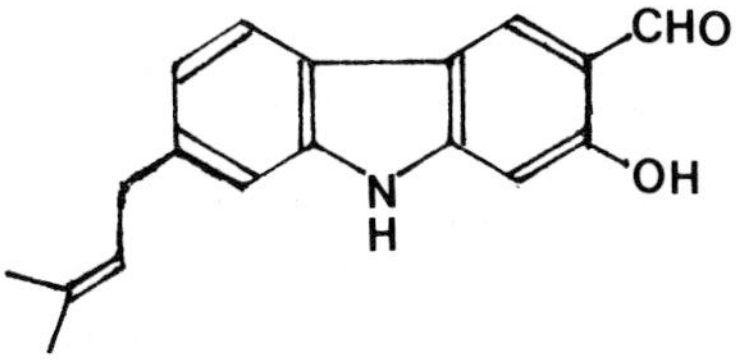

Clausanitin

The C-23 compounds continue to provide interest in the alternative cyclisations possible between the terpenoid unit and the carbazole. (+)- Mahanimbine (Vol. IV B, p.69) is racemised at 150°C, presumably by reversible opening of the dihydropyran ring and at 200°C is converted into cylomahanimbine and mahanimbidine, a conversion previously promoted by acid catalysis (N.S. Narosimhan and S.L. Kelkar, Indian J. Chem., 1976,14B,430). Exozoline, from *M. exotica* is merely dihydrocyclomahanimbine (N. Ganguly and A. Sarkar, Phytochem.,1978,17,1816).

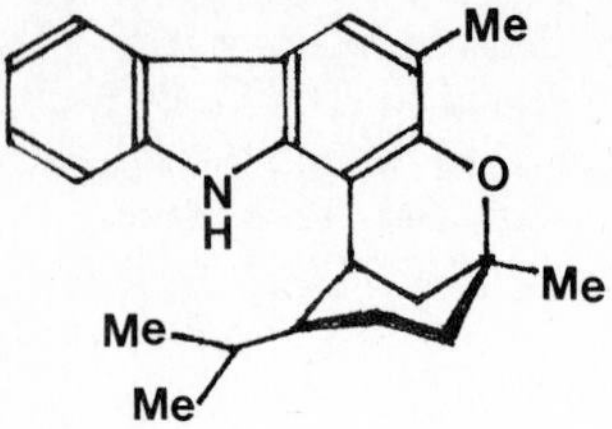

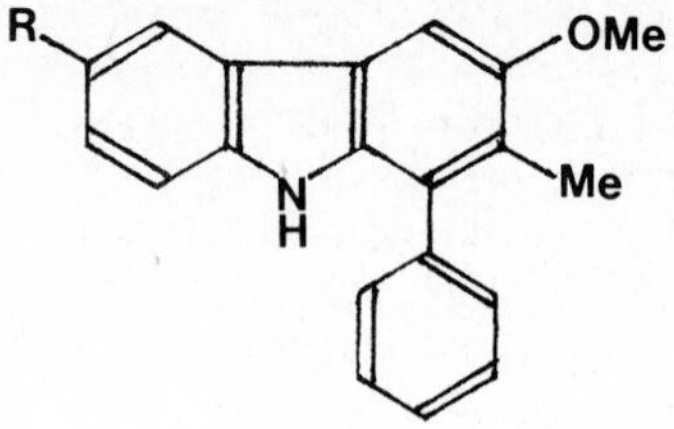

Exozoline

Hyellazole R = H

Chlorohellazole R = Cl

A new source of carbazoles has been found in the form of the blue-green algae *Hyella caespitosa* (J.H. Cardellina *et al*., Tetrahedron Letters, 1979, 4915). Hyellazole and chlorohyellazole have been characterised and synthesised. (S. Kano, E. Sugino and S. Hibino, Chem. Comm., 1980, 1241.)

2. *Alkaloids containing a tryptamine unit*

(a) *Compounds without an isoprene moiety*

(i) *Simple tryptamine and tryptophan derivatives*

Simple tryptamines and substituted tryptamines have been isolated from a number of differing sources.

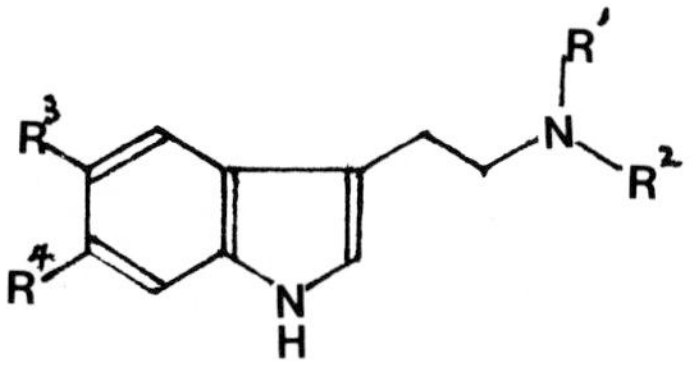

Tryptamine
$R^1=R^2=R^3=R^4=H$

Tryptamine, *N*,*N* - dimethyltryptamine (R-1 = R-2 = Me; R-3 = R-4 = H) and bufotenine (R-1 = R-2 = Me; R-3 = OH, R-4 = H) have all be isolated from the roots and stems of *Desmodium caudatum* (A. Ueno *et al*., Chem. and pharm. Bull (Japan), 1978, 26,2411). The same source also supplies bufotenine - *N* - oxide. The various mono and dimethyl analogues, dipterine (R-1 = Me; R-2 = R-3 = R-4 = H), serotonine (R-1 = R-2 = H ; R-3 = OH; R-4 = H) and 5 - hydroxy - *N* - methyltryptamine (R-1 = Me; R-2 = R-4 = H; R-3 = OH) have been isolated from the red - violet octocoral *Paramuricea chamaeleon* (G. Cimino and S. De Stefano, Comp. Biochem. Physiol. Sect.C, 1978,61, 361). Other marine sources, the sponges *Smenospongia echina* and *S. aurea,* have yielded 5, 6 - dibromo - (R-1 = R-2 = Me; R-3 = R-4 = Br) and 5- bromo (R-1 = R-2 = Me; R-3 = Br; R-4 = H) - *N*, *N* - dimethyltryptamines respectively (P. Djura, J. org.Chem.,1980,45,1435). *N*(b) - acyl tryptamines are known from a number of sources. The Argentinian tree *Prosopis nigra* has been shown to contain *N* - acetyl tryptamine (R-1 = Ac; R-2 = R-3 = R-4 = H), the first time this simple compound has been isolated from a natural source (D. der Santos Filhao and B. Gilbert, Phytochem.,1975,14,821) and the benzoyl derivative (R-1 = COC_6H_5; R-2 = R-3 = R-4 = H) has been obtained from the New Caledonian member of the *Rutaceae* family, *Myrtopsis myrtoidea* (M.S. Hifnawy *et al.*,

1977,16,1635). The novel formamide derivative (R-1 = Me; R-2 = CHO; R-3 = R-4 = H) occurs in the leaves of *Acacia simplicifolia* (C. Coupat, A. Ahond and T. Sevenet, ibid., 1976,15,2019). Mature kernels of sweet corn, *Zea mays*, contain the p-Coumaryl derivative (6) (A. Ehmann, *ibid* ., 1974,13,1979) and the novel thiazolic acid derivatives (7) has been isolated from *Thermoactinomyces* species (Y. Konda *et al.*, Chem. and pharm. Bull (Japan), 1976,24.92).

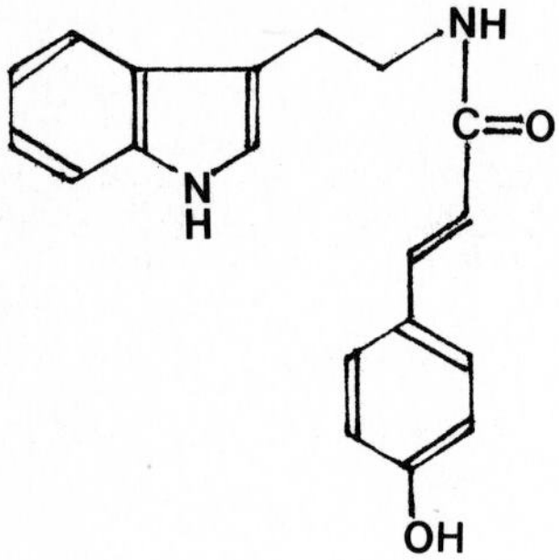

6

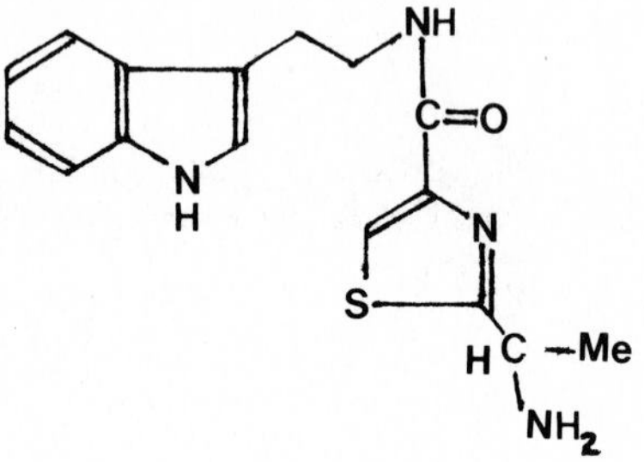

7

A tryptamine derivative which has experienced oxidation is borreline, obtained from several Borreria species of Guyana (A. Jossany *et al.*, Tetrahedron Letters 1977,1219). The biosynthetic origin of this compound is obscure.

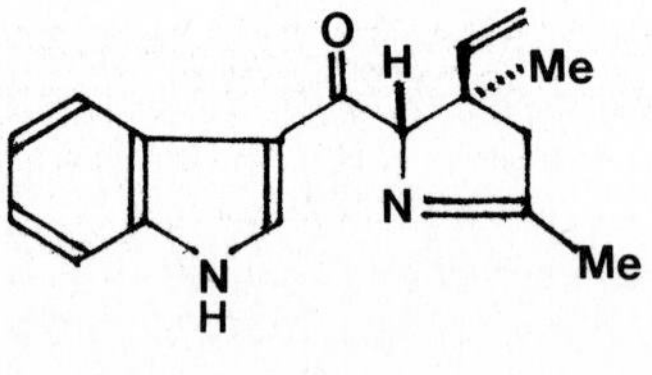

Borreline

A similarly oxidised derivative is dendrodoine, a cytotoxic thiadiazole derivative which has been isolated from the tunicate *Dendrodoa grossularia* (S. Heitz, *et al.*, Tetrahedron Letters, 1980,21,1457). Differently oxidised tryptamine derivatives are the two oxindoles donaxarine and donaxaridine, the alkaloids of *Arundo donax* (K.A.Ubaibullaev, R. Shakirov and S. Yu. Yunusov, Khim.prirod.Soedinennii, 1976,553). Donaxarine may be obtained from donaxaridine in the laboratory by condensation with acetaldehyde.

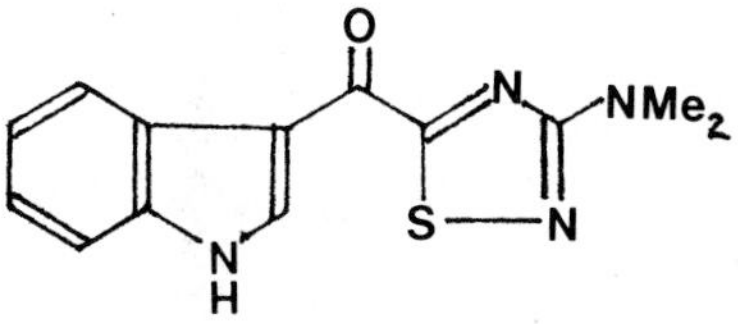

Dendrodoine

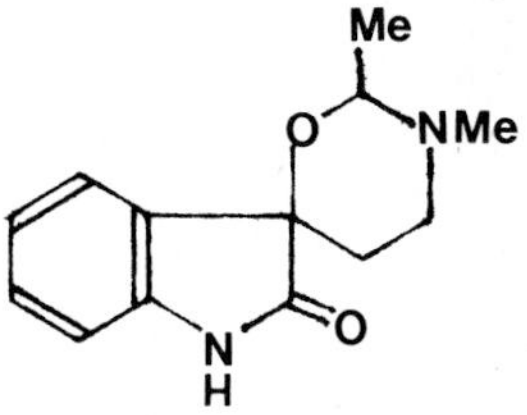

Donaxarine

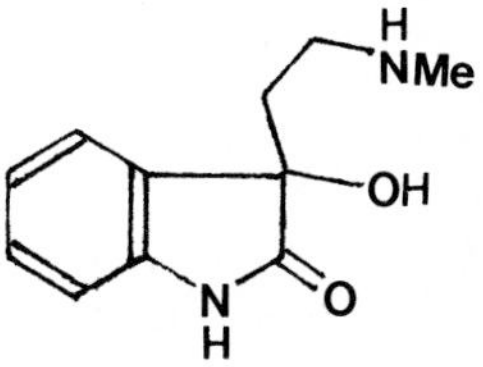

Donaxaridine

Compounds retaining the tryptophan carboxyl carbon atom have been isolated. *N - Acetyl - L - tryptophan* has been found in *Claviceps purpurea* (H - J.L. Liang and J.A. Anderson, Phytochem., 1978,17,597), and *N,N* - dimethyl tryptophan methyl ester has been reported from jequirity seeds *(Abrus precotorius)* (N. Mandava, J.D. Anderson and S.R. Dutky, *ibid*, 1974,13,2853). A number of more complex systems have been reported where esterification, amide formation or cyclisation has occurred. Hypaphorine, for example, has been known for many years and has been shown to occur in lentil seedlings, but has now been found elaborated as its esters of erysorine and erysodine in the compounds erysophorine and erysodinophorine respectively, isolated from *Erythrina arborescens*. (M. Hofinger, X. Monseur, M. Pais and F.X. Jarreau, *ibid.*, 1975, 14, 475, K.P. Tiwari and M. Masood, *ibid.*, 1979, 18, 704).

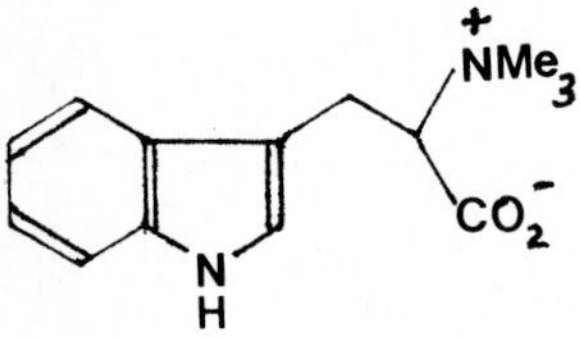

Hypaphorine

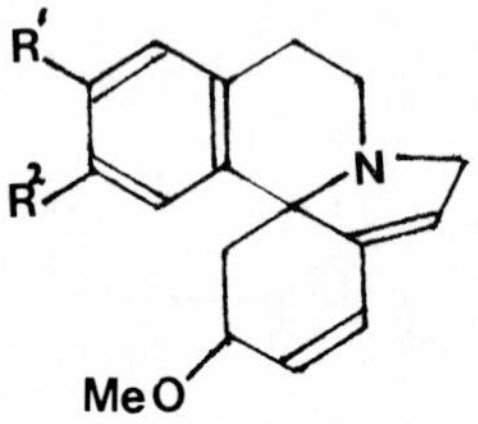

Erysophorine
R^1 = MeO
R^2 = hypaphorine

Erysodinophorine
R^1 = hypaphorine
R^2 = MeO

An example of amide formation is the alkaloid (9) isolated from *Evodia rutacaecarpa* (B. Danieli, G. Lesma and G. Palmisano, Experientia, 1979, 35, 156). The compound is readily synthesised from the optically active dihydro - β - carboline (8) and *N* - methylanthranilic acid.

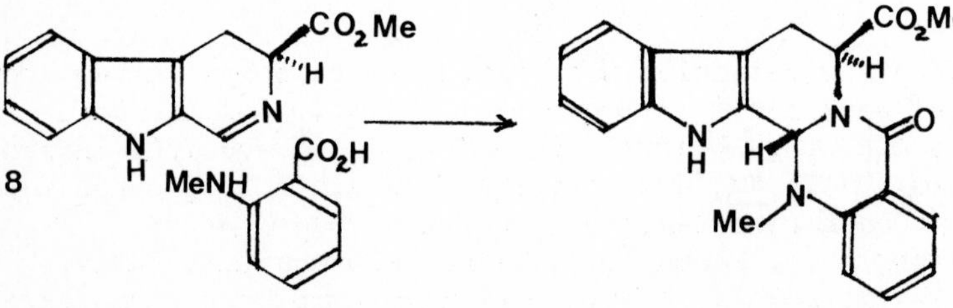

8

9

Reagents : PPh_3, CBr_4, THF, pyridine.

The carbomethoxyl group in (9) is clearly axial from its nmr spectrum and this is to be expected since the equatorial disposition would be disfavoured by interaction with the amide carbonyl group.

A more simple example of cyclisation-amide formation is exhibited by the marine metabolite (10) isolated from *Smenospongia aurea* (P. Djura *et al.*, J.org.Chem.1980,45, 1435): the similarity to neoechinulin E is obvious.

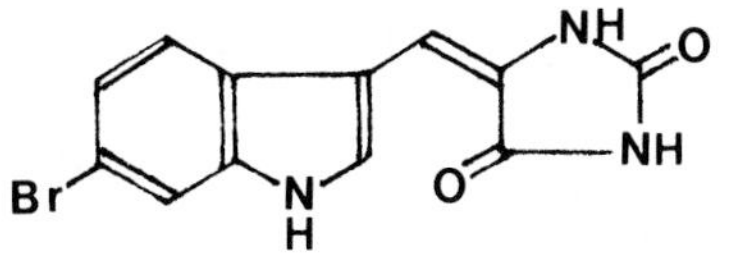

10

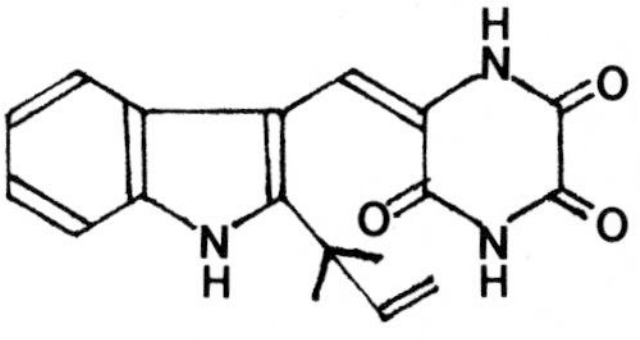

Neoechinulin E

Related 6 - brominated indole derivative is surugatoxin which has been obtained from the guts of the gastropod *Babylonia japonica,* found in Suruga Bay, Japan. The toxin is peculiar to species obtained from this bay and is not obtained from animals of this species found elsewhere. The structure was established by X-ray diffraction studies of the heptahydrate (T. Kosuge *et al.*, Tetrahedron Letters 1972,2545).

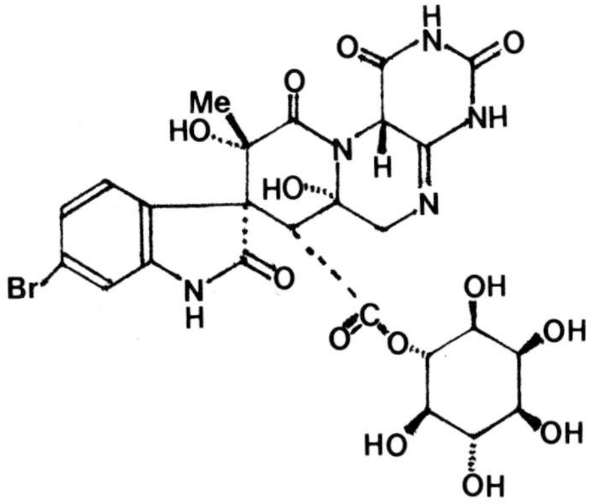

Surugatoxin

(ii) *Eserine types*

Interest in the alkaloids of the Calabar bean has been largely concerned with synthesis, with little structural investigation other than absolute sterochemical deductions. Two variants of the eserine type have been isolated from a marine source *Flustra foliacea*, a moss animal. These are flustramines A and B which contain a bromine atom at the indole C - 6 position. They differ in the attachment of an isopentenyl unit at C - 3a; the *cis* - fusion of rings B and C follows from nuclear Overhauser enhancement difference (J.S. Carle and C. Christophersen, J. Amer.chem.Soc, 1979,101,4012; J. org.Chem,1980,45,1586).

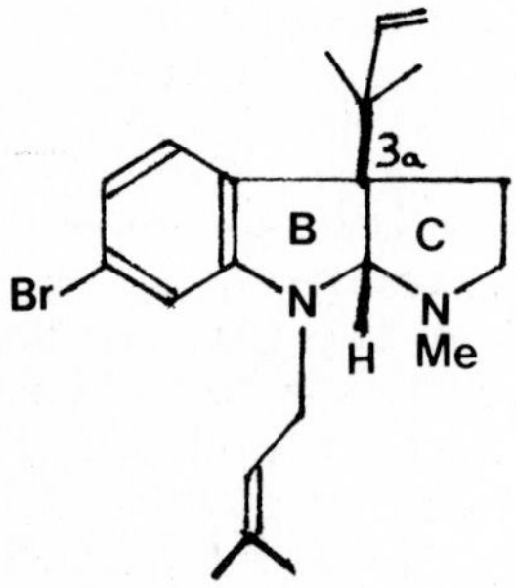

Flustramine A

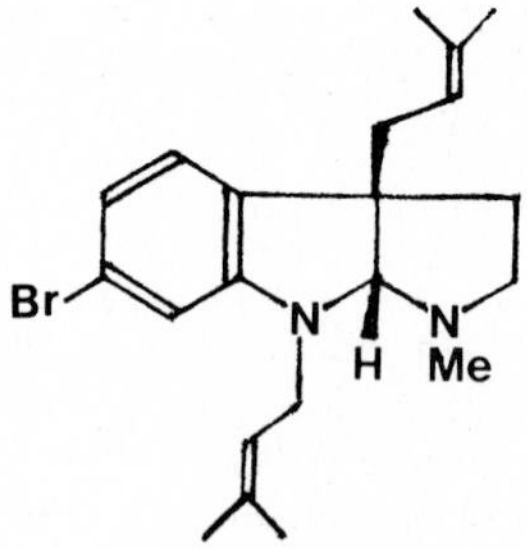

Flustramine B

(b) *Compounds containing an isoprene (but not terpene) derived moiety*

(i) *The ergot alkaloids*

The ergot alkaloids attract attention largely on account of the pronounced physiological activities of these compounds, with the resultant poisonings of humans and domesticated animals. An example of this is provided by the investigation of *Penicillium islandicum* found growing on freshly dug green peanuts. These contain two chlorine-containing toxins together with their non-chlorinated analogues and these have been identified as the rugulovasines A and B and the corresponding 8-chloro derivatives, (R.J. Cole *et al.,* Tetrahedron Letters, 1976, 3849). These compounds had previously been isolated from P.Concavo - Rugulosum. The structures have been proved by X-ray crystallography (M. Abe *et. al.,* Agric. and Giol Chem.(Japan) 1970,34,485). They are interconverted by ring opening; the chlorinated isomers appear to be the first halogenated ergot alkaloids reported.

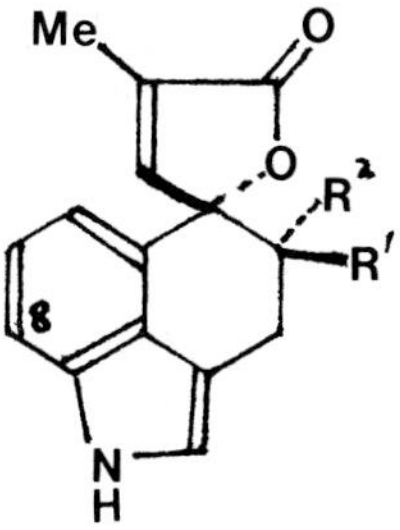

Rugulovasine A

R^1=H R^2 = NH Me

Rugulovasine B

R^1 = NH Me R^2 = H

The isolation of (11) from cultures of *Claviceps purpurpea* J.A. Anderson and M.S. Saini, Tetrahedron Letters, 1974,2107) represents yet another missing link in the biosynthesis of these compounds from tryptophan.

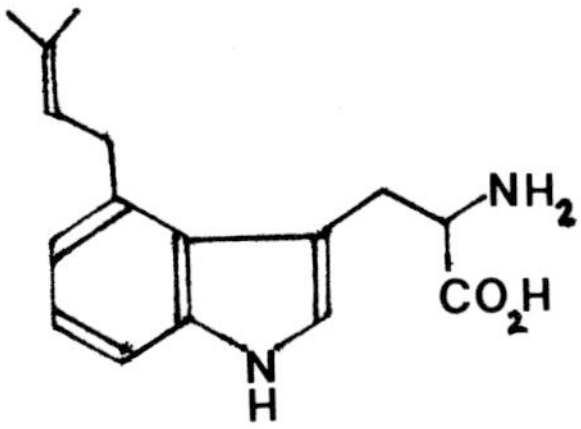

11

A new structural type is represented by two new clavine alkaloids isolated from *Claviceps paspali* (H. Tscherter and H. Hauth, Helv. 1974,57,113). These are paspaclavine and paliclavine. Treatment of the former with dilute acid hydrolyses the amino - acetal function to give paliclavine. The structure and relative stereochemistries were deduced from their ^{1}H nmr spectra; the depicted absolute stereochemistry is deduced by analogy with other ergot alkaloids.

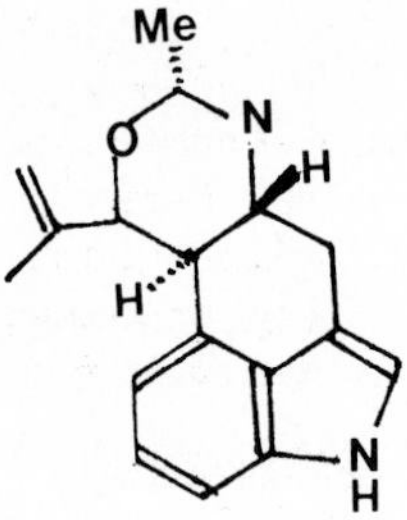

Paspaclavine 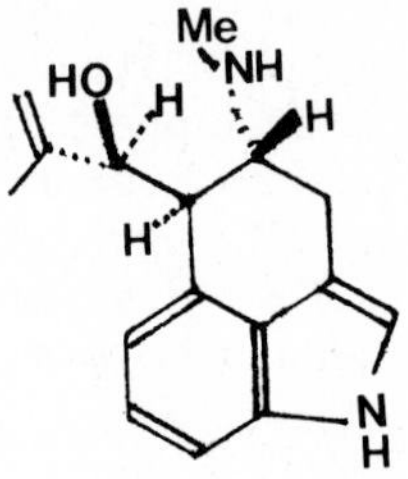 Paliclavine

Ipomeoea species continue to provide alkaloids of this type; especially rich in ergolines is *I. Violacea,* the notorious "Morning Glory".

(ii) *Mould Metabolites*

A whole range of mould metabolites has been obtained containing a diketopiperazine unit derived from tryptophan and amino acids incorporating isopentenyl units, attached either normally or in reverse manner. These in the form of the brevianamides and echinulins were reviewed earlier (2nd Edn. IVB pp.79-83) and since that time there have been many additions to the types known.

Probably the most simple example containing all of these features is compound (12) isolated from proteinase casein hydrolysate (T. Shiba and K. Numanii, Tetrahedron Letters, 1974,509). In this case the C_5 unit almost certainly originates from leucine. The structure follows from an unambiguous synthesis from leucine and tryptophan.

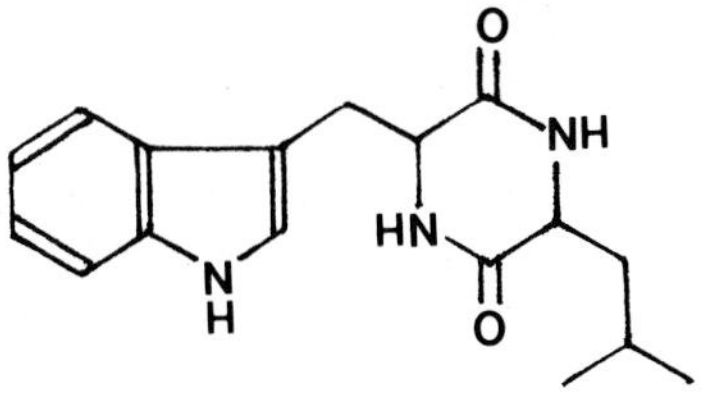

12

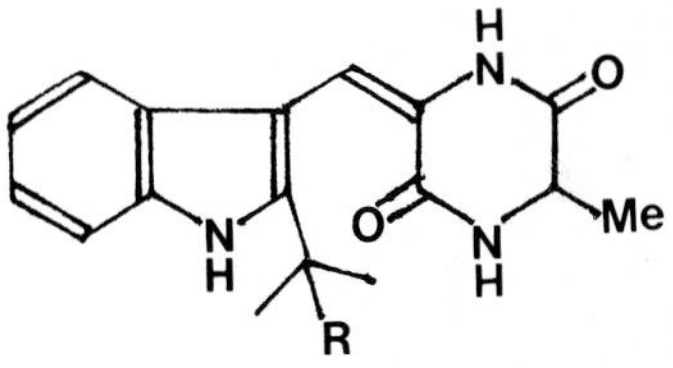

13 R = $CH{=}CH_2$
14 R = Et

The more usual substitution pattern, related to echinulin, is demonstrated by compound (13) obtained, together with its dihydro compound (14) from *Aspergillus rubus* infecting oil cakes (H. Itokawa *et al.*,Yakugaku Zasshi, 1973,93,1251).

Brevianamide E, together with several minor variants, and its biosynthetically close relatives of the austamide group, have been shown to be more widely distributed. Deoxybrevianamide E has been shown to be the major metabolite isolated from *Penicillium italicum,* the blue mould found on citrus fruits; this is accompanied by the 12,13 - dehydro derivative and the indole equivalent of austamide (15) (P.M. Scott, *et al.,* Applied Microbiol., 1974,28,892).

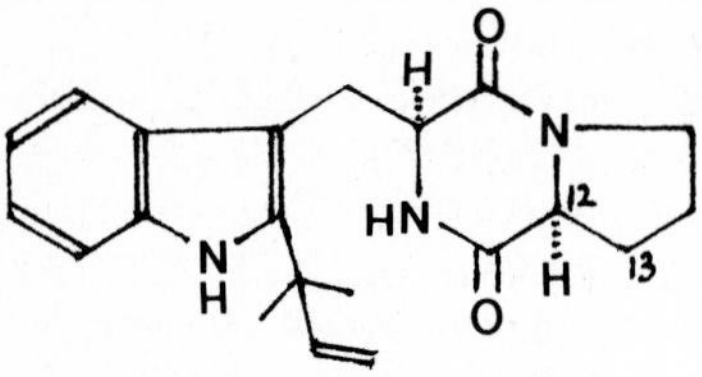

Deoxybrevianamide E

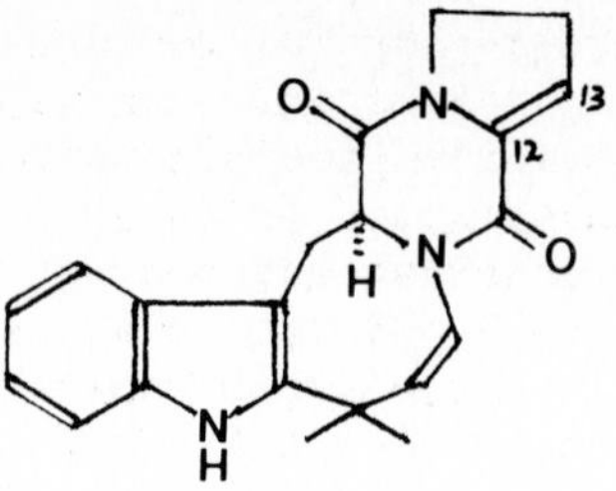

15

There are two successful, similar syntheses of deoxybrevianamide E, (R. Ritchie and J.E. Saxton, Chem.Comm., 1975,611; T. Kametani, *et al.*, J. Amer.chem.Soc.,1980,102 3974). The later workers have converted their product into a mixture of brevianamide E and the stereoisomeric *cis*-fused product. By comparison of the ^{1}H nmr spectra the stereochemistry of brevianamide E has been defined.

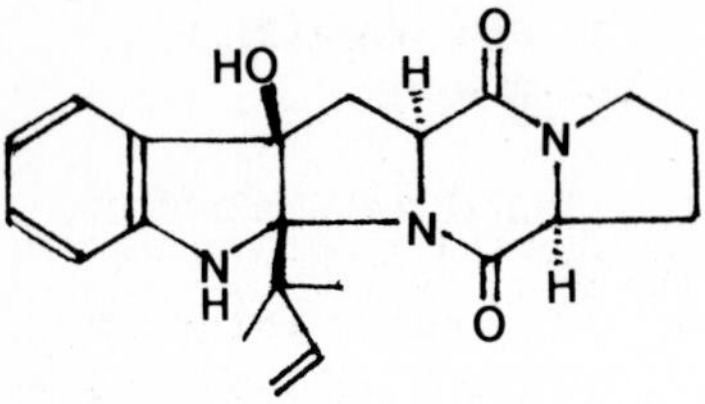

Brevianamide E

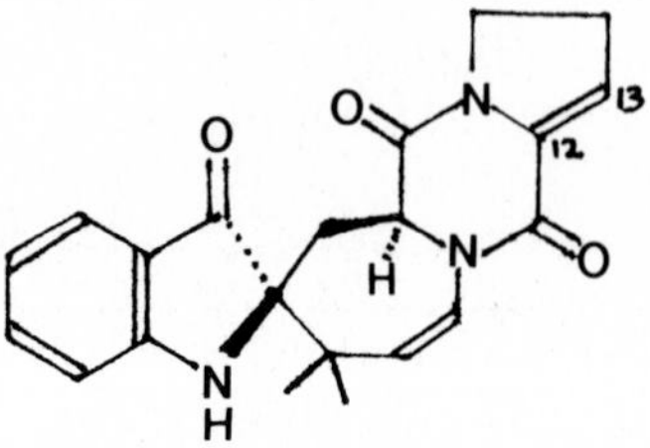

Austamide

Brevianamide A has been isolated from *P. ochraceum*(J.E. Robbers, J.W. Straus and J. Tuite, Lloydia, 1975,38,355). Austamide has been synthesised (A.J. Hutchinson and Y. Kishi, J. Amer.chem.Soc.,1979,101,6786) and the penultimate step in the synthesis was the formation of 12,13 dihydro - 12 - hydroxyaustamide. This was shown to be identical to a minor metabolite from *Aspergillus ustus* (P.S. Steyn and R. Vleggar, Phytochem.,1976,15,355).

Compounds containing the diketopiperazine unit and a dehydrotryptophan skeleton have been found to be widespread and have led to a confusion of nomenclature. They have mostly been isolated from *Aspergillus amstelodami* but have been obtained from other *Aspergillus* species. The members of the neoechinulin group are characterised by having the dehydrotryptophan bridge, the reversed isopentenyl group at C - 2 of the indole unit and variously oxidised diketopiperazine systems. Neoechinulin itself, together with neoechinulins C and D have isopentenyl groups at C - 6 of the indole whereas neoechinulins A and B do not have this substituent. Recently the unsubstituted analogue of neoechinulin has been isolated and termed neoechinulin E (R. Marchelli *et al.*, J.chem.Soc. Perkin I, 1977,713). The structure follows from comparison of the ^{1}H nmr with that of neoechinulin and by isolation of 2 - (1,1-dimethylallyl) indole from the alkaline hydrolysate.

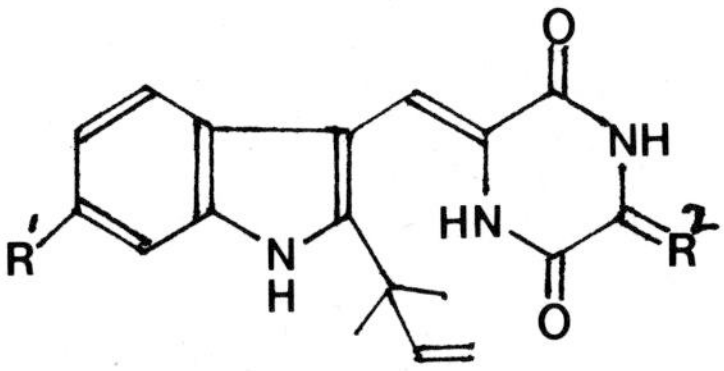

Neoechinulin
$R' = CH_2 \cdot CH = CMe_2$ $R^2 = O$
Neoechinulin E
$R' = H$ $R^2 = O$
Neoechinulin C
$R' = CH_2 \cdot CH = CMe$ $R^2 = CH_2$
Neoechinulin B $R = H$
$R = CH_2$

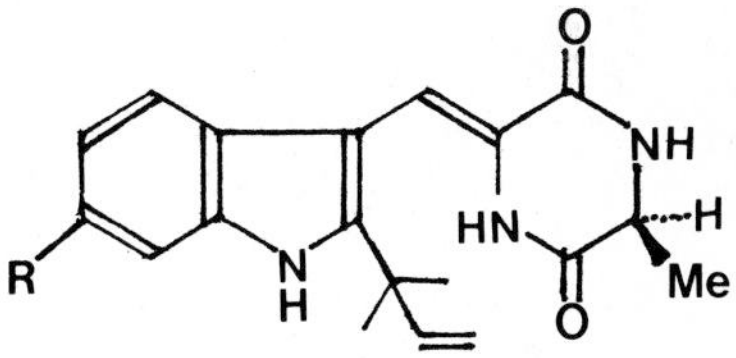

Neoechinulin A
$R = H$

Neoechinulin D
$R = CH_2 - CH = CMe_2$

The isoechinulins differ from the neoechinulins in having the aromatic isopentenyl unit substituted at C - 5 rather than C - 6. Isoechinulin A is the isomer of neoechinulin D and occurs in *A. amstelodami* and *A. ruber* where it is believed to be responsible for growth inhibition in the silk worm larvae (H. Nagasawa, *et al.*, Tetrahedron Letters, 1976,1601).

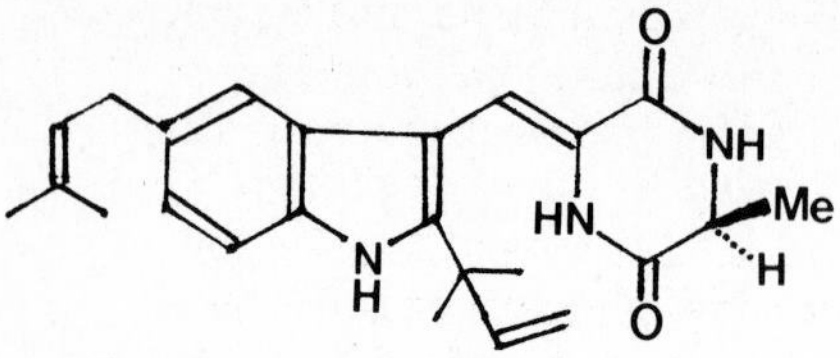

Isoechinulin A

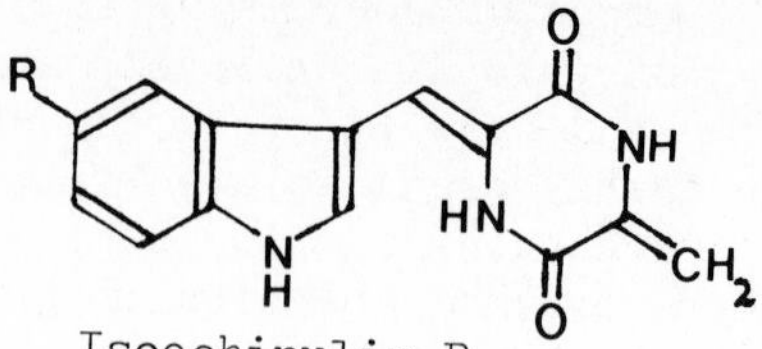

Isoechinulin B
R = CH_2- CH =CMe_2
Isoechinulin C
R = CH_2- CH - CMe_2 (epoxide, O bridging CH and CMe_2)

Other constituents of the same mycelium are isoechinulins B and C. Isoechinulin B is related to neoechinulin C whereas isoechinulin C is the epoxide of isoechinulin B.

There is not, as yet, any simple relationship in the cryptoechinulin series, although it may well be that since some of these compounds are identical with others the nomenclature may be discontinued. Cryptoechinulin A is identical with neoechinulin C and cryptoechinulin C is identical with neoechinulin E. Cryptoechinulins B and D are, however, new types and are related to each other. They may be considered as derived from auroglaucin (also found in *A. amstelodami*) and neoechinulins C and B respectively by regiospecific Diels-Alder condensation of the terminal diene system of auroglaucin with the exocyclic methylene of the neoechinulin. These have been isolated from *A. amstelodai* (R. Cardillo *et al.*, Chem.Comm.,1975,778) and have been synthesised (*idem.*, *ibid.*, 1976,435) Aurechinulin has been isolated and shown to be identical to cryptoechinulin B (Y. Kishi, *et al.*, Yakugaku Zasshi, 1977, 97,582). The stereochemistry in the Diels-Alder adduct is described as *cis*, not unreasonably, but the configuration at the spiro centre is not known.

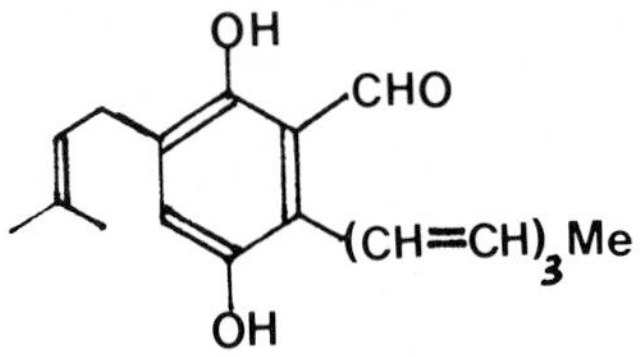

Auroglaucin

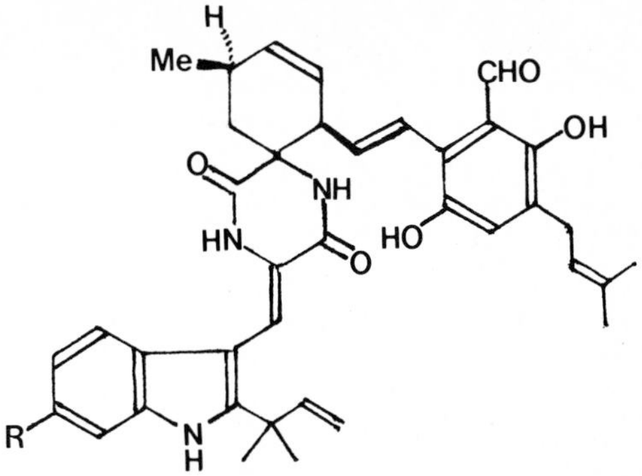

Cryptoechinulin B
R = H

Cryptoechinulin D
R = CH_2-CH = CMe_2

Cryptoechinulins E and F have as yet not been characterised. Cryptoechinulin G represents yet a further variant in that it contains two isopentenyl units attached to the aromatic ring. (G. Gatti, R. Cardillo and C. Fuganti, Tetrahedron Letters, 1978, 2605). These are shown to be *ortho* to each other from the ^1H nmr spectrum and the substitution pattern shown is favoured on the grounds of the ^{13}C nmr spectrum and on the restricted rotation to be expected if there is an isopentenyl unit at C - 4. The compound was isolated from *A. ruber.*

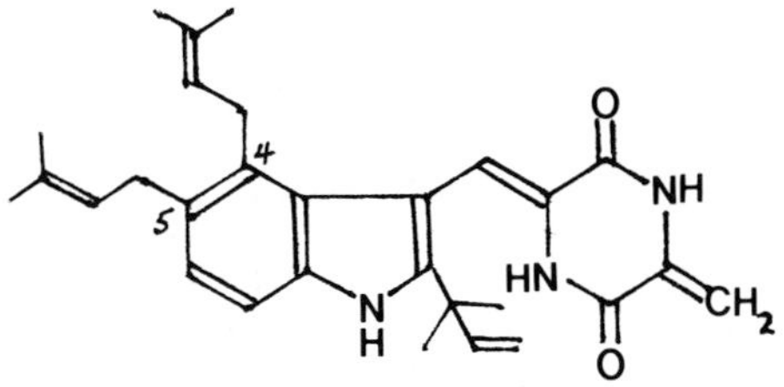

Cryptoechinulin G

Verrucologen, from *Penicillium verruculosum,* is a very powerful mycotoxin and is related to fumitremorgin B (2nd Edn., Vol. IVB, p.83) as its peroxide which has undergone cyclisation. The structure and relative configuration have been deduced from X-ray diffraction studies (J.W. Kirksey *et al.*, J.Amer.chem.Soc.,1974,96,6785). It is also found as a congener of FTB in *Aspergillus caesopitosus*. *A. fumigatus* contains a number of metabolites of related structures which are termed the *fumitremorgins*(FT) and labelled FTA to FTJ. FTB was the first structure to be proven by X-ray methods and FTA is simply the *O*-dimethylallyl ether of verrucologen (Kirksey *et al.*, Tetrahedron Letters, 1975,1051, M. Yamazaki *et al.*, *ibid.*, 1241).

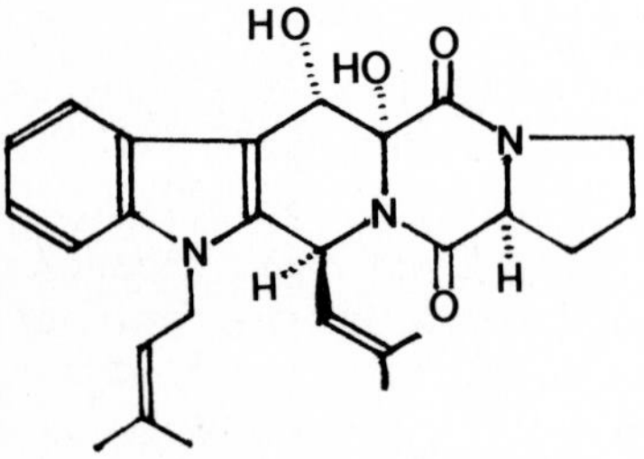

Fumitremorgen B

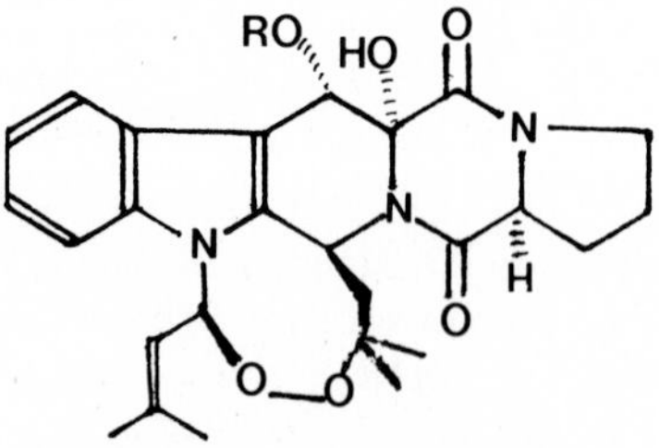

Verrucologen R = H

FTA R = CH_2–CH = CMe_2

Compounds of similar structures have been isolated from a strain of *A. clavatus* collected from mould-infested rice in Thailand. (G. Buchi *et al.*,J.org.Chem, 1977,42,244). These compounds have been termed the tryptoquivalines (and names derived from this) and some of these have been found to be identical with the fumitremorgins. A number of these are unusual in containing hydroxylamine units. The structures of the compounds are shown as below and these, with their stereochemistries have largely been derived from correlation with one another (Yamazaki *et al.*, Tetrahedron Letters 1976,2861; *idem,* Chem. and Pharm. Bull (Japan), 1977,25,2554; 1978,26,111).

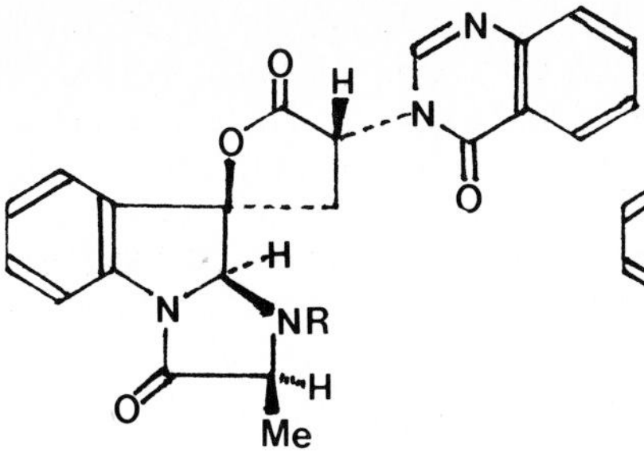

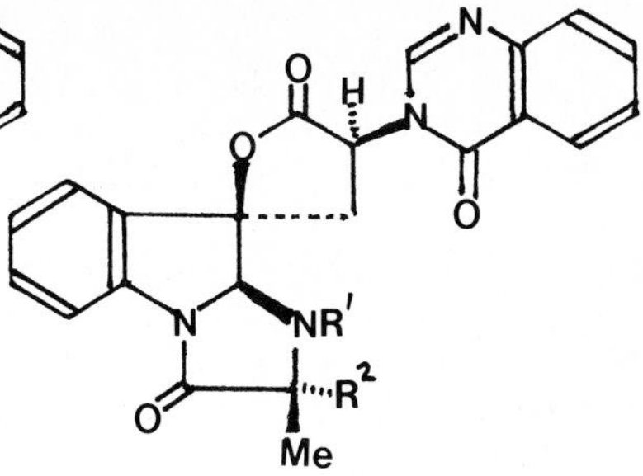

FTH R = OH

FTF R = H

	R^1	R^2
FTE	OH	H
FTJ	H	H
FTG	OH	Me

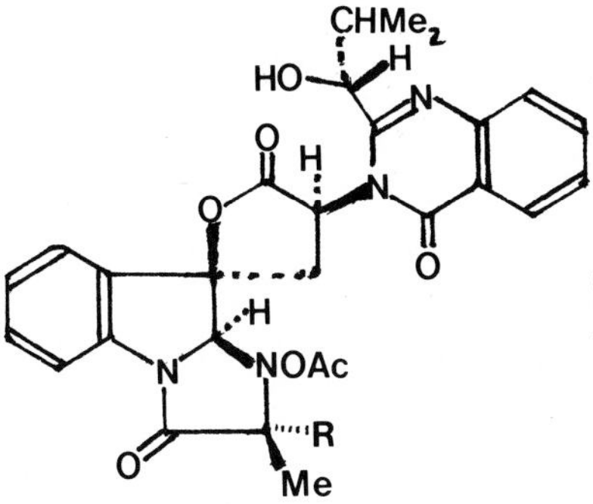

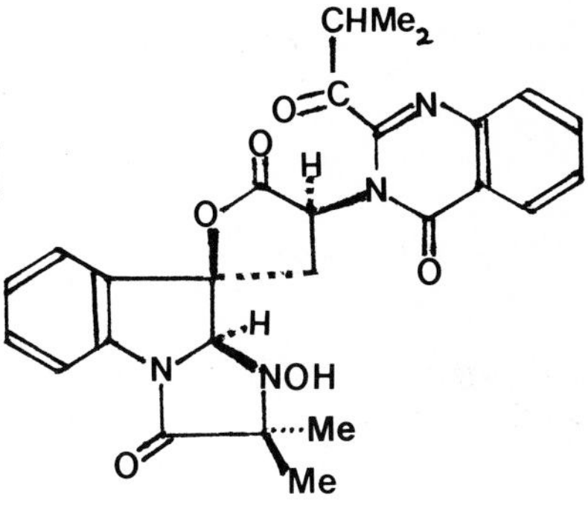

FTC = Tryptoquivaline C
R = Me
FTD = Tryptoquivaline D
R = H

FTI

The absolute stereochemistries shown for FTC and FTD follow from single crystal X-ray diffraction studies (J.P. Springer, Tetrahedron Letters, 1979,339).

A compound showing an indolic N-O-Me group is oxaline, the major metabolite of a toxigenic strain of *Penicillium oxalicum* (P.S. Steyn *et al.*,chem.Comm.,1974,1021: Tetrahedron, 1976,32,2625). This crystalline compound, m.p. 220-1°, has been shown by X-ray diffraction studies of a single-crystal to have the structure depicted. Unusual features of this structure are the coupling of the tryptophan unit to hystadine, the attachment of three nitrogen functionalities at C-2 and the presence of a reverse isopentenyl unit at C-3 (this is attached at C-2 in the brevianamide-austamides and in the neoechinulins). Similar features are shown by roquefortine, a metabolite of *P. roqueforti,* the essential fungus in the production of blue cheeses (P.M. Scott, M.A. Merrien and J. Polonsky, Experientia, 1976,32,140). Roquefortine, m.p. 195-200° has molecular formula C-22 H-23 N-5 O-2 and ^{13}C nmr studies showed the presence of 2CH-3, 1CH-2, 2CH, 1CH-2 =, 8-CH =, 2Sp-3 and 4Sp-2 fully substituted C and 2C = O, accounting for 20 protons. Two of the three remaining protons were accounted for as exchangeable NH protons in the ^{1}H nmr spectrum. Catalytic reduction gives a dihydro compound. The presence of $^m/e$ 69 in the mass spectrum of roquefortine is attributable to an inverted isopentenyl group by comparison of the ^{1}H and ^{13}C nmr spectra of roquefortine and its dihydro compound. This group was placed at the indoline β - position by comparison of the ^{13}C nmr spectrum of roquefortine with that of other indoline alkaloids. Reduction of the alkaloid with zinc in acetic acid gives an amorphous 3,12 - dihydro compounds which on hydrolysis gives histidine. This led to comparison of the ^{13}C nmr spectrum with that of oxaline and allowed assignment of the structure.

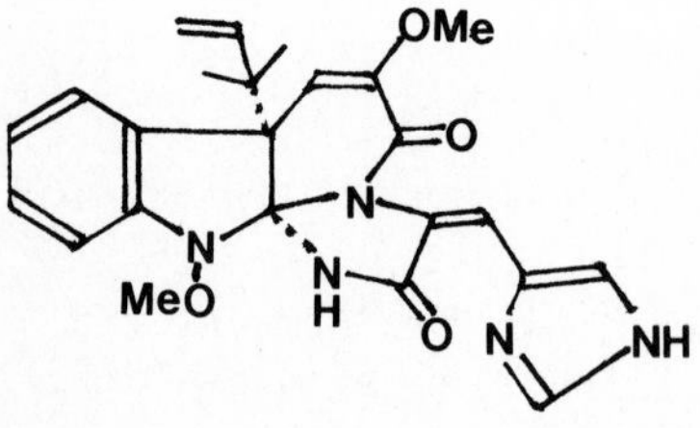

Oxaline

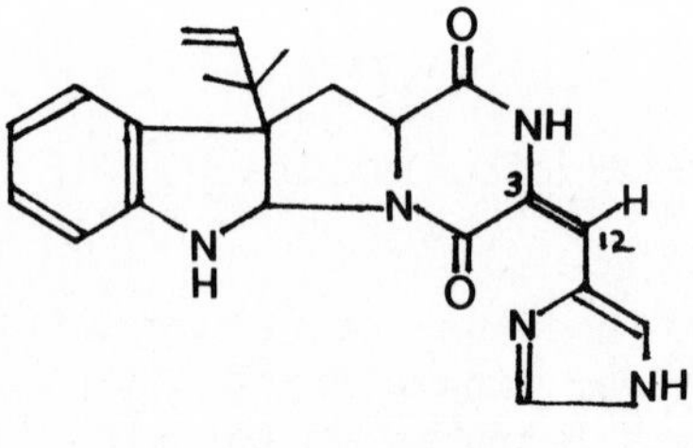

Roquefortine

The stereochemistry about the 3,12 double bond follows from a study of the ^{1}H nmr spectrum of the product obtained by photoisomerisation of roquefortine. In the spectrum of the isomer, the proton attached to C-12 has moved to lower field as a result of shielding by the oxo-group at C-4 (*idem.*,J.Agric.Food Chem,1979,27,201). Treatment of roquefortine by hot methanolic hydrochloric acid leads to loss of the isopentenyl group at C-3, as shown below.

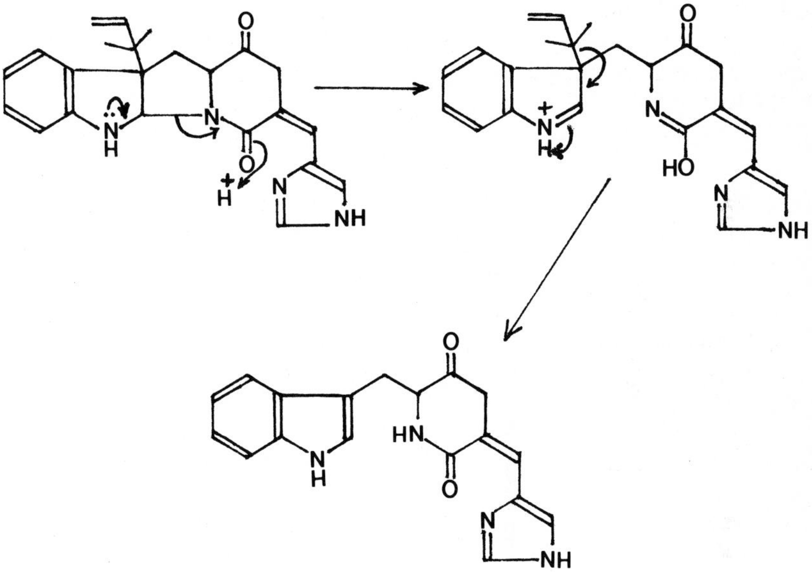

Comparison of the ^{13}C nmr spectrum of roquefortine with that of oxaline supports the assignment of the stereochemistry to the former (R. Vleggar and P.L. Wessel, chem. Comm., 1980,160). 3,12 - dihydro roquefortine has been shown to be a minor metabolite of *P. roqueforti* (M. Abe *et.al.*,Agric.Giol.Chem.,1977,41,2097; 1979,42,2387).

The close biosynthetic relationship between oxaline and roquefortine has been supported by their co-occurrence in *P. oxalicum*(R. Vleggar, *loc. cit)*. Roquefortine also occurs in *P. commune*(R.E. Wagener, N.D. Davis and U.L. Diener, Appl. Environ. Microbiol., 1980,39,882).

A novel structure has been assigned to marcfortine A, a new metabolite of a strain of *P.roqueforti*(J. Polonsky *et al.*, Chem. Comm., 1980,601). The ^{1}H and ^{13}C nmr data indicate a monoketopiperazine with two isoprene units with a substituted oxindole system; the full structure has been deduced from X-ray analysis. It is probable that this compound is obtained from a diketopiperazine with loss of the tryptophan carbonyl oxygen. The presence of the reversed isopentenyl unit a C-3 relates this to oxaline and roquefortine while the presence of the seven membered ring linking the two phenolic oxygen atoms is unique.

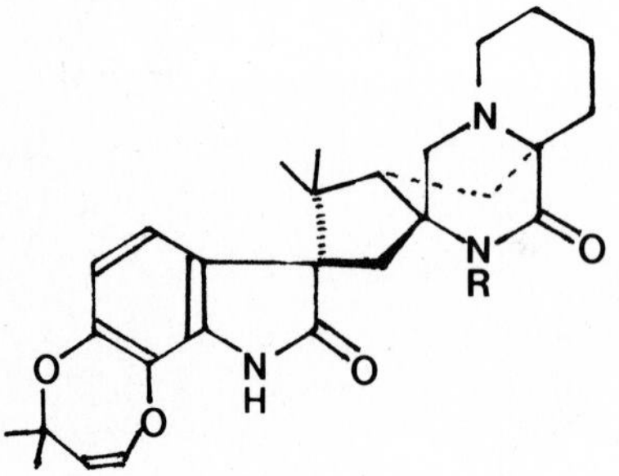

Marcfortine A R = Me

Marcfortine B R = H

Two congeners have subsequently been characterised, Marcfortine B is desmethyl marcfortine B and Marcfortine C differs from Marcfortine B in having the isopentyl unit joined directly to the 7-position of the indole. (T. Prangé, M.A. Billion, M. Vuilhorgne, C. Pascard and J. Polonsky, Tetrahedron Letters, 1981, 22, 1977).

(c) Compounds containing a terpene derived moiety

(i) *Alkaloids with the Corynanthe-Strychnos unit*

(1) *Glycoalkaloids*

The considerable interest in the elaboration of the monoterpene derived unit has led to extensive investigation of the biosynthesis of indole alkaloids derived from loganin. As noted earlier (2nd Edn. Vol IV.B, p.88 *et seq.*) the incorporation of the secologanin unit onto tryptamine leads to two possible stereoisomers at C-3, strictosidine and vincoside having 3α-H and 3β-H respectively. Initially it was believed that strictosidine was not a significant precursor of the alkaloids but this was subsequently found to be untrue.

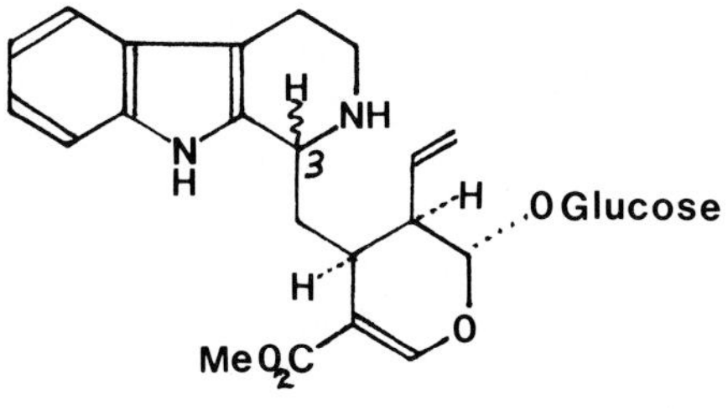

Strictosidine 3α-H

Vincoside 3β-H

Initial experiments using enzyme preparations from *Rhazya stricta* showed that strictosidine was converted into the various C-19 and C-20 isomers of the heteroyohimbine system (J. Stöckigt and M.H. Zenk, F.E.B.S. Letters, 1977, 79, 233) and this was then supported by studies in whole plants of *Catharanthus roseus (idem.,* chem.Comm,1977,646) where the three major skeletal types were obtained from doubly-labelled strictosidine. Significantly vincoside was not incorporated into these alkaloids; this has been confirmed independently (A.R. Battersby, N.G. Lewis and J.M. Tippett, Tetrahedron Letters, 1978, 4849).

Conversion to alkaloids having the 3α-H would be expected to occur without loss of radioactive label from this position. This was found to be the case when C-3 tritium-labelled strictosidine was fed to *Rauwolfia canescens* and *Mitrogyna speciosa* (Zenk *et al.,ibid.*, 1978, 1593). The general importance of this observation is emphasised by the fact that alkaloids having 3β-H can also be obtained from strictosidine, with loss of tritium from C-3. Hence strictosidine occupies a central position in the major biosynthetic pathways leading to the monoterpenoid indole alkaloids. 5-Carboxystrictosidine (2nd Edn., Vol. 1VB, p.94), which is naturally occurring, is found not to be incorporated into a whole range of alkaloids (Stöckigt, *ibid.*, 1979, 2615). However, it is reasonable to assume that 5-carboxystrictosidine is involved in the biosynthesis of adirubin.

5-Oxostrictosidine has been isolated from *Adina rubescens* heartwood (R.T. Brown and A.A. Charalambides, Experientia, 1975,31,505).

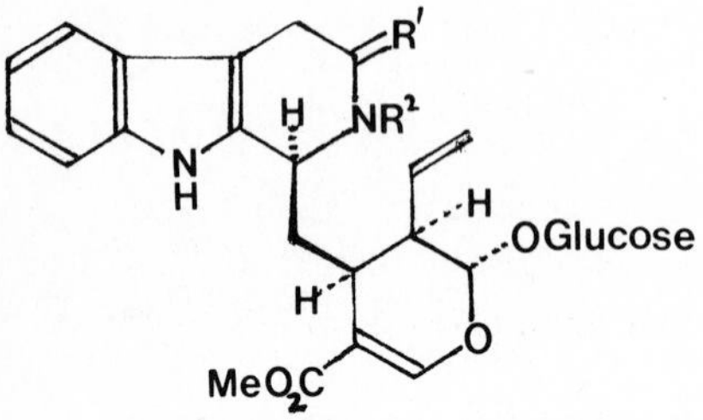

R′ NR² H N H H OGlucose H MeO$_2$C O

5-oxostrictosidine $R^1 = O$ $R^2 = H$

Dolichantoside $R^1 = H_2$ $R^2 = Me$

N_b-Methylstrictosidine, dolichantoside, occurs in the roots of *Strychnos gossweileri* from Zaire. (C. Coune and L. Angenot, Planta Med., 1978,34,53; Plant Med. Phytother, 1978, 12, 106).

A number of carbazole derivatives are known which clearly arise from dehydrogenation of strictosidine or vincoside. Desoxycordifoline has been isolated from the heartwood of *Adina rubescens* (Brown and B.F.M. Warambwa Phytochem, 1978,17,1686). *Lyaloside,* from *Pauridiantha lyalli,* is the C-20 epimer of desoxycardifoline (L. Levesque, J.L. Pousset and A. Cave, Compt. Rend., 1975,280,C593);

the stereochemistry follows from a detailed analysis of the ^1H coupling constants in the tetrahydropyran ring and presumably epimerisation occurs because of ring opening and subsequent recyclisation involving the C-21 aldehyde function. 14-Oxolyaloside, pauridianthoside, has been obtained from the same source (*idem.,* Fitoterapia, 1977, 48, 5).

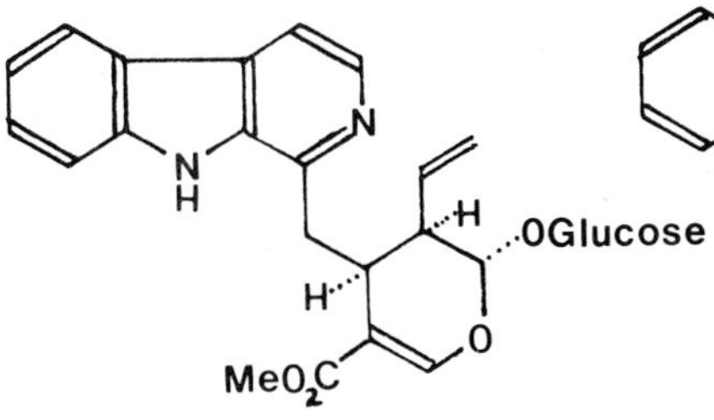

Desoxycordifoline

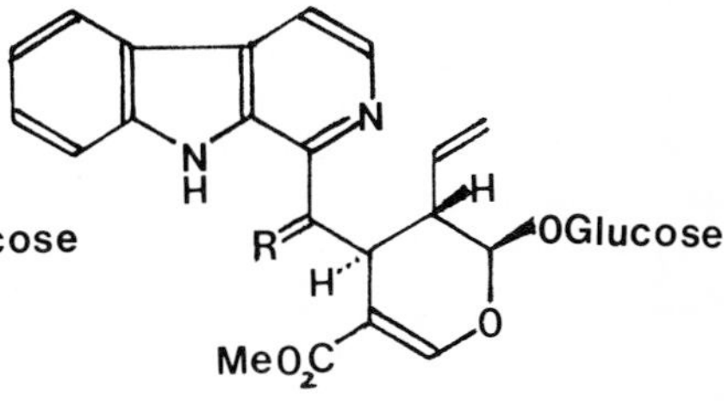

Lyaloside R = H_2

Pauridianthoside R = O

(2) *Yohimbine, heteroyohimbine and secoyohimbine types and related oxindoles.*

An important heteroyohimbine alkaloid is cathenamine. This has been isolated by standard methods from the leaves of *Guettarda eximia* (H.P. Husson, C. Kan-Fan, J. Sevenet and J.P. Vidal. Tetrahedron Letters 1977, 1889) and also from the incubation of secologanin and tryptamine in a cell-free extract of *Catharanthus roseus* (Stockigt *et al.,* F.E.B.S. Letters, 1976, 70, 267; *idem.,* chem.Comm., 1977, 164). It is 20,21-dehydroajmalicine and is reduced to tetrahydroalstonine with sodium tetrahydroborate. It may be obtained from Nb, 21-dehydro-geissoschizine, another alkaloid from *G. eximia*, by treatment with dilute aqueous alkali (Kan-Fan and Husson, chem.Comm., 1979, 1015). Treatment of this precursor with 2% hydrochloric acid gives the hemiacetal 17-hydroxydihydro cathenamine (16).

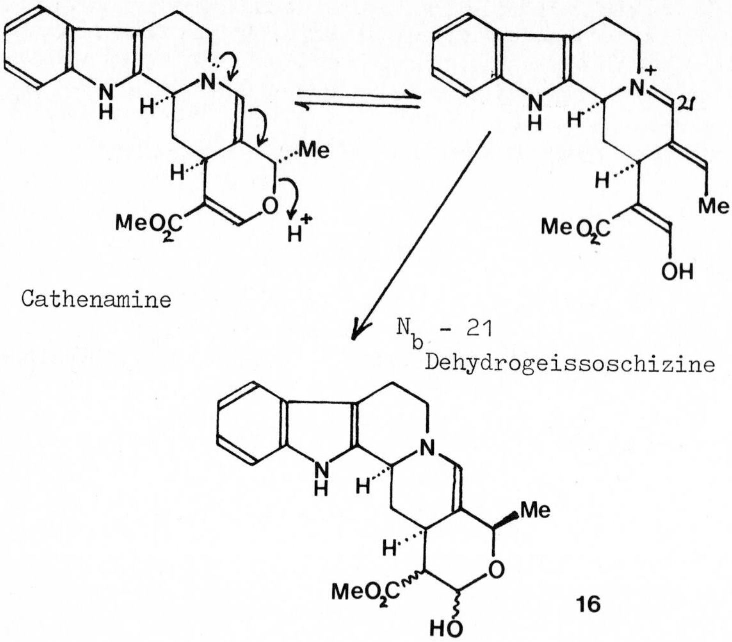

Cathenamine

N_b - 21 Dehydrogeissoschizine

16

This inter-relationship has allowed biomimetic syntheses of heteroyohimbine type alkaloids, the whole sequence of reactions being carried out in a single reaction vessel. These so-called "one-pot" syntheses allow remarkable variation with simple changes of conditions. For example if tryptamine is condensed with secologanin in the presence of β-glucosidase (to remove the sugar) and sodium cyano-borohydride the initially formed imine is reduced to 2,3-secoakuammigine. However if the glucosidase and cyanoboro-hydride reagents are added after the initial condensation the same compound is obtained together with akuammigine, tetrahydroalstonine and a very small amount of ajmalicine. All of these products show the same stereochemistry at C-19 and this suggests a intramolecular proton transfer in some conjugated enamine intermediate with stereospecific β-attack by the enolate anion at C-19 to give the pentacyclic products by way of cathenamine (Brown, Leonard and S.K. Sleigh, Chem. Comm., 1977, 636).

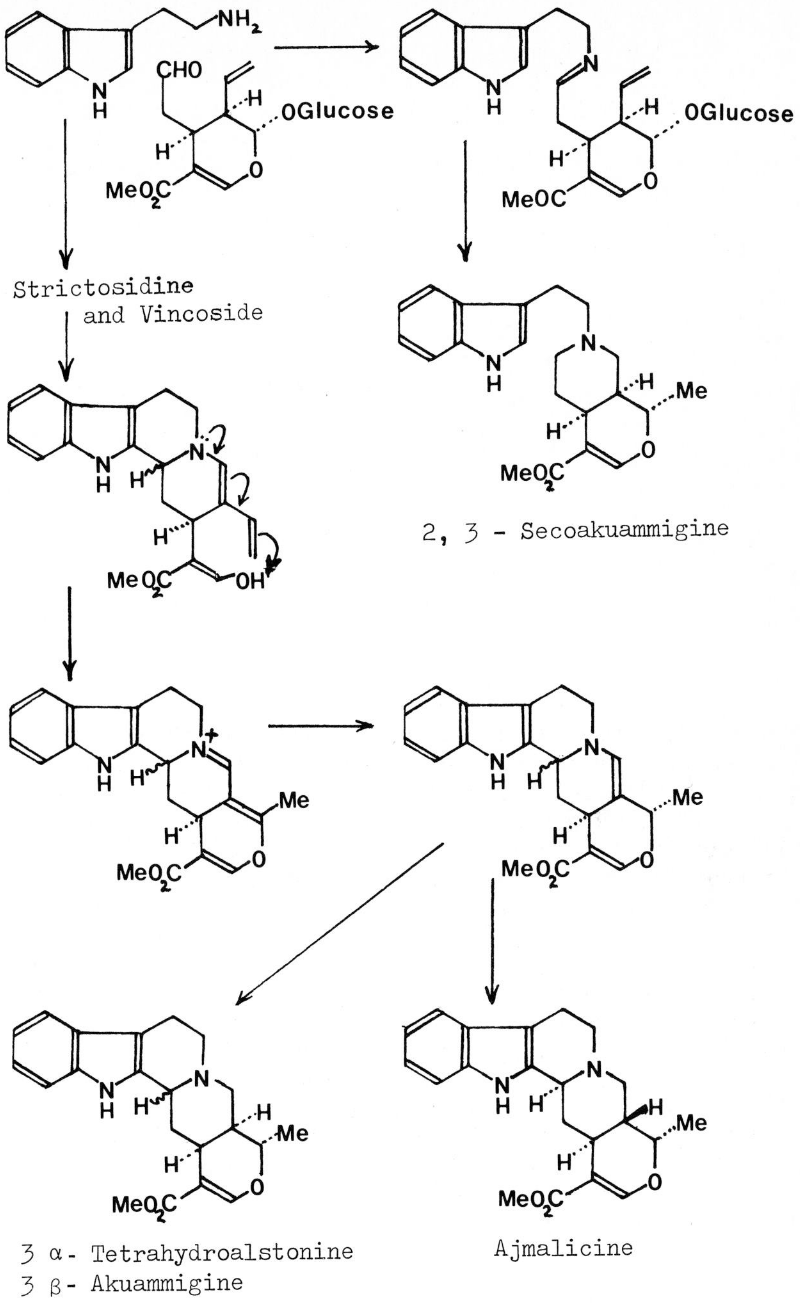

NH$_2$ CHO H OGlucose N H H MeO$_2$C O

N H N H OGlucose H MeOC O

Strictosidine
and Vincoside

N H N H H MeO$_2$C H Me O

2, 3 - Secoakuammigine

N H N H H MeO$_2$C OH

N H N$^+$ H Me H MeO$_2$C O

N H N H Me H MeO$_2$C O

N H N H H Me H MeO$_2$C O

3 α- Tetrahydroalstonine
3 β- Akuammigine

N H N H H Me H MeO$_2$C O

Ajmalicine

The synthesis of yohimbine types by a similar process has been achieved (Brown and S.B. Pratt, chem.Comm., 1980 165). In this case the secologanin moiety is modified before condensation with the tryptamine. The dicarboxylic ester related to secologanin is converted to a mixture of two aldehyde esters (17) before the condensation and the final cyclisation effected by standard means. A mixture of yohimbine and β-yohimbine resulted.

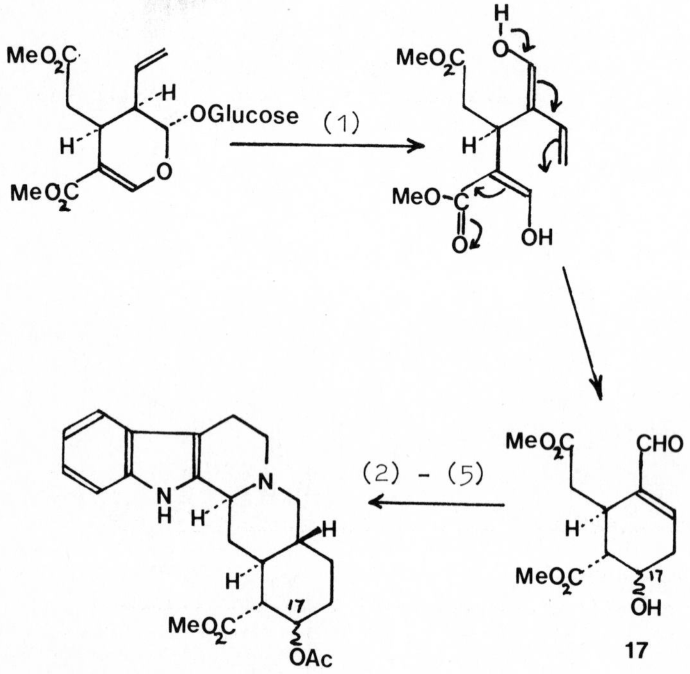

β- Yohimbine acetate C-17H α
Yohimbine acetate C-17H β

Reagents : (1) β-glucosidase (2) Acetic anhydride
(3) Tryptamine (4) $NaBH_3CN$
(5) $POCl_3/NaBH_3CN$

The absolute configurations at C-3, C-15 and C-20 in sitsirikine and at C-3 and C-15 in isositsirikine follow from the correlation of these alkaloids with corynantheine (2nd Edn. Vol.IV, pp.99-101). The stereochemistry at C-16 follows from the following observations. Hydration of the double bond in sitsirikine using mercuric acetate and sodium tetrahydroborate gives a diol which cyclises to the ether, cyclositsirikine. The stereochemistry at C-3 remains unaltered in this process as shown by comparison of the c.d. spectrum. The stereochemistry at C-16 follows from the 300 MHz ^{1}H nmr spectrum. The proton at this position appears as a triplet of doublets which shows two large diaxial couplings and one small axial-equatorial. Hence this proton must be axial and the stereochemistry at C-16 must be as shown (Brown and J. Leonard, Tetrahedron Letters 1979, 1805).

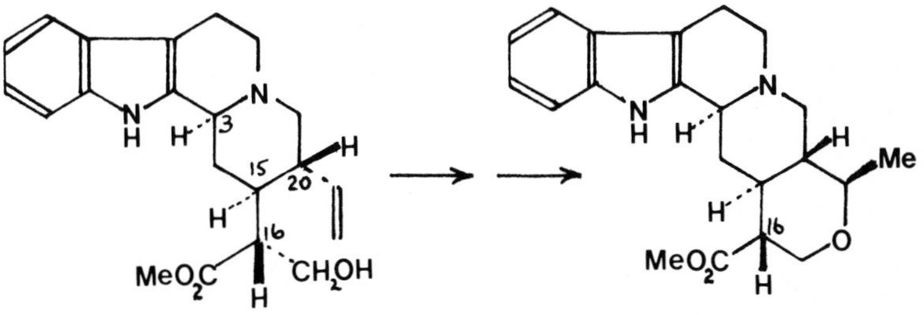

Sitsirikine Cyclositsirikine

The compound enantiomeric at C-16 may be similarly converted to a cyclic ether and in this case the proton is equatorial. Epimerisation with sodium methoxide converts this compound to the thermodynamically more favoured cyclositsirikine. Hydrogenation of iso-sitsirikine to the two epimers at C-20 allows correlation of this alkaloid with sitsirikine at C-16 (T. Hirata, S.L. Lee and A.I. Scott, chem.Comm., 1979, 1081).

The 2,3-bond in a variety of yohimbine and heteroyohimbine alkaloids may be cleaved by heating the compound with anhydrous formic acid in formamide (J. Le Men Tetrahedron Letters, 1978, 2153; S. Sakai *et al.*, Chem. and pharm. Bull (Japan) 1978, 26, 678). Presumably the C-3, N-4 immonium ion is produced by reverse Mannich reaction and this is reduced by the formic acid. Cleavage of the N-4, C-5 bond may then be used to release the rings D and E of the molecule which may then be condensed with tryptophyl (or substituted tryptophyl) bromide. The 2,3-bond may be regenerated by mercuric acetate oxidation. By this method reserpine has been converted to deserpidine (Sukai *et al.*, Heterocycles, 1978, 67).

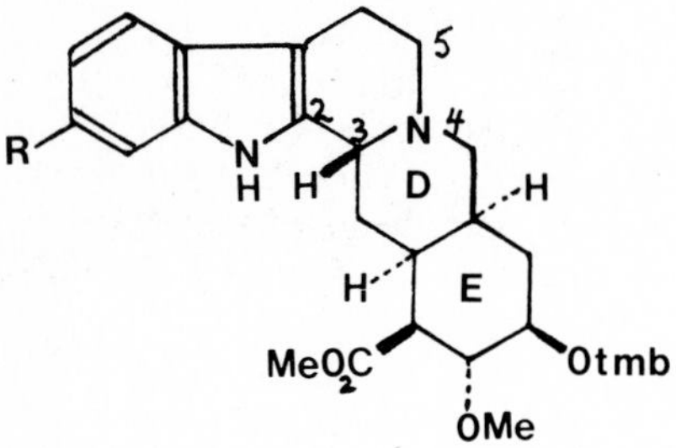

tmb = 3,4,5-$(MeO)_3$ C_6H_2CO

Reserpine R=MeO
Deserpidine R=H

The formic acid - formamide reagent has obvious applications for correlation between the oxindole and hetero-yohimbine groups of alkaloids (Le Men *et al.*, *ibid.*, 1977, 1129).

An interesting observation in the oxindole series is the production of only one isomer of the oxindole from venanatine and alstovenine (differing stereochemistry at C-3). These give a compound which can be coverted to its stereoisomer only with considerable difficulty; reflux in 30% aqueous acetic acid for 48h gave only 30% of the other isomer and this reverts to the original compound on attempted chromatography (P.L. Majumder, S. Joardar and T.K. Chanda, Tetrahedron, 1978, 34, 3341). Spectroscopic analysis of the products suggests the structures to be of the *allo* series (18) and (19); the destabilisation of the less readily obtained isomer may be attributed to considerable non-bonded interaction between the 9-methoxy-group and the *N*-4 lone pair. The two epiallo-isomers of the two suggested structures would be even less stable.

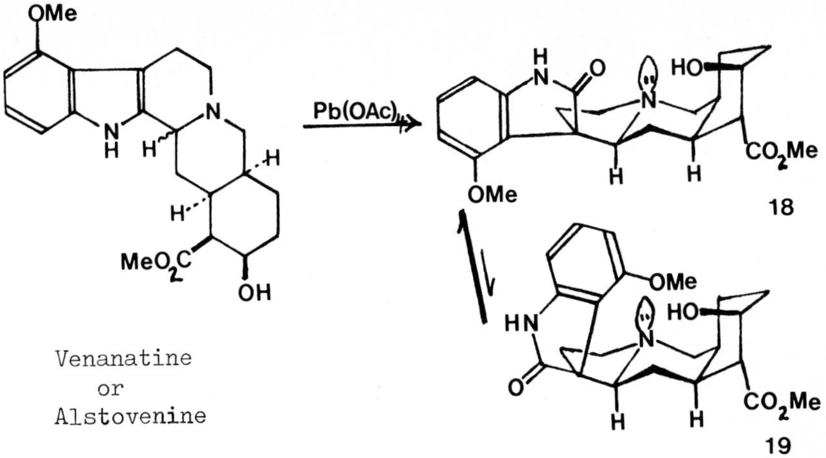

Venanatine
or
Alstovenine

$Pb(OAc)_4$

18

19

The widely held belief that oxindoles derived from the *pseudo*-yohimbine series are too unstable to exist because of steric crowding (see, for example, 2nd Edn. Vol IVB p.103), has been disproved (R.T. Brown and R. Platt, Chem.Comm.,1976. 357). When the oxindole analogue of dihydromancunine (20) are reduced with sodium tetahydroborate three diols, separated and characterised as their acetates, are obtained. Two of these are identical with the oxindoles obtained from dihydrositsirikine diol diacetate and this confirms the stereochemistry at C-20. The third product,given the *pseudo*-stereochemistry (21), is equilibrated in glacial acetic acid to a mixture of the other two. The possibility of inversion at C-20 is excluded by the incorporation of one deuterium atom (at C-16) when the reduction is carried out in deuteriomethanol. Under more mild borohydride reduction the hydroxy-ester is also isolated in the *pseudo*-series, as a mixture of C-16 epimers.

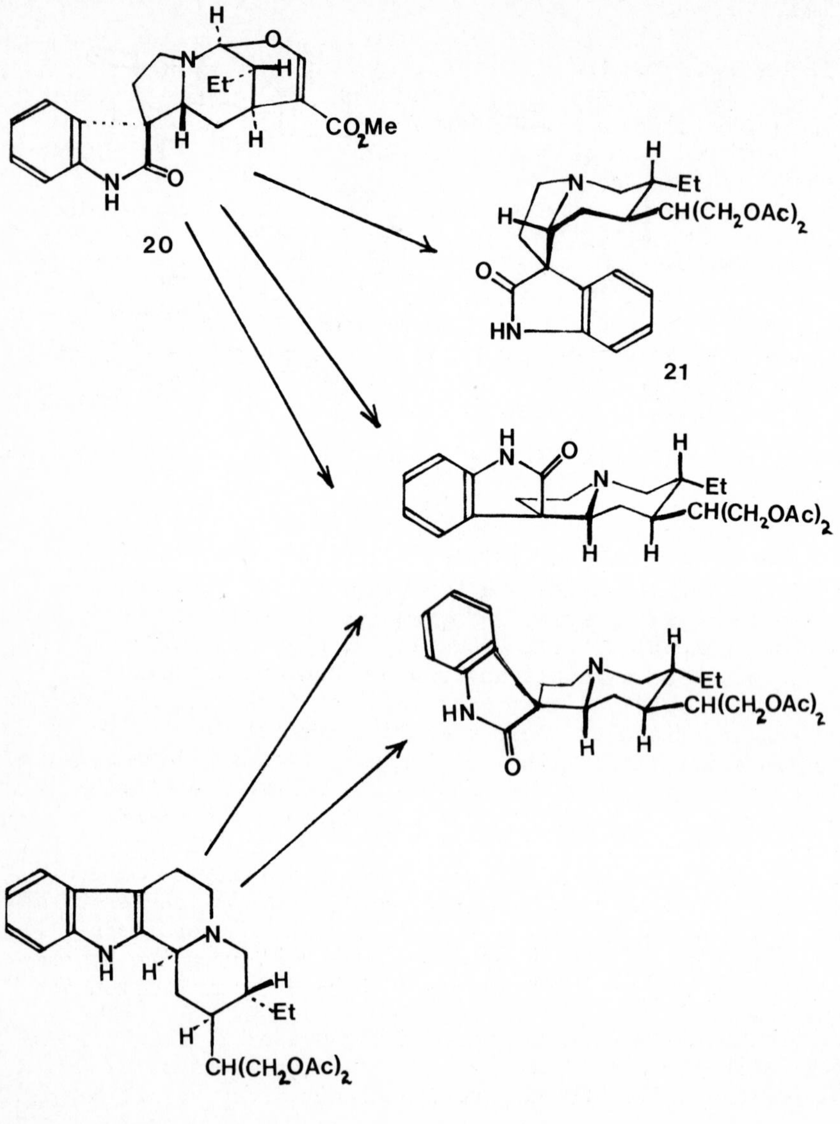

20

21

Dihydrositsirikine diol diacetate

(3) *Picraline-type alkaloids*

Three closely related alkaloids, from the aerial parts of *Alstonia lanceolifera*, provide an interesting variant on the picraline groups. These have been extensively oxidised and the N-4, C-5 bond has been cleaved with the generation of a lactone. The cinnamoyl group may be removed by methanolysis and on reaction with base there is loss of formaldehyde by reverse-aldol reaction. Diazomethane converts the methanolysis product to a methyl ester and generates an indoline chromophore in the product (G. Lewin, N. Kunesch and J. Poisson, Compt. Rend., 1975, 280, C, 987; J. Indian Chem. Soc. 1978, 55, 1096). The compounds are related to lanciferine, the other two being the 10-hydroxy and 10-methoxy compounds (R=OH; R=OMe).

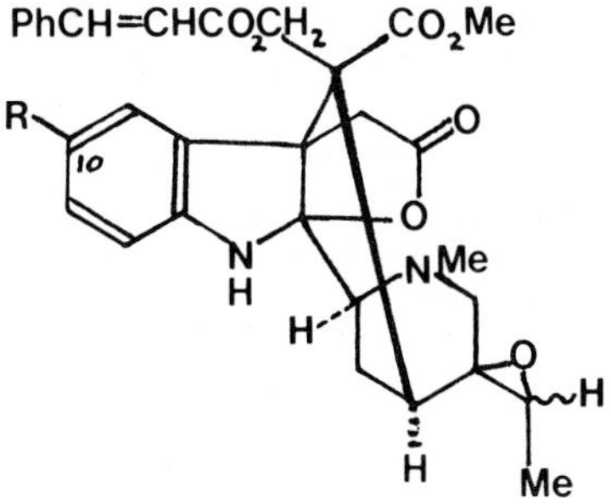

PhCH=CHCO$_2$CH$_2$ CO$_2$Me

Lanciferine R = H

The structure and stereochmistry of lanciferine have been deduced by careful correlation of spectra with degradation products from desacetyldesformo picraline (q.v.). Final correlation has not been achieved because of difficulty with epoxidation of the double bond in the ethylidene side-chain. (G. Lewin and J. Poisson, Bull. Soc.chim.Fr.,Part 2,1980,400).

The New Caledonian species *A.quaternata* provides a number of inter-related members of this group. Cathofoline, and its 19,20- epoxide, quaternoxine, are clearly related to deformo-akuammiline and quaternine is similarly related to picraline.

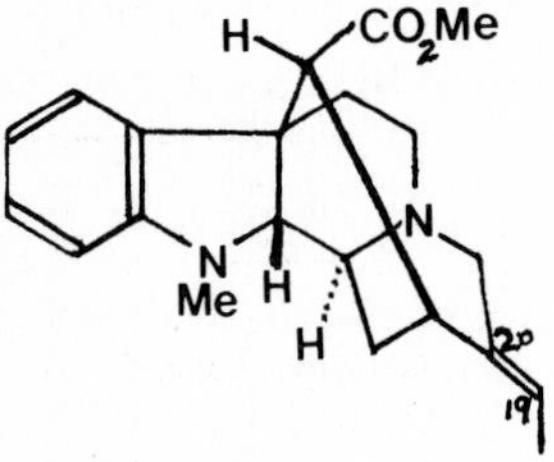

Cathofoline

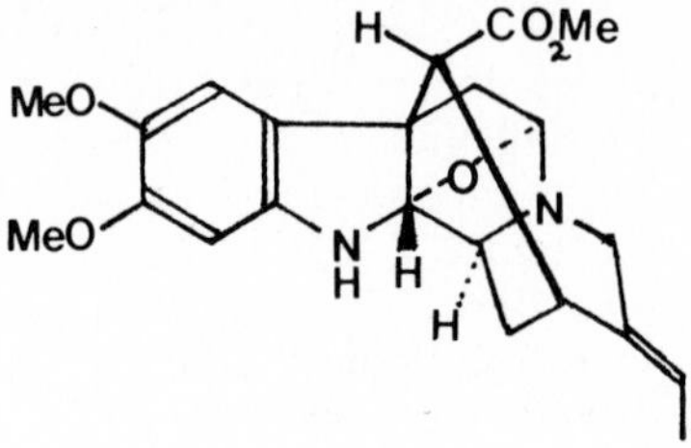

Quaternine

Quaternoline

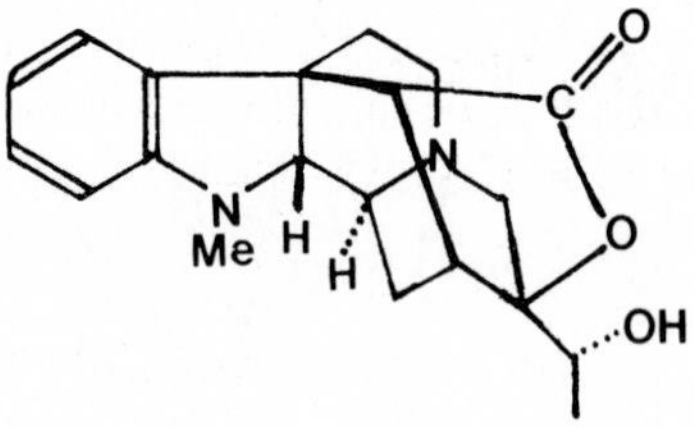

Quaternoline is clearly derived from quaternoxine (P. Potier *et al.*, Phytochem., 1975, 14, 1849). *Rhazya stricta* contains a number of picraline type alkaloids. Strictamine has been shown to be as depicted by X-ray crystallographic studies; strictalamine must have the same stereochemistry since both of these alkaloids are reduced to the same alcohol by complex metal hydrides (Y. Ahmad *et al.*, J. Amer. chem. Soc., 1977, 99, 1943). Rhazinaline has been shown to be 16-formyl-16-epistrictamine (A. Chatterjee et al., Bull chem. Soc. (Japan), 1976, 49, 2000).

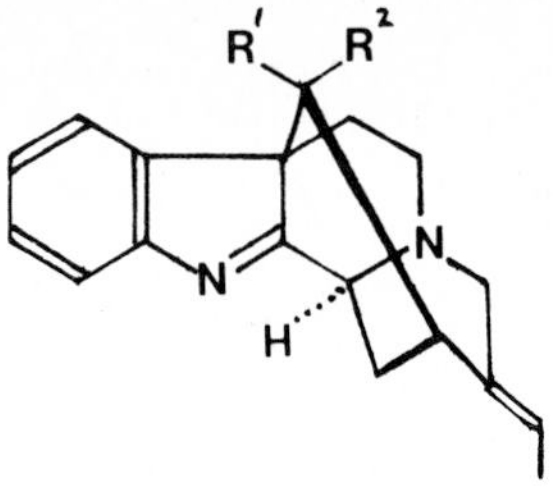

Strictamine R^1=H R^2=CO_2Me

Strictalamine R^1=H R^2=CHO

Rhazinaline R^1=CO_2Me R^2=CHO

The seeds and leaves of the Senegalese plant *Hunteria elliottii* and *H. congolana*, from Zaire, contain alkaloids of yet a further variant of the picraline system. In these compounds N-4 is bonded to C-2 rather than C-3 (L. Le Men Olivier, Plant. Med. Phytother, 1978, 12, 173). They are elaborations of corymine and are 3-*O*-acetyl- and 3,17-di-*O*-acetyl-3-epicorymine. Further derived from the corymine system is isocorymine, where N-4 has become detached from C-18 with the formation of a hemiacetal ring and a lactone from C-17 to C-3.

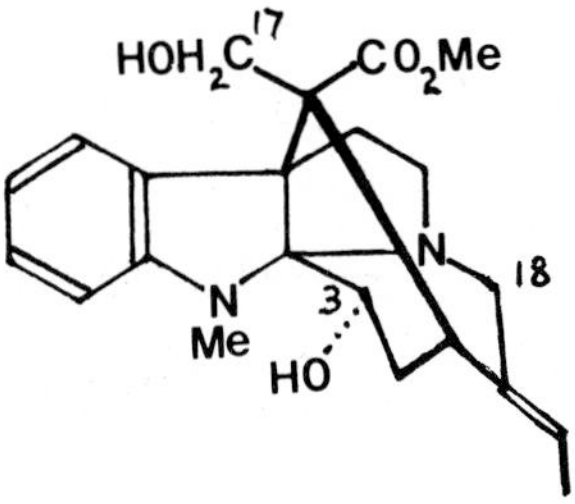

3-epi-Corymine

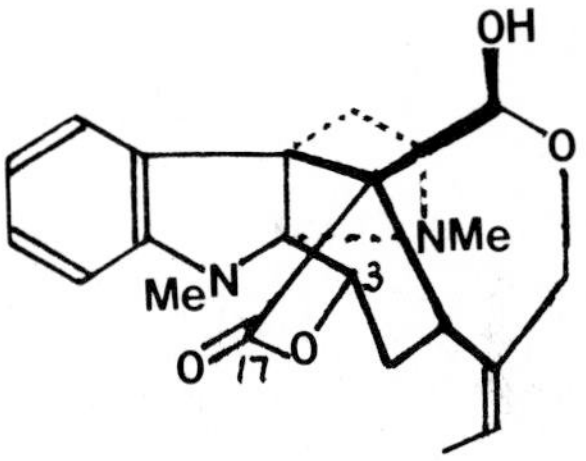

iso-Corymine

Yet a further variant is present in the form of eripinal where the oxidation level is as shown. Reduction of this with sodium tetrahydroborate gave the bis-lactone erinine.

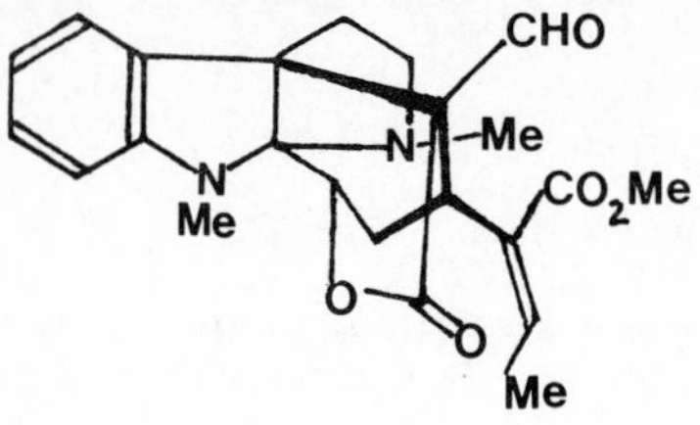

Erpinal

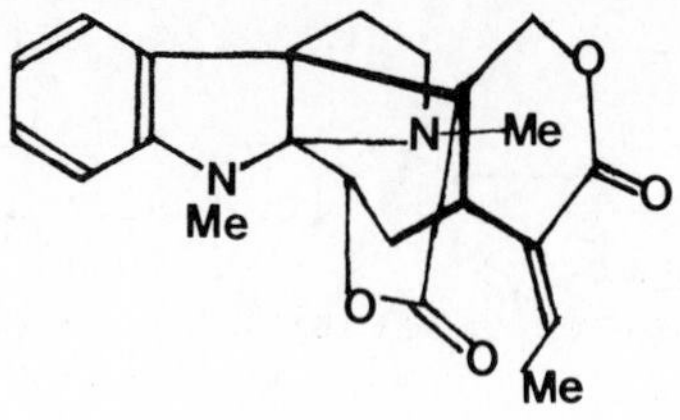

Erinine

An extremely complex variation of the picraline skeleton is seen in naureline (M. Hesse *et al.*, Helv. 1977, 60, 1419). The alkaloid, from *Alstonia scholaris*, could be formed from picraline, as shown, or from strictamine; the structure follows from single-crystal X-ray diffraction studies.

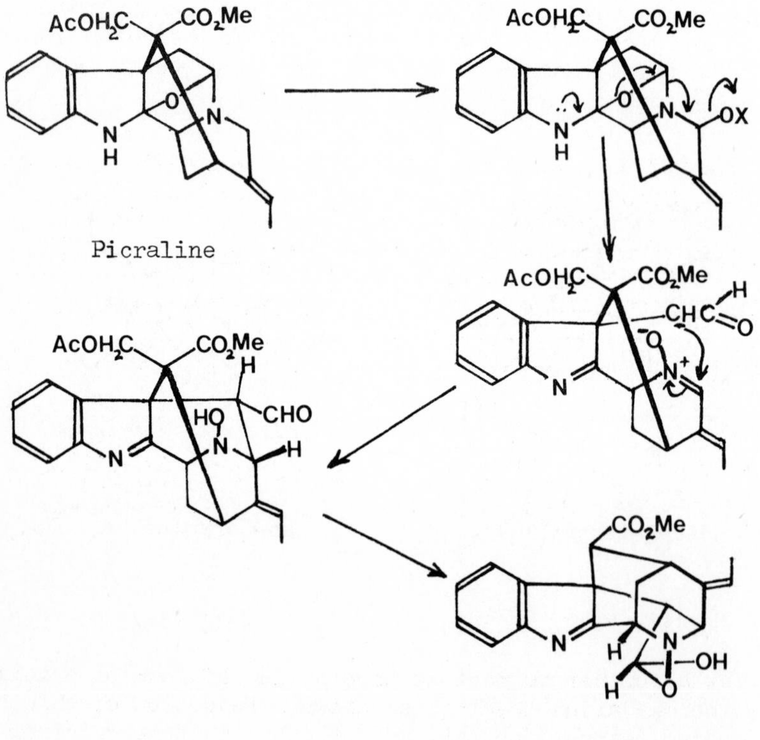

Picraline

Naureline

(4) *Ajmaline - Sarpagine type alkaloids*

Alstonia lanceolifera has been found to contain ajmaline derivatives which are esters of 3,4,5-trimethoxy-cinnamic acid (J. Poisson *et al.*, Phytochem., 1975, 14, 2067). These are derivatives of vincamajine containing hydroxy or methoxy groups at C-10 (R=OH or R=OMe). The similarity in the oxygenation pattern and the presence of the $C_6 - C_3$ unit to their congener lanciferine (q.v.) is significant.

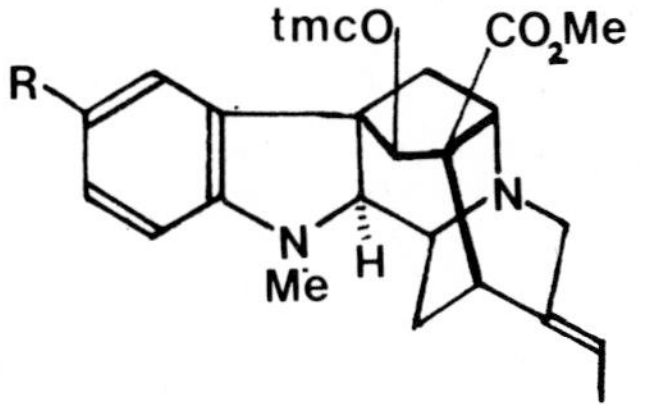

tmc = 3,4,5 - $(MeO)_3C_6H_2CO$

Vincamajine R = H

Purpeline (rauflexine) and reflexine are two alkaloids isolated from *Rauwolfia reflexa*. They are related as ketone and secondary alcohol and may be interconverted. (A. Chatterjee *et al.*, Experientia, 1976, 32, 1236). The correct substitution pattern in the aromatic ring follows from a detailed examination of the ^{1}H and ^{13}C nmr spectra (*idem.*, Tetrahedron Letters 1978, 3879).

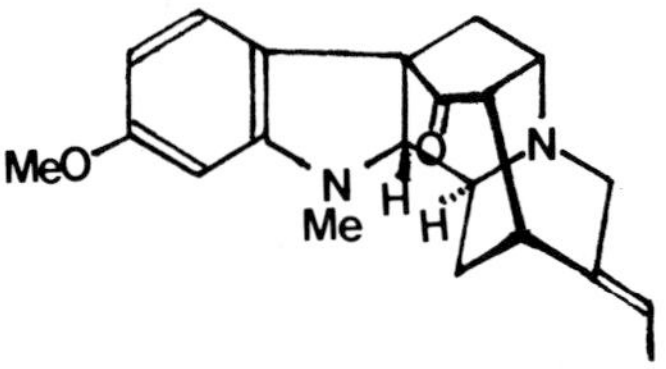

Rauflexine

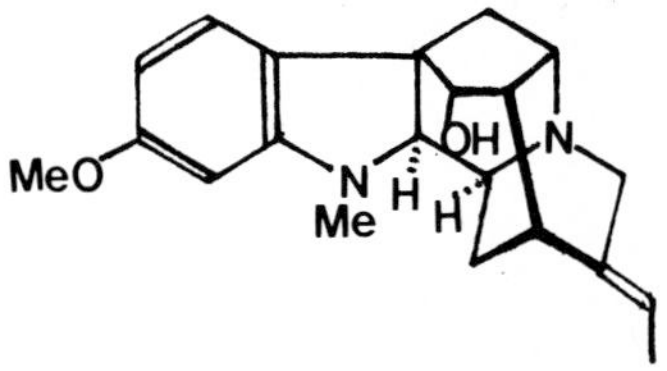

Reflexine

An alkaloid which is most probably related to the sarpagine group is ervitsine, a 2-acyl indole from the root bark of the Madagascan plant *Pandaca boiteau*. The structure follows from X-ray studies and correlated with methuenine. Ervitsine is a Mannich base and as such will undergo ring opening in dilute acid to give an intermediate which may re-cyclise to a compound which may be hydrogenated to give the hydrogenation product from methuenine.

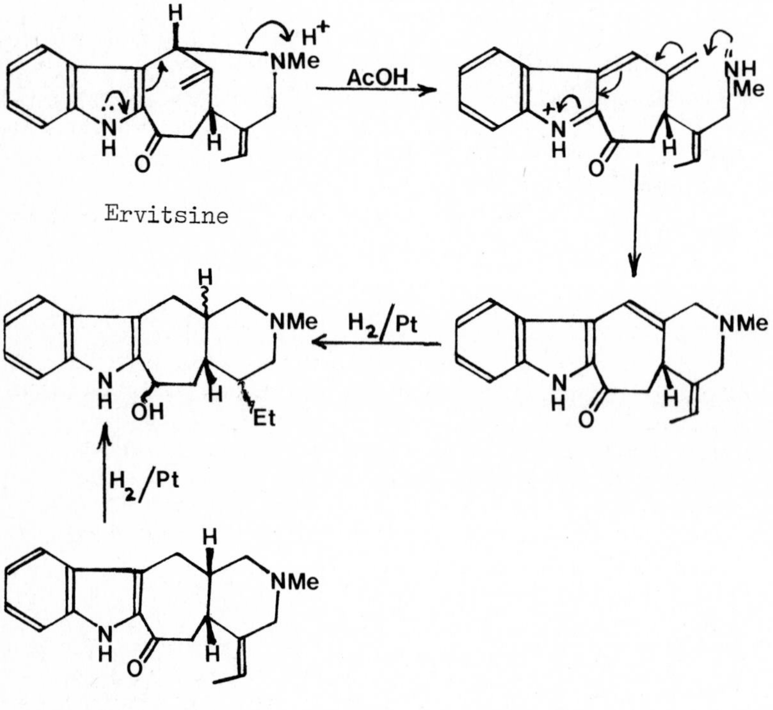

Ervitsine

Methuenine

Ervitsine may be obtained from vobasine *N*-oxide or its biochemical equivalent. Similar conversions of *N*-oxides of vobasine and its two 19,20-dihydrocompounds, dregamine and tabernaemontainine (differing in stereochemistry at C-20) has been achieved in the laboratory (A. Husson *et al.* Tetrahedron, 1973,29,3095; P. Mangeney *ibid.*, 1978,34,1359).

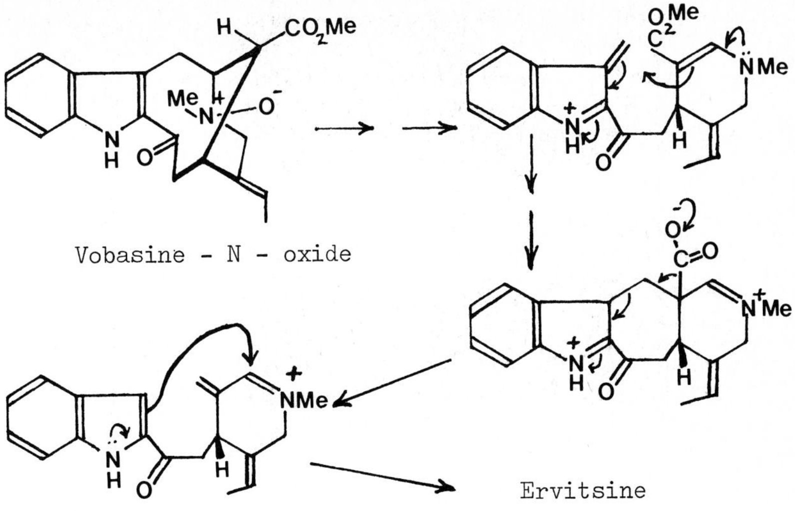

Vobasine - N - oxide

Ervitsine

Gardneria multiflora continues to provide oxindole and related alkaloids which are clearly derived from sarpagine type precursors (S. Sakai *et al.*, Yakugaki Zasshi, 1977, 97,399). These are all related to gardneramine and may be obtained from this alkaloid or easily converted into it. The structures are tabulated below.

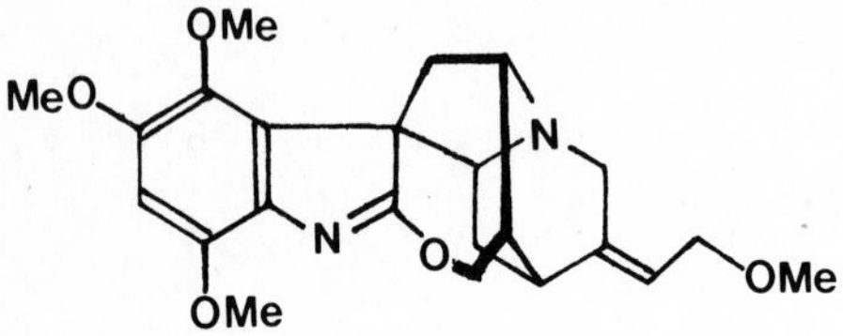

Gardneramine

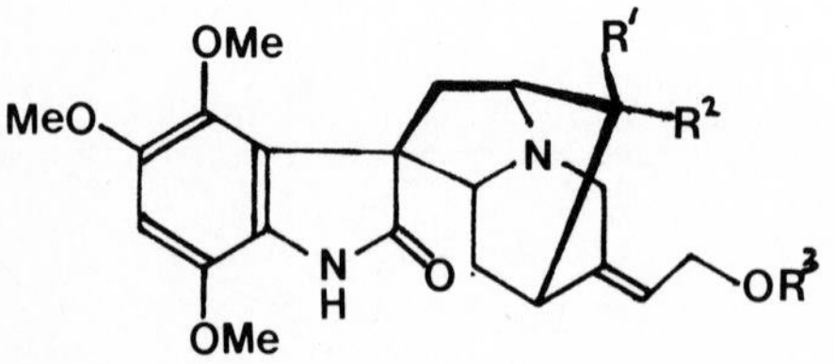

Alkaloid J
R^1=H R^2=CH_2OH R^3=Me

Alkaloid L
R^1=H R^2=CH_2OH R^3=H

Alkaloid I
R^1=CH_2OH R^2=H R^3=Me

Alkaloid M
R^1=CH_2OH R^2= R^3=H

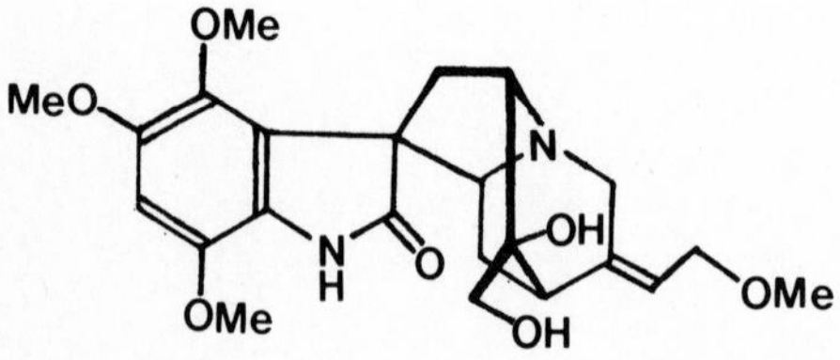

Alkaloid N

(5) *Carboline derivatives not containing a C-21 to N-4 bond*

The structure assigned to akagerine contains the novel N-1 to C-17 bond (2nd Edn. Vol. IV B p.111); however a number of other alkaloids related to this compound have been obtained from *Strychnos* species. *S. dale* and *S. elaecarpa* both contain akagerine and kribine, the hemiacetal from 17-*epi* akagerine (W. Rolfsen *et al.*, Planta Med., 1978 34 264). Kribine is a mixture of two isomers and the methyl ethers (i.e. the acetals) of these co-occur although they may be artefacts produced in the methanol extraction. *S. decussata* contains a variety of 10-hydroxy-akagerine and kribine acetals (*ibid.*, J. Nat. Prod., 1980, 43, 97).

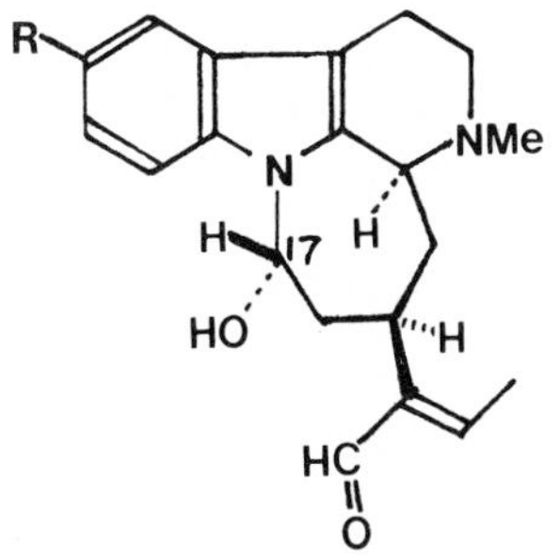

Akagerine R=H
10-Hydroxyakagerine R=OH

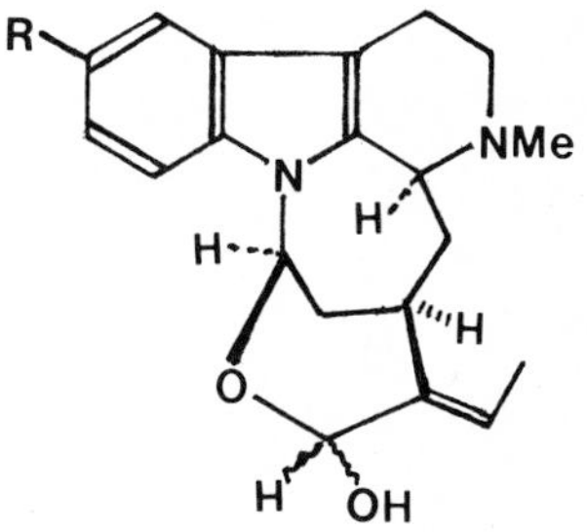

Kribine R=H
10-Hydroxykribine R=OH

(6) *Strychnos alkaloids*

A number of alkaloids of Strychnos holstii and S. henningsii have been shown to be *seco*-strychnine types with the C-3 to N-4 bond cleaved and differing from strychnine in having the O to C-18 bond also cleaved. These are holstiine, rindline and hostiline and the structures are deduced from mass spectral and nmr data. The mass spectrum shows m/e 124 and $M^{+\bullet}$=57 ($M^{+\bullet}$=71 for rindline and hostiline) which indicate the fragmentations a and b respectively (G. Spiteller *et al.*, Phytochem, 1975,14,1411).

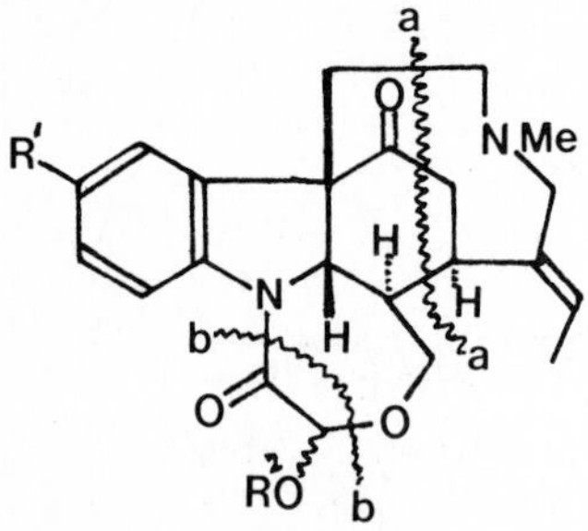

Holstiine (R^1=R^2=H)

Rindline (R^1=OMe R^2=Me)

Hostiline (R^1=H R^2=Me)

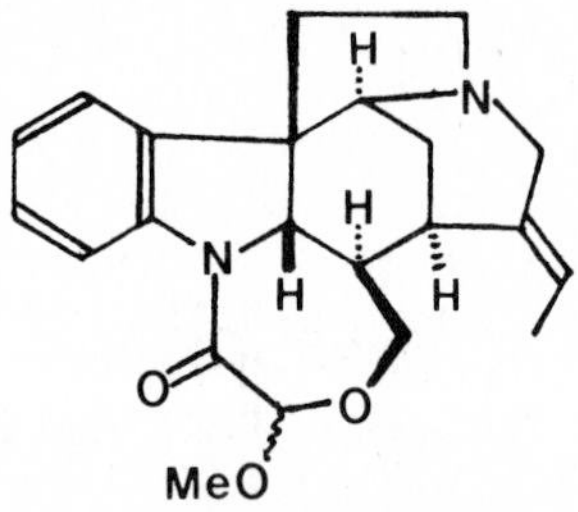

Tsilanine

A feature of these compounds is the unusual amide-hemiacetal function on the indole nitrogen atom. They may be seen as related to tsilanine, which is more closely related to strychnine, and is also found in *S. henningsii*, taking its name from the Madagascan name for the plant, 'tsilanimboana'. A number of rather similar skeletal types have also been reported. These show the oxidation at C-3 at the carbonyl and carbinolamine levels. *S. fendleri* provides strychnofendlerine where the indole *N*-acetyl group evident in strychnine biosynthesis is attached although the oxidation level of an aldehyde in a Wieland-Gumlich aldehyde type intermediate is missing (G. Galeffi *et al.*,Gazz., 1976,106,773). Strychnofendlerine is the C-19 epimer of the more well-known isosplendine (from *S.Splendens*).

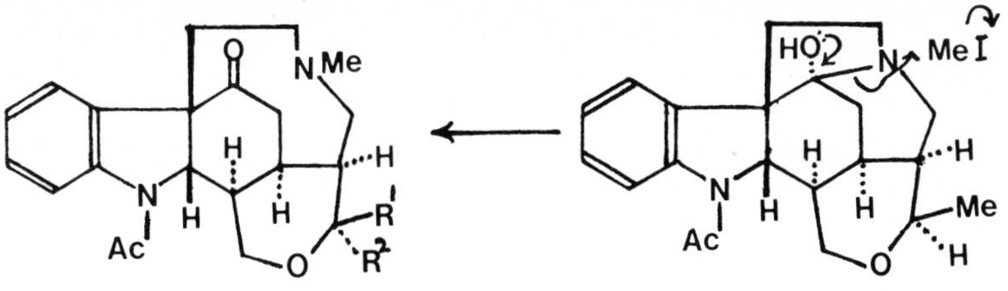

Strychnofendlerine
R^1 = Me R^2 = H.
Isosplendine R^1=H R^2=Me

N-1-acetyl-
Strychnosplendine

Strychnosplendine contains a carbinolamine unit and when N-1-acetylstrychnosplendine is obtained from the same plant it is possible to convert this alkaloid to strychofendlerine by methylation. *S. icaja* has provided a number of new strychnine related species which are in a highly oxidised state. (N.G. Bisset and A.A. Khalil, Phytochem., 1967,15,1973). Typical examples are provided by (22) and (23).

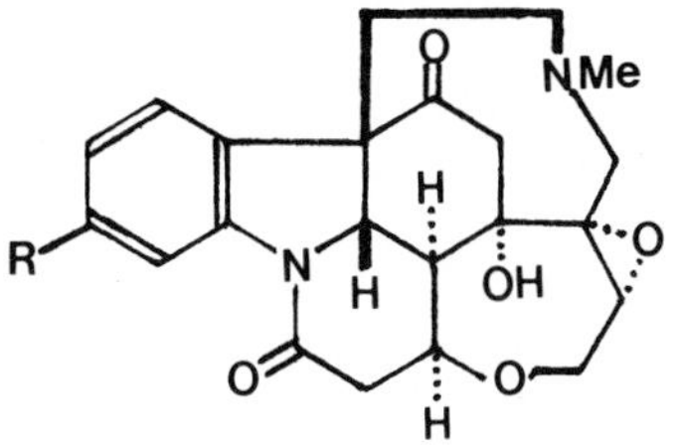

22 R=H
23 R=OMe

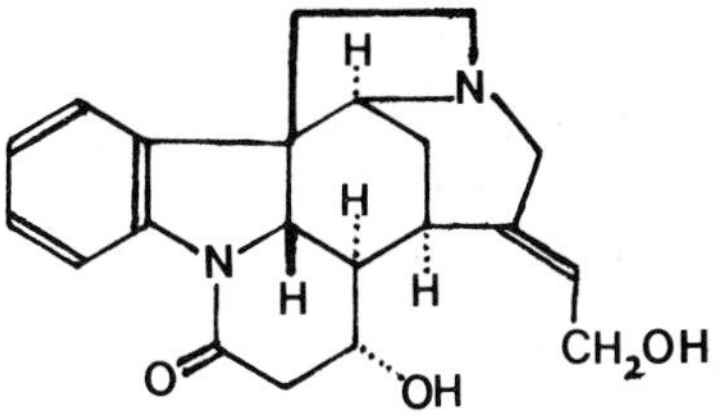

Protostrychnine

A compound which is probably the immediate precursor of strychnine has been obtained from *S. nux-vomica* root bark. (K.H.C. Baser, N.G. Bisset and P.J. Hylands, Phytochem. 1979, 18,512). Given the name protostrychnine, this minor alkaloid is converted to strychnine by converting it into its *O*-18 ester of tiglic acid and reacting this with phosphorus oxychloride followed by saponification and treatment with dilute hydrochloric acid.

A remarkably stable compound having an unusual structure is hunteracine from *Hunteria eburnea* (R.H. Burnell, A. Chapelle and M.F. Khalil, Canad. J. Chem., 1974,52,2327). It is stable to high temperature even though it is quaternary salt; in nature it probably arises from a hydroxyindolenine related to stemmadenine.

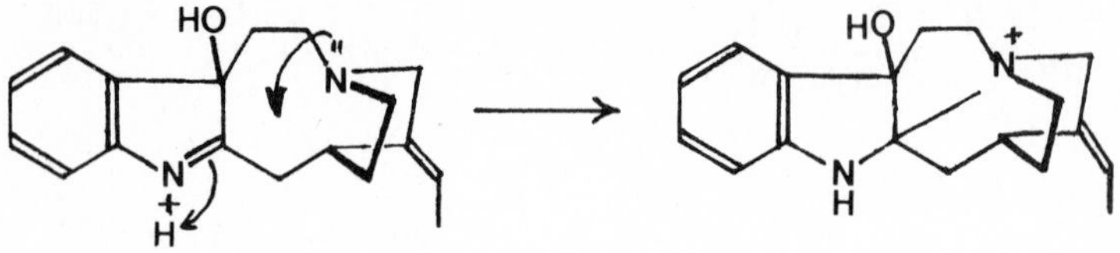

Hunteracine

An alkaloid not containing an indole related unit but found as a congener of indole alkaloids is goniomine, m.p. 223°, from *Gonioma malagasy* (H.P. Husson *et al.*, J. Amer. chem. Soc., 1980, 102, 5920). The ir spectrum indicates the presence of a carbonyl group and the most important features of the ^{1}H nmr spectrum is the presence of a methyl doublet (δ 1.42) and two protons showing a mutual small coupling in the alkene proton region. The mass spectrum gives the molecular formula as C-19 H-22 O-2 N-2 and reduction with sodium tetrahydroborate in methanol gives a dihydrocompound which still displays carbonyl absorbtion in the ir spectrum. The ^{1}H nmr spectrum of this derivative shows an additional methyl doublet with the disappearance of the alkene proton signals. Hence it follows that the alkaloid contains an exocylic methylene group conjugated with the carbonyl group which is reduced rather than the carbonyl function. The structure has been confirmed by a single-crystal X-ray diffraction study of the dihydrocompound. This structure is a new system which is considered to be derived from a strychnos skeleton, possibly related to precondylocarpine, as shown.

HOH$_2$C CO$_2$Me

Precondylocarpine

Goniomine

It is questionable if goniomine is a genuine alkaloid or an artefact; the analogy with rhazinilam (q.v.) is noteworthy.

(ii) *Alkaloids with a seco-type unit*

There have been no further reports of monomers related to secodine although there have been several syntheses directed towards this compound. Secodine itself has been obtained by two independent routes (Kutney *et al.*, Canad. J. Chem., 1979, 57,289 ; S. Raucher, *et al.*, J. Amer. chem. Soc., 1981, 103, 2419). The compound obtained dimerises rapidly to compounds of the presecamine series.

(iii) *Alkaloids with the Aspidosperma unit*

This class of alkaloids occurs with probably the largest number of variations in oxidation level and oxygenation patterns. The possibility of cyclisations of these oxidised compounds allows for a wide range of structures. Lochnericine and hörhammericine have been isolated from *Catharanthus lanceus* and *C. trichophyllus*. (G.A. Cordell and N.G. Farnsworth, J. pharm. Soc., 1976, 65, 366). It is believed that the cytotoxic and anti-tumour propertimes of the latter may be attributed to their presence. Cathovalinine, from *C. ovalis*, is an isomer of hörhammericine differing in stereochemistry at positions 14/15 and/or 19; the stereochemistry follows from X-ray crystallographic studies. (Potier, *et al.*, Tetrahedron, 1976, 32, 1899). The stereochemistry of hörhammericine is unknown.

R = H Lochnericine
R = OH Hörhammericine

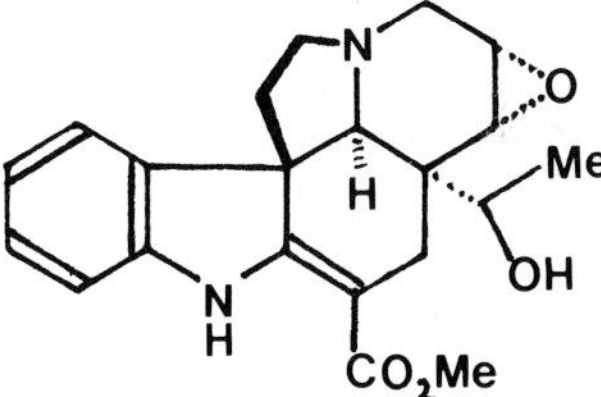

Cathovalinine

Cathophylline is a new type of compound from this source. Its mass spectrum shows that two oxygen atoms must be present in the angular sidechain and since it does not show the properties of either an acid or of an aldehyde it must be hydroxyketone.

Cathophylline

The two epimers of 19-hydroxytabersonine have been isolated from *C. ovalis*(R.Z. Andriamialisoa, N. Langlois and P. Potier, Tetrahedron Letters, 1976, 163) and their structures established by synthesis from 19-iodotabersonine (q.v.). They are accompanied in the plant by two further alkaloids whose spectral properties resemble vincoline (2nd Edn. Vol.IV B, p. 132) except for the presence of an additional carbon atom as a methoxy group. These have been named kitramine and kitraline (*idem.*, Phytochem., 1979, 18, 467) and may be obtained from the two epimeric 19-hydroxytabersonines by oxidation with lead tetra acetate followed by methanolysis. Clearly they are the 2-methoxy compounds since melobaline and vincoline are obtained by hydrolysis instead of methanolysis.

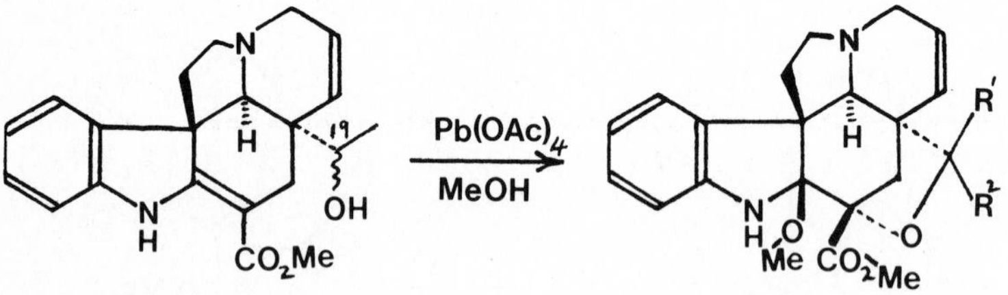

$Pb(OAc)_4$ / MeOH

19-Hydroxytabersonine

Kitraline R^1=H R^2=Me

Kitramine R^1=Me R^2=H

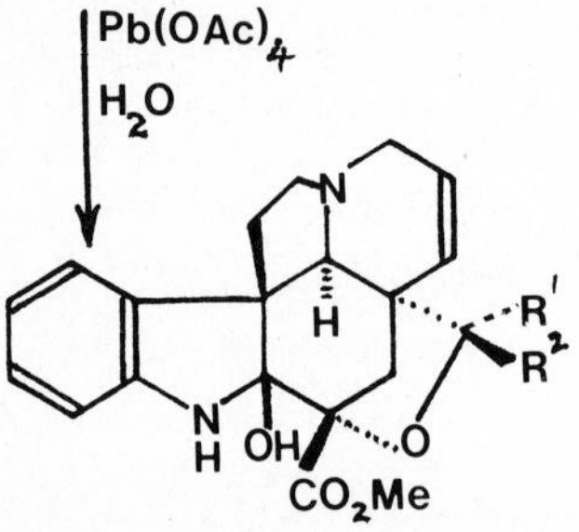

$Pb(OAc)_4$ / H_2O

Vincoline R^1 = H
R^2 = Me

Melobaline R^1 = Me
R^2 = H

Aspidosperma alkaloids oxidised at C-3 have been isolated from *Amsonia elliptica* (N. Aimi *et al.*, Chem. and pharm. Bull (Japan), 1978, 26, 1182). In particular 3-oxotabersonine and the corresponding 14,15 epoxide have been isolated and characterised.

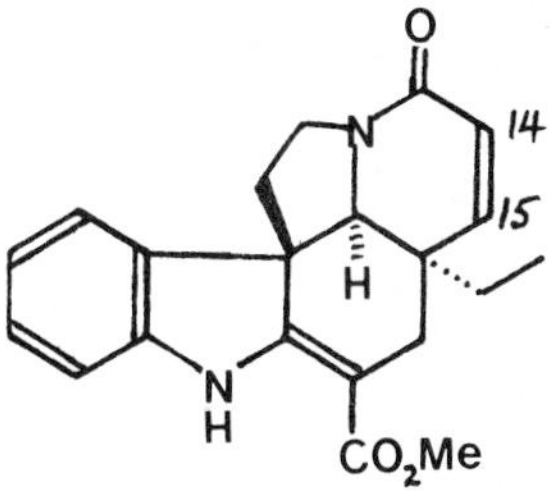

3 - Oxotabersonine

Two further oxygen cyclised Aspidosperma units have been isolated. *Melodinus celastroides* contains buxomeline, (A. Rabaron *et al.*, Phytochem., 1978, 17, 1452). If this structure is in fact correct, this is a remarkable alkaloid since the presence of carbinolamine and carbinolamine ether units in such close proximity (at N-4) would suggest the compound to be readily hydrolysed. Alalakine from *Aspid-dosperma album* has an ir-spectrum similar to that of obscurinervine (2nd. Edn. Vol.IVB p.127) (P. Potier Compt. Rend., 1978, 287C, 63), and its mass spectrum indicates a similarity to the aromatic part of the same alkaloid. The structure follows from a detailed analysis of its 400 MHz ^{1}H nmr spectrum.

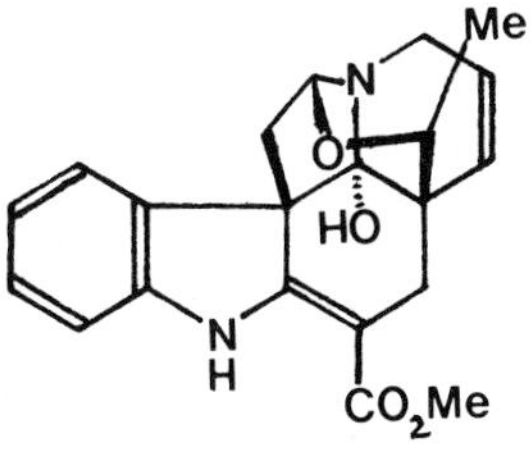

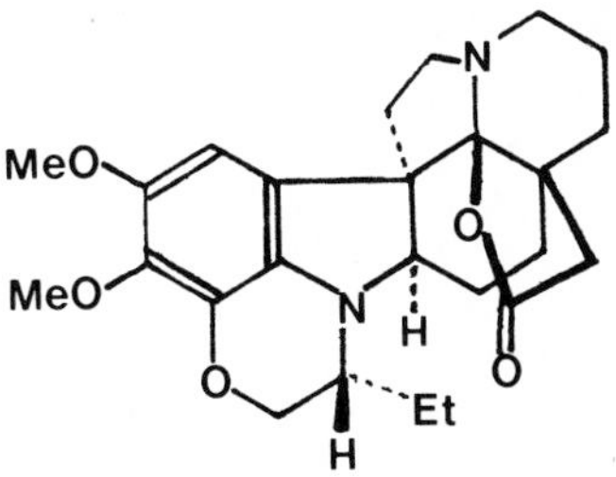

Buxomeline

Alalakine

The relationships of rhazinilam to an oxygenated Aspidosperma type has been described (2nd. Edn., Vol. IVB, p.125) and there is some laboratory analogy for this conversion. When tabersonine is oxidised with *m*-chloroperbenzoic acid and rearranged with triphenylphosphine in acetic acid the major products are vincamine derivatives (J. Le Men *et al.*,Tetrahedron, 1980, 36, 511). It is possible to convert one of these products to the eburnamine alkaloid criocerine. A minor product from the oxidation-rearrangement is a pyrrole-*N*-oxide (24) which can be hydrogenated, hydrolysed and decarboxylated to give a hydroxy-rhazinilam which on reduction with lithium tetrahydridoaluminate gives rhazinilam (25, R = H).

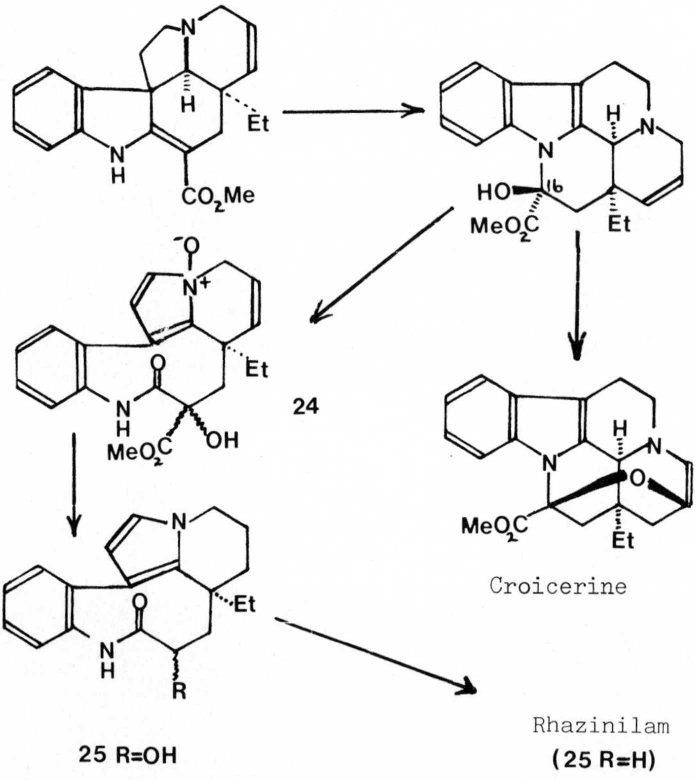

N H Et CO$_2$Me

H N N HO 16 MeO$_2$C Et

$^-$O N$^+$ Et O N H MeO$_2$C OH

24

H N N O MeO$_2$C Et

Croicerine

N Et O N H R

25 R=OH

Rhazinilam
(25 R=H)

The reaction of vindolinine with iodine in weakly alkaline solution gives the two epimers of 19-iodotabersonine, a compound which has proved invaluable in preparation of other Aspidosperma systems. Elimination of hydrogen iodide gives Δ^{18} tabersonine which on heating gives andranginine (2nd. Edn. Vol. IV B p.145) together with an *allo catharanthine* derivative. Andranginine may be formed via a secondine type intermediate.

Vindoline

19 - Iodotabersonine

Andranginine

15-Methoxy-14,15-dihydro-Δ^{18}-allocatharanthine.

Andranginine and the C-21 epimer are obtained as racemates whereas the *allo*-catharanthine derivative is optically active. This suggests that the cyclisation to the andranginine system is not concerted but since the *allo*-catharanthine derivative is optically active the C-1?/C-20 bond is not broken in the reaction.

An alternative rearrangement occurs when 19-iodotabersonine is heated with sodium acetate in DMF (*idem.*, *ibid.*, 1976, 32, 2839). A high yield of the eleven-membered ring compound (26) is obtained, presumably by the mechanism shown. Since this unit is found in some bisindoles this is an important observation.

19 - Iodotabersonine $\xrightarrow[\text{DMF}]{\text{NaOAc}}$

26

When 19-iodotabersonine is oxidised using silver tetrafluoroborate in DMSO (Kornblum oxidation) the 19-oxo compound is obtained which may be reduced to minovincine.

This in turn may be converted by dry methanolic hydrogen chloride into 16-*epi*-19-oxo kopsinine, a member of the aspidofractine group. (N. Lauglois and R.Z. Andriamialisoa, J.org. Chem., 1979, 44, 2468).

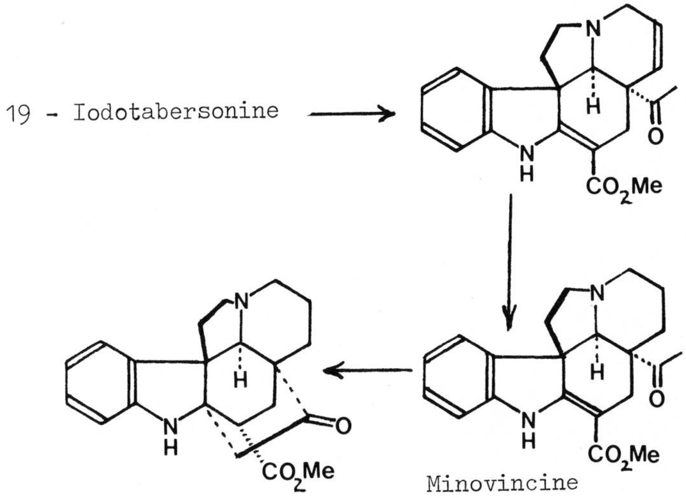

19 - Iodotabersonine

Minovincine

(iv) *Alkaloids containing the Iboga unit*

The variants in this small group of alkaloids remain few although investigations of the *Pandaca* species show these plants to contain such compounds. Hydroxylated derivatives include 3-hydroxycoronaridine and the ether formed by joining this oxygen atom to C-6, i.e. eglandine. These have been isolated from the root bark of *Gabunia eglandulosa* (J. Le Men *et al.*, Bull Soc. Chim. France, 1974, 1369). A compound from the same source but with a higher oxidation level is eglandulosine; on reduction with lithium tetrahydroaluminate all three compounds are reduced to coronaridinol, the product of metal hydride reduction of coronaridine. This allows correlation of the structures.

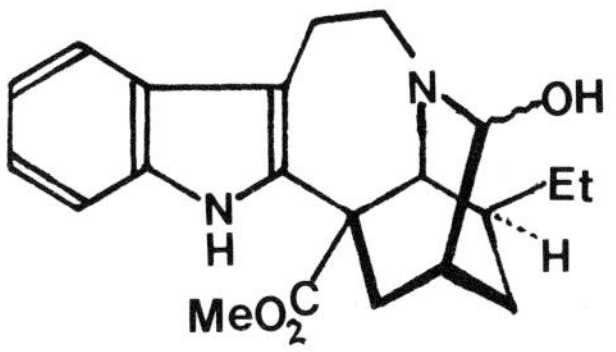

3 - Hydroxycoronaridine

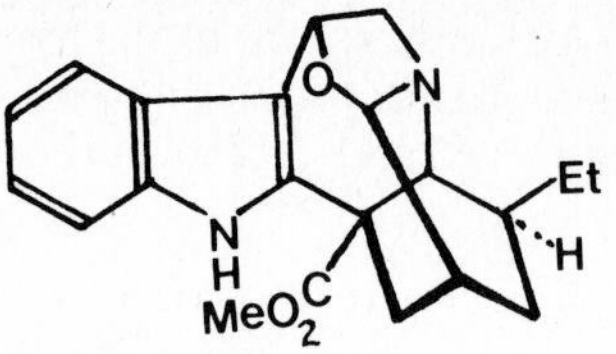

Eglandine

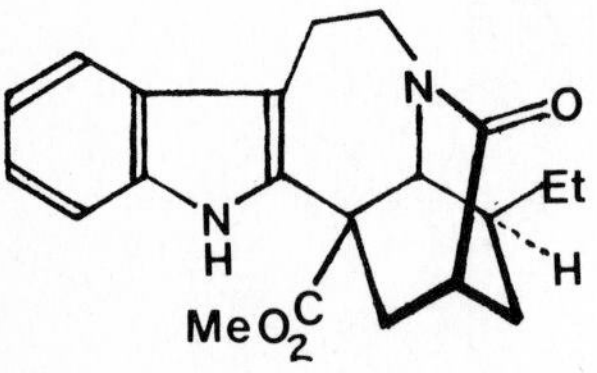

Eglandulosine

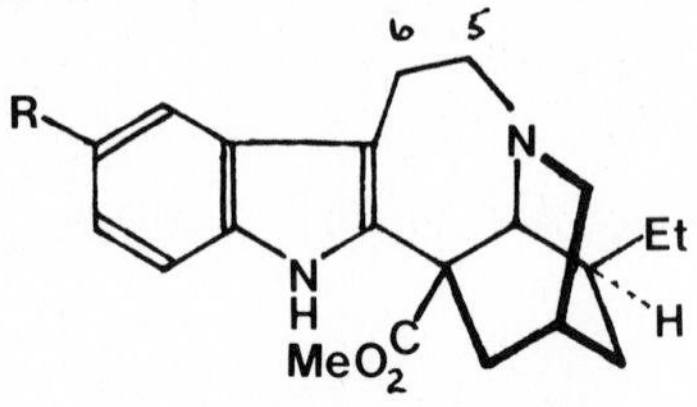

Coronaridine R = H
Voacangarine R = OMe

Highly oxygenated compounds have been isolated from *Tabernaemontara divaricata* (K. Ratogi, R.S. Kapil and S.P. Popli, Phytochem., 1980, 19, 1209); these include 5-oxo and 6-oxo-coronaridines. Yet a different cyclisation of an oxygenated form is displayed by heyneatine from *Ervatamia heyneana* (S.P. Gunasekera, G.A. Cordell and N.R. Farnsworth, *ibid.*, 1213). The presence of the carbinolanime ether is obvious from the ^{1}H nmr spectrum and the reduction with sodium tetrahydroborate gives 19(*S*) - voacangarine. Also obtained from the same source is 10-methoxyeglandine *N*-oxide.

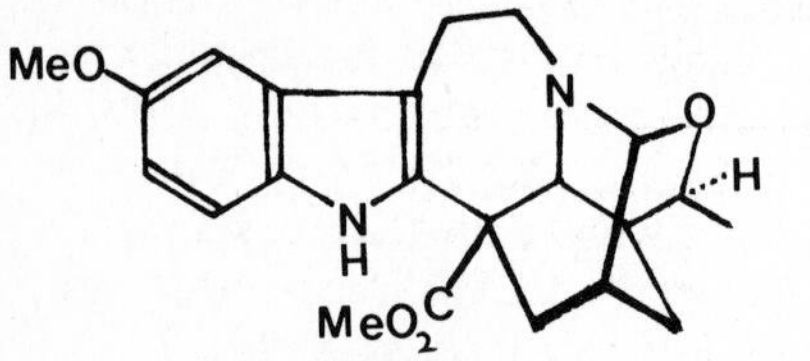

Heyneatine

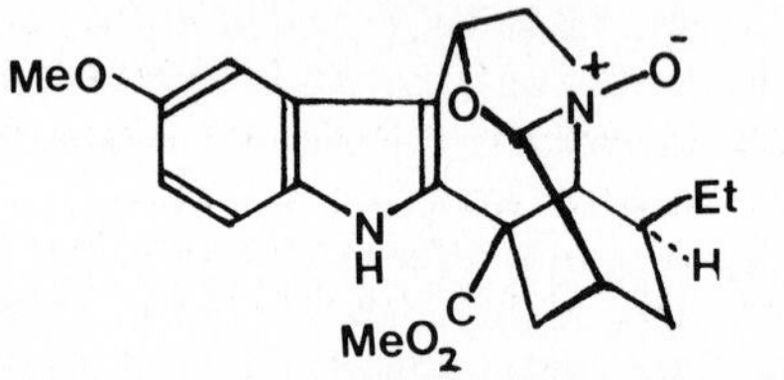

10 - Methyoxyeglandine-N-oxide

Further compounds having the *pseudo*-Aspidosperma skeleton have been obtained. *Capuronetta elegans* provides 14,15-anhydrocapuronidine and 14,15-anhydro-1, 2-dihydrocapuronidine (H.P. Husson *et al*., Phytochem, 1978, 17, 1605).

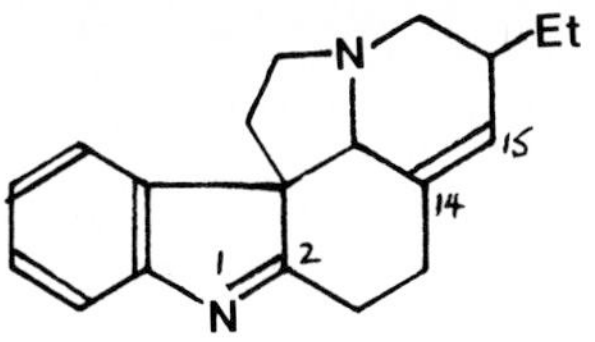

14,15 - Anhydrocapuronidine

Two further variants on the *pseudo*-Aspidosperma skeleton (which in turn are probably derived from the Iboga by way of a cleavamine type) are shown by ibophyllidine and iboxyphilline. These have been obtained from *Tabernaemontana iboga* and *T. subsessilis*,the bark of both of these species being used as stimulants (in small quantities) or as hallucinogens by the natives of Gabon. Ibophyllidine shows a β-anilinoacrylic ester chromophore but shows a prominent peak at m/e 110 (as opposed to m/e 124 in vincadifformine) and this suggests a lower homologue. An ethyl group is obvious from the ^{1}H nmr spectrum and the ^{13}C nmr indicates that this is attached to one of three methine groups in the molecule, two of which, including the one bearing the ethyl group, are attached to nitrogen. There is only one quaternary carbon atom, namely C-7, and this allows the structure shown to be deduced. The structure of iboxyphylline has been deduced by X-ray crystal analysis (R. Goutarel *et al.*, Tetrahedron, 1976, 32, 2539). These compounds could arise in nature from pandoline by oxidative fission of the 20,21-bond followed by Mannich recyclisation (to iboxyphylline) or by loss of C-21 followed by cyclisation and reduction.

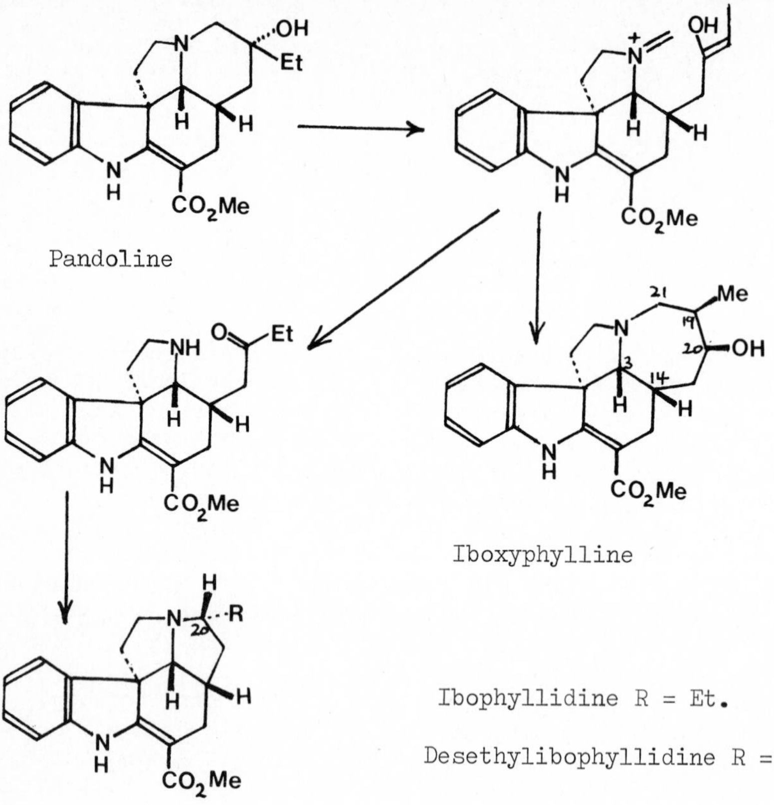

Pandoline

Iboxyphylline

Ibophyllidine R = Et.

Desethylibophyllidine R = H

The C-20 epimer of ibophyllidine has been obtained from the trunk bark of *T. albiflora* together with the desethylibophyllidine. (Husson *et al.*, Tetrahedron Letters, 1980,55).

Desethylibophyllidine has also been obtained from Anacampta disticha (Husson *et al., loc. cit*). This is the first example of an alkaloid from this group which has lost the ethyl side-chain; this, presumably, occurs in the contraction of the piperidine ring during biosynthesis.

The stereochemistry of an indoxyl alkaloid in this group has been deduced using ^{13}C nmr spectroscopy using lanthanide shift reagents. Iboluteine has the R-configuration at C-2 with the six-membered ring of the spirocyclic linkage adopting a twist boat conformation. This is attributed to the proximity of the carbonyl group to the axial hydrogen atoms at C-5 and C-21 when the ring is in the chair conformation (E. Wenkert and H.E. Gottlieb, Heterocycles, 1977, 753).

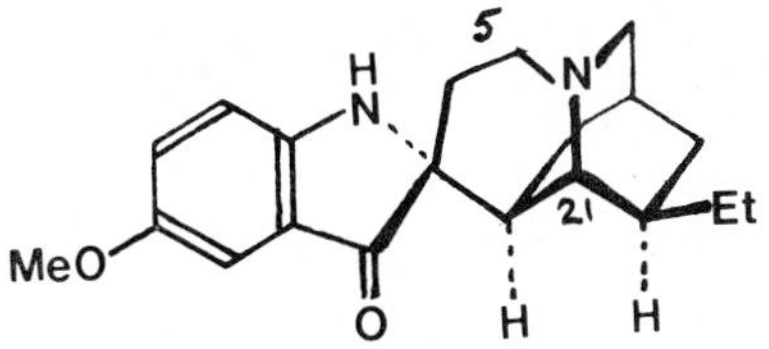

Iboluteine

The synthetic activity in this area is largely concerned with the attempts to prepare bisindoles related to the vinblastine group.

(v) *Novel types*

The number of alkaloids related to aristoteline has increased, although the sole source has been Aristotelia species, where they occur in very small quantities. They are characterised by having a C-9 or C-10 unit of standard monoterpenoid origin but which does not suggest the intermediacy of loganin.

Aristoteline has been isolated from *A. Serrata,* the New Zealand wine-berry, and the structure based on chemical and spectral analysis supported by X-ray crystallography (B.F. Anderson *et al.*, Chem. Comm., 1975,511). It also occurs in the Chilean species, *A.chilensis*,together with a minor alkaloid aristotelone, which exhibits the uv and ir spectra of *pseudo*-indoxyl (P.G. Sammes *et al.*, Phytochem., 1976,15,574). Because of the minute quantities of material available the structural assignment as a *pseudo*-indoxyl analogue of aristoteline relies solely upon mass spectral data.

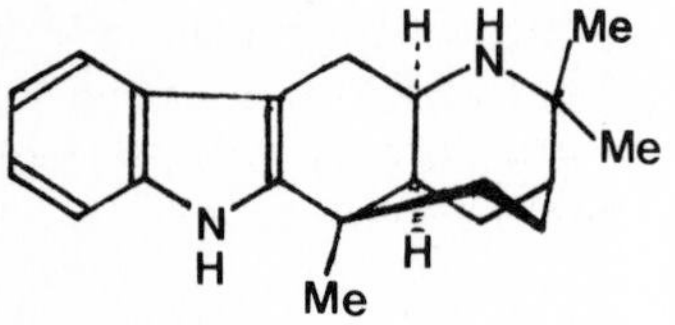

Aristoteline

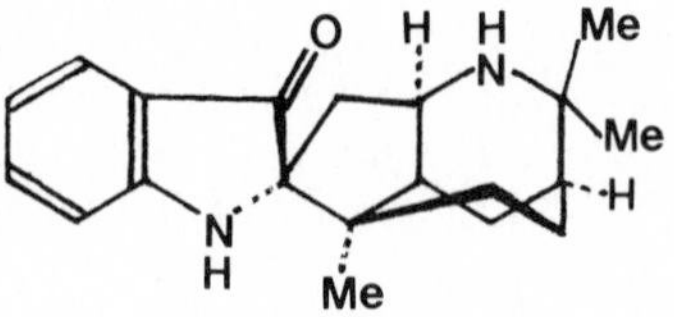

Aristotelone

Aristotelinone, from *A. serrata,*is a 3-acylindole which may be reduced to aristoteline by lithium tetrahydroaluminate, and must have the structure shown.(I.R.C. Bick *et al.*, Tetrahedron Letters, 1980, 545). Serratoline, a congenor, has a remarkably similar ^{1}H nmr spectrum to that of one of the β-hydroxy-compounds obtained in the reduction of aristotelinone but also has an extra singlet at δ 7.30 p.p.m. and shows the uv absorption of an indolenine. On reduction with sodium tetrahydroborate the alkaloid gives an indoline believed to be (27, R^1=H, R^2=OH) which has undergone rearrangement and which on acid catalysed dehydration gives aristoteline. This allows the structure for serratoline to be postulated.

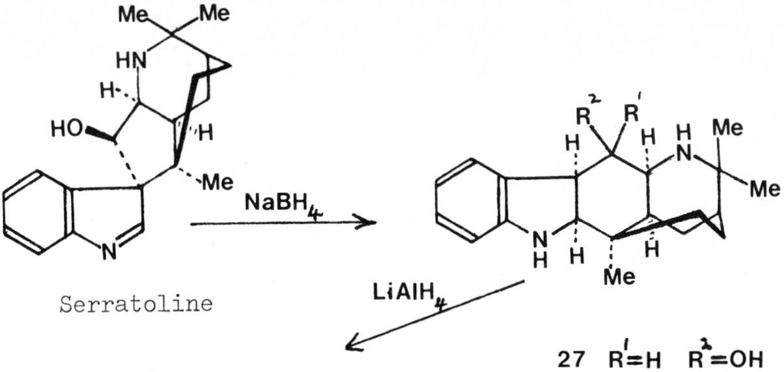

$NaBH_4$

$LiAlH_4$

Serratoline

Aristoteline

27 R^1=H R^2=OH

A.chilensis yields extremely small amounts of two further alkaloids of this type; because of the small quantities available (9mg. of both from 10kg. of leaves) it is possible that they are artefacts due to aerial oxidation. Their structures have been deduced by X-ray crystallography (Sammes *et al.*, chem. Comm., 1978,79). Aristotelinine is a hydroxyindolenine derivative of a hydroxyaristoteline, and the other, aristone, may be derived from this by rearrangement.

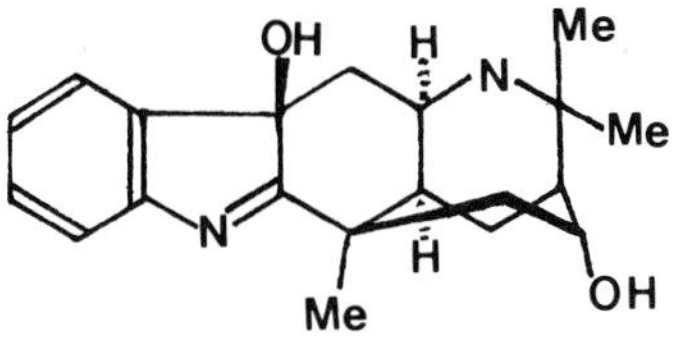

Aristotelinine

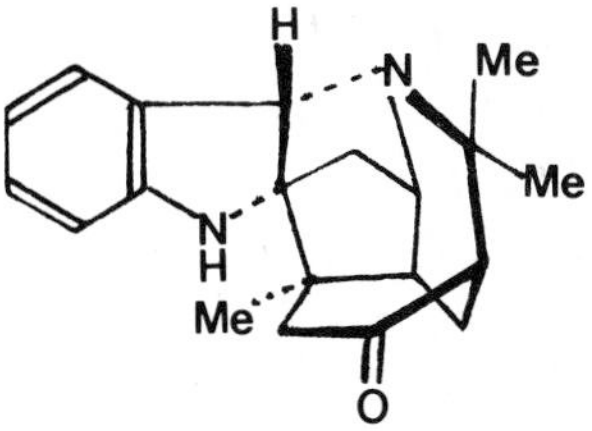

Aristone

The structure originally postulated for peduncularine, a compound isolated from *A. peduncularis* (Bick *et al.*, chem. Comm., 1971,1155) has been shown to be incorrect. The currently accepted structure follows from a more detailed study of mass spectral, ^{1}H and ^{13}C nmr spectroscopic and chemical data (H.P. Ros *et al.*, Helv., 1979 62, 481). The structure does not appear to be directly derived from a simple monoterpenoid unit but obviously it originates from a similar source.

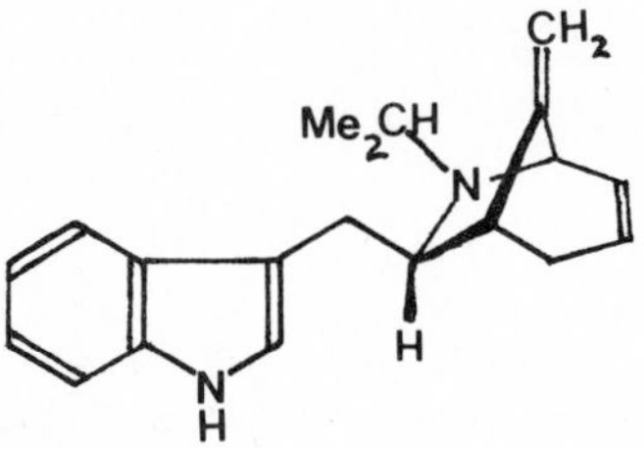

Peduncularine

Two further minor alkaloids from this source, which is the Tasmanian variety, are sorelline and hobartine. Their structures, which follow from a detailed analysis of their nmr and mass spectra, contain an unrearranged monoterpenoid unit.

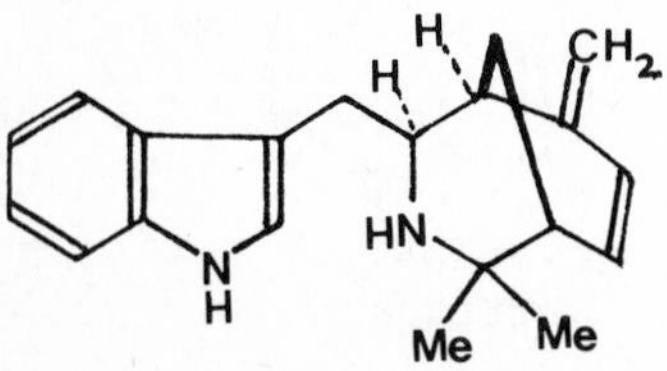

Sorelline

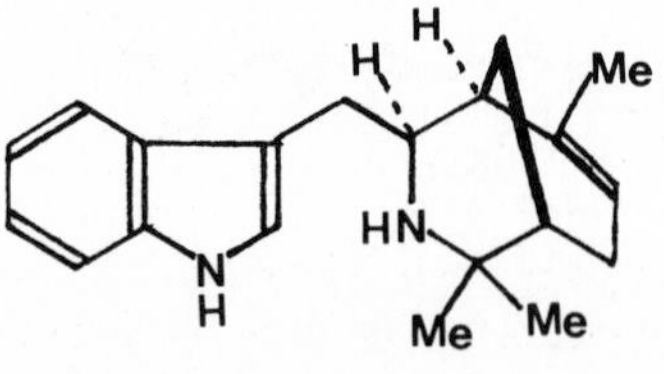

Hobartine

3. *Bis-indole alkaloids*

(a) *Compounds containing two identical "halves" linked symmetrically.*

Full details of the structural investigation of the quadrigemines have been published (K.P. Parry and G.F. Smith, J. chem. Soc. Perkin Trans. I, 1978, 1678). Related to these alkaloids but not strictly a bisindole is psychotridine from the shrub *Psychotria beccarioides,* found in New Guinea (J.A. Lamberton *et al.*, Aust. J. Chem., 27, 1974,639). This compound is a pentamer of tryptamine. The number of similarities between the ^{13}C nmr spectrum of this compound and that of hodgkinsine simplified the structural assignment.

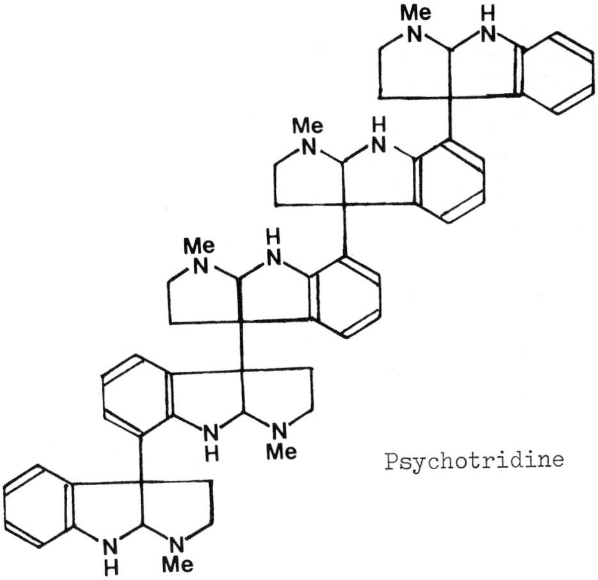

Psychotridine

A compound which may be identified as a dimer of diketopiperazine alkaloid origin is ditryptophenaline which has been isolated from *Aspergillus flavus*. The structural assignment follows from X-ray studies and may be seen as two molecules of tryptophan and two molecules of *N*-methylphenylalanine (G. Buchi *et al.*, Tetrahedron Letters, 1977, 2403.

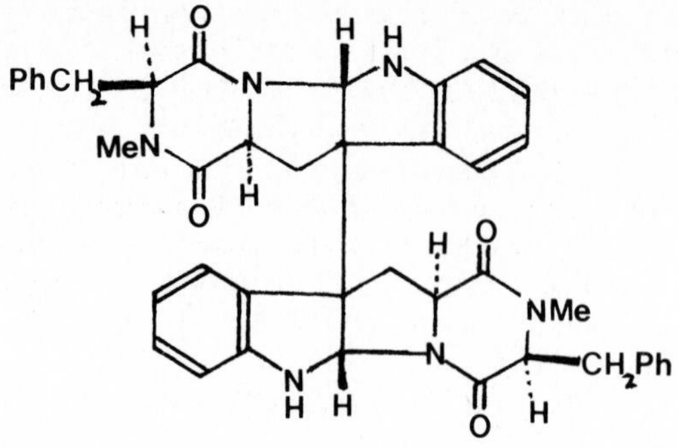

Ditryptophenaline

Other bisindoles of a similar origin biosynthetically are the antibacterial agents, the melinacidins. These antibiotics are produced by *Acrostalagmus cinnabarimus Var. melinacidinus* and are clearly related to chaetocin and verticillin (2nd Edn. Vol.IVB p.147) (A.D. Argoudelis and S.A. Mizsak, J. Antibiotics, 1977, 30,468). They only differ in their oxygenation pattern.

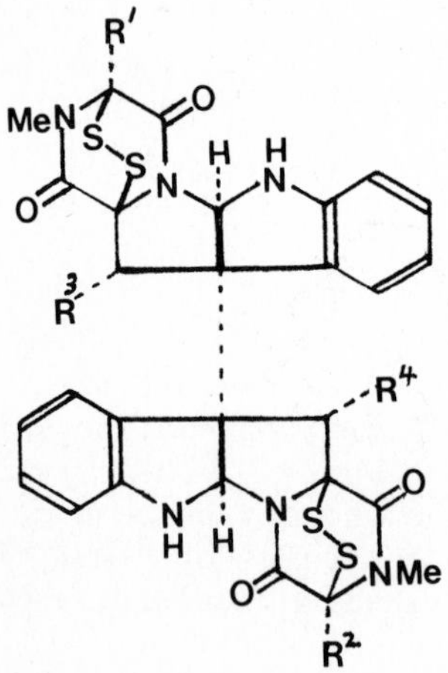

Melinacidin II
R^1 = Me
R^2 = CH_2OH
R^3 = H or OH
R^4 = OH or H

Melinacidin III
R^1 = R^2 = CH_2OH
R^3 = H R^4 = OH

Melinacidin IV
R^1 = R^2 = CH_2OH
R^3 = R^4 = OH

A further variant on this skeleton is displayed in chetomin, from a culture of *Chaetomium cochliodes*. The structure follows from an elegant combination of ^{15}N and ^{13}C nmr spectroscopy (A.G. McInnes, A. Taylor and J.A. Walter, J. Amer. chem. Soc., 1976, 98, 6741). It differs principally from the melinacidins in that one tryptamine unit has not become involved in an esorine-type unit. The compound was initially isolated more than 30 years previously (S.A. Waksman and E. Bugie, J. Backteriol., 1944, 48, 527).

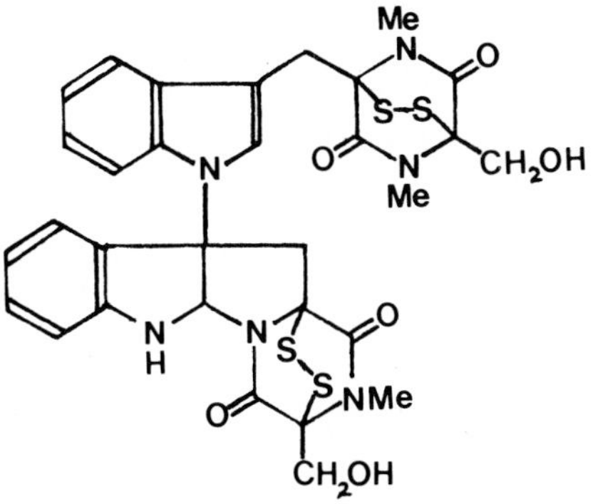

Chetomin

A compound containing two "*Iboga*" units has been found to occur in the leaves of *Bonafousia tetrastachya* (M. Damak C. Poupat and A. Lond, Tetrahedron Letters, 1976, 3531). This is 12,12'-bis-11-hydroxycoronaridine, (28).

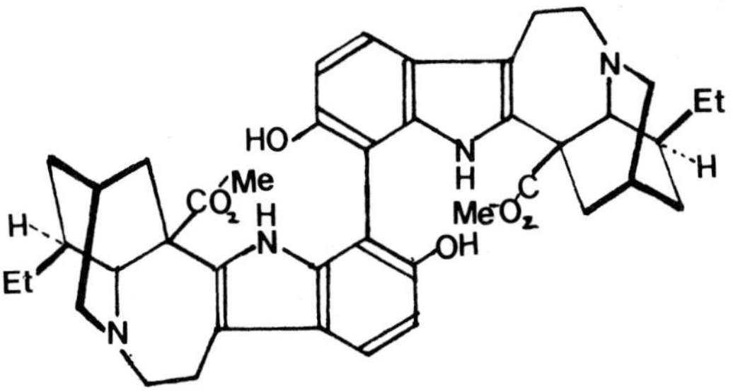

28

(b) Compounds not composed of identical halves nor linked symmetrically.

(i) Sesquimeric compounds

The structure and absolute stereochemistry of unsambarensine has been established by X-ray crystallography (O. Dideberg, L. Dupont and L. Angenot, Acta. Cryst., 1975, B31, 1571.

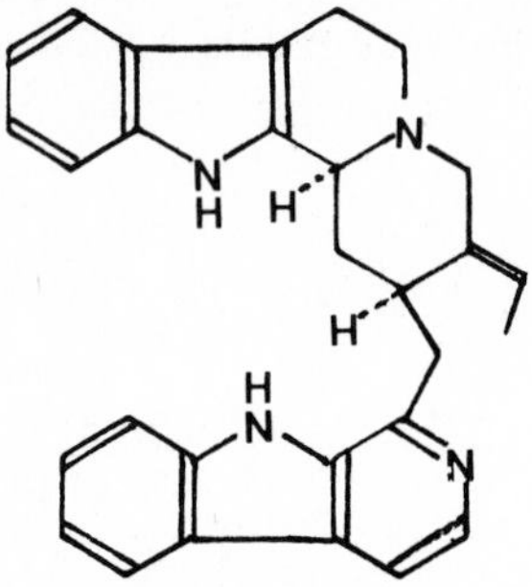

Usambarensine

A number of oxygenated analogues related to the tetrahydrocarbazole, usambarine have been obtained from *Strychnos usambarensis*, (Angenot, C. Coune and M. Tits, J. Pharm. Belg., 1978, 33, 11; 284). These differ in the pattern of the oxygenation on the aromatic ring and in the position of the double bond in the side-chain from usambarensine. They are called usambaridines Br and Vi.

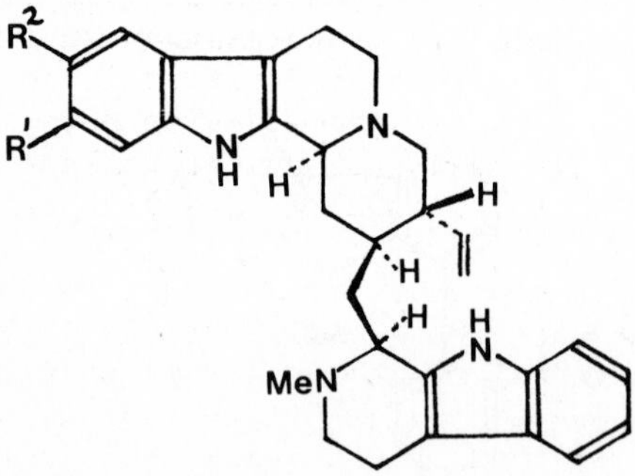

Usambaridine Br

R^1 = OH R^2 = H

Usambaridine Vi

R^1 = H R^2 = OH

Congeners of these compounds are strychnobaridine, which has the structure of usambaridine with a phenolic hydroxyl group in each aromatic ring (position not known), and the interesting variant, strychnopentamine. The presence of a pyrrolidine substituent on the aromatic ring is unusual; the structure has been determined by X-ray methods (*idem.*, Acta.Cryst.,1977,B33,1801). A number of oxindoles derived from usambaridine types have been isolated; the only one to which a definite structure has been assigned is strychnofoline (*idem.*, *ibid.*, 1796). These are probably the only oxindole alkaloids isolated from *Strychnos* species to date.

The structure was assigned from X-ray crystallographic studies (Angenot, Plant, Med.Phytother., 1978,12,123).

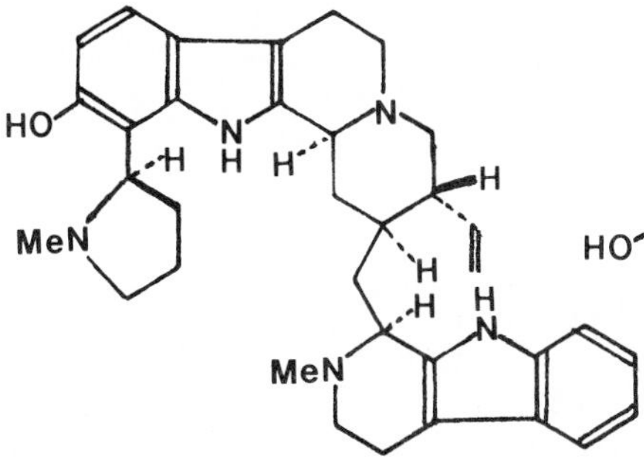

Strychnopentamine 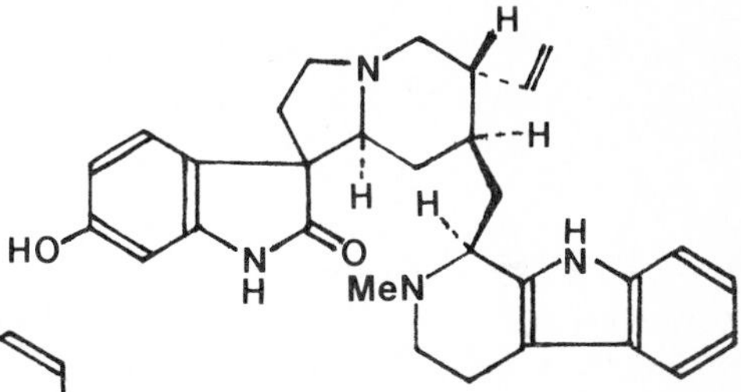   Strychnofoline

The stereochemistry of the roxurghines has been reinvestigated (2nd Edn. Vol. IVB p.152); the assignments to roxburghines C, D and E have been supported but it is now believed that the stereochemistry at C-3 is β-H. The methods used were 300 MHz ^{1}H nmr and ^{13}C nmr spectroscopy (E. Wenkert *et al.*,Helv. 1976,59,2254). A further stereoisomer roxburghine X has been isolated from *Uncaria elliptica*.

A dihydrocarbazole structure is found in tchibangsenine from *S.tchibangensis*(J. Le Men *et al.*,Phytochem, 1978,17,539); this is intermediate in oxidation level between usambarensine (a carbazole) and the usambaridines (tetrahydrocarbazole). The alkaloid may be obtained by condensation of *geissoschizoic acid* with tryptamine followed by cyclisation promoted by phosphorus oxychloride (*idem.*,Heterocycles, 1979,12,1409). The analogy with a possible biosynthetic pathway is obvious.

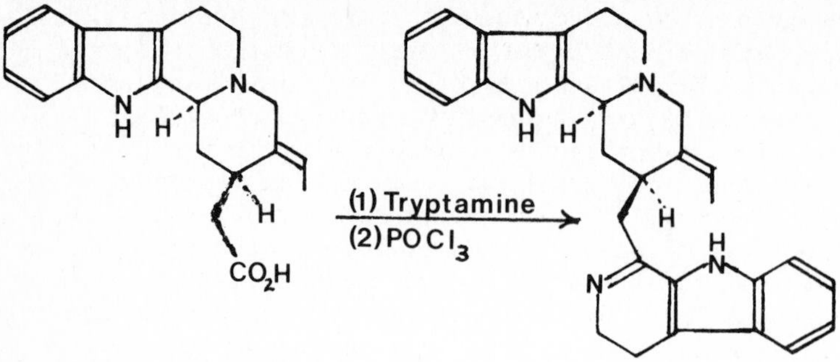

(1) Tryptamine
(2) $POCl_3$

Geissoschizoic acid Tchibangsenine

A series of bisindoles containing a ten carbon unit of isoprenoid but not monoterpenoid origin have been isolated from *Borreria verticillata* (J.L. Pousset, A. Cave, A. Chiaroni and C. Riche, Chem. Comm., 1977, 261) and *Flindersia fournieri* (F. Tillequin, R. Rousselet, M. Koch, M. Bert, and J. Sevenet, Ann. Pharm. Fr. 1979, 37, 543; Phytochem., 1979, 18, 1559; 2066). *B. verticillata* contains borreverine; the structure has been determined by X-ray methods and the compound, which is a congener of borrerine, may be considered as a condensation product of tryptamine and esorine units with two isopentenyl moieties.

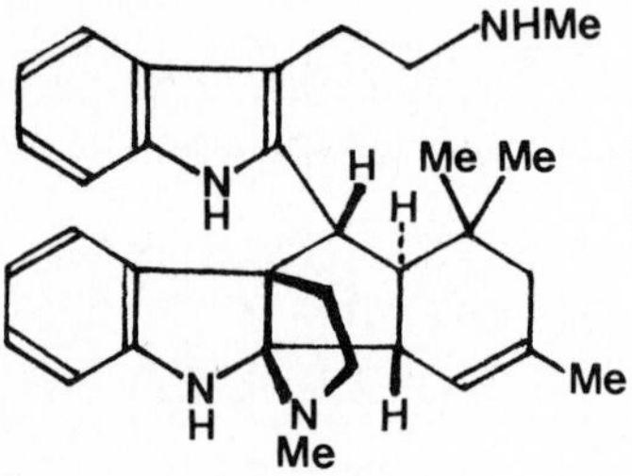

Borreverine 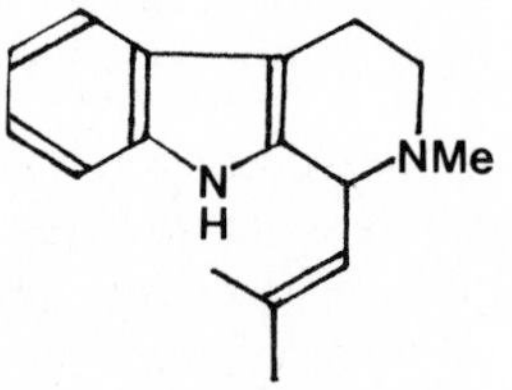 Borrerine

(ii) *The Secamines and presecamines*

There have been no further variants on these compounds isolated although their occurrence in other species has been observed. Tetrahydropresecamine has been isolated from *Pandaca mintiflora*(Le Men *et al.*, Phytochem; 1975, 14, 1648) and from the leaves of *Hunteria elliottii (idem., ibid.*, 1978,17,167). The relative stereochemistry of the secamines and the presecamines have been discussed (G.A. Cordell, G.F. Smith and G.N. Smith, J. Indian Chem.,Soc.,1978,55,1083). The rearrangement from presecamines to secamines would appear to be kinetically controlled since the diastereoisomer at C-16 is produced (see 2nd Edn. Vol. IVB, p.156).

(iii) *Bis-indoles from* Vinca rosea

Catharine was first isolated from *Catharanthus roseus* as early as 1961 (G.H. Svoboda, M. Gormann, N. Neuss and A.J. Barnes, J. pharm.Sci., 1961,50,409) and it was soon realised its structure contained a vindoline component. The other component has been shown to be an unusual enamide, presumed to be derived by oxidative fission of the piperidine ring in a velbanamine derivative. Hence this compound is closely related to vinblastine (2nd. Edn. Vol.IVB p.157); the structure of the alkaloid follows from X-ray crystallography (P. Potier *et al.*,Compt. rendu, 1974,279,C,75; J. Guilhem *et al.*, Acta. Cryst., 1976, B32, 936). These *N*-formyl compounds have found considerable application in cancer therapy.

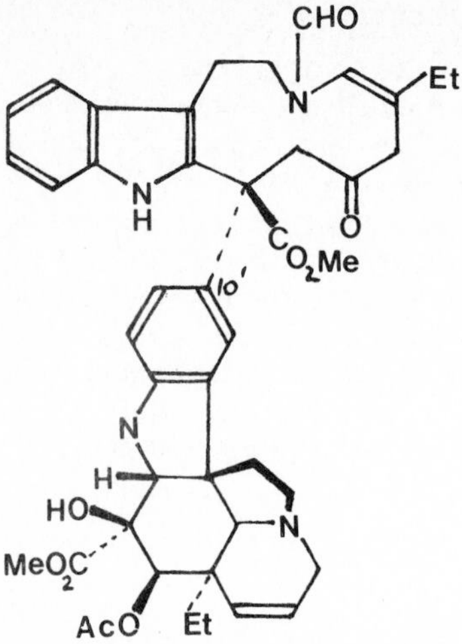

Catharine

A number of alkaloids related to leurosine have been isolated. The diol related to the epoxide leurosine is known as vincadioline and an isomer of this is leurocolombine. The structure of the latter alkaloid follows from a detailed analysis of its ^{13}C nmr spectrum; all of the vindoline carbon atom could be identified and all of the velbanamine unit carbon signals are unchanged from those of vinblastine with the exception of C-14. Hence the additional hydroxyl group is placed at this position (Svoboda *et al.*, J.pharm. Sci., 1975, 64,1953). The vincamidine structure has been assigned from the ^{13}C nmr spectrum. The lack of identity with dihydrocatharine supports the location of the ketone carbonyl in a position not predicted from biosynthetic theory. It was shown by X-ray methods (Potier *et al.*,Tetrahedron, 1978, 34,677) that in fact the D ring of the velbanamine moiety has been cleaved. This is quite plausible biosynthetically.

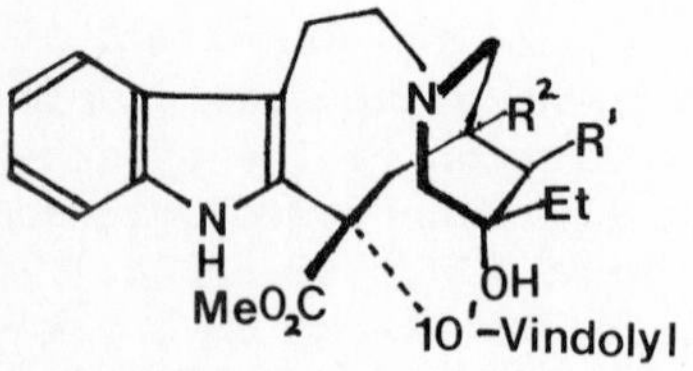

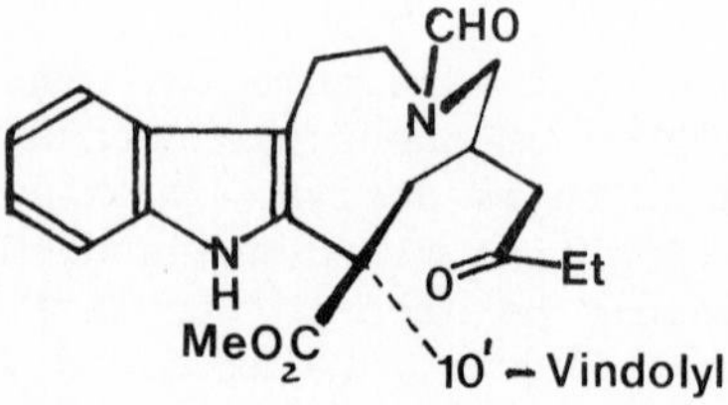

Vincadioline R^1 = OH R^2 = H

Leurocolombine R^1 = H R^2 = OH.

Vinamidine

A new series of compounds related to these as simple stereoisomers appears to have been isolated from *C. ovalis* (Potier *et al.*, Tetrahedron Letters, 1976,2849).

Synthesis of systems related to the vinblastine groups have developed with the application of the Polonovski reaction by Potier and by Kutney. Formation of the series with the natural stereochemistry at C-16 is remarkably sensitive to the exact experimental conditions; it would appear that the fragmentation and coupling occurring as a concerted reaction gives the natural stereochemistry but the stepwise process gives the enantiomer at C-16 (Potier *et al.*, Chem.Comm.,1975,670, Kutney *et al.*,Heterocycles 1975,3,639).

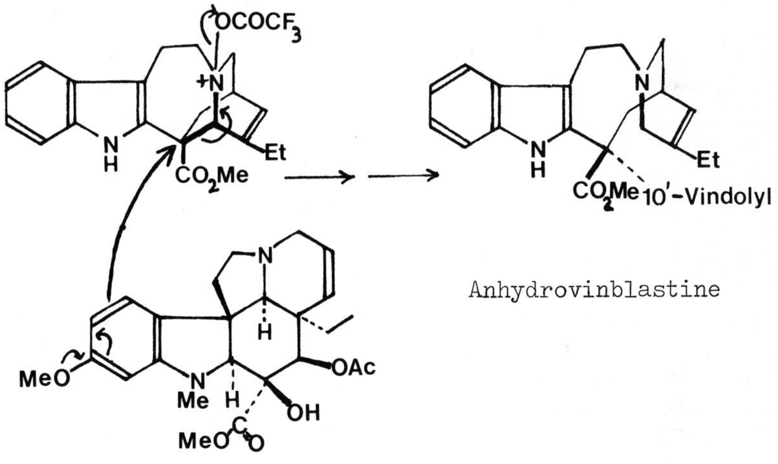

Anhydrovinblastine

Vindoline

The reaction is complicated by cleavage of the tryptamine bridge in the catharanthine-*N*-oxide half during the Polonovski reaction leading to some coupling of the vindoline moiety to these positions, i.e. giving compounds of the type (29).

OCOCF3 +N N H CO2Me Et +H N +N Et CO2Me Vindoline

29 N N Et CO2Me 10′–Vindolyl

The success of this coupling process has allowed the synthesis of a number of analogues differing in the vindoline unit. However the value of this method for the synthesis of anhydrovinblastine is that this serves as a useful precursor to a whole range of alkaloids of this type, e.g. leurosine, many of which are found in minute quantity in nature but are of considerable value in chemotherapy. Vinblastine itself has been synthesised by this method (Potier *et al.*, J. Amer.chem.Soc., 1979,101,2243).

(iv) *Other representative bis-indole alkaloids*

Gonioma malagasy has been shown to contain dihydropycnanthine which contains decarbomethoxydihydrovindoline, and pleiocarpamine components (P. Rasoanaivo and G. Lukacs, J.org. Chem., 1976, 41, 376). Pycnanthine has previously been characterised as containing the vindolinine unit but with the correction of the structure of this alkaloid it is necessary to revise the structure of the bis-indole. By careful comparison of the ^{13}C nmr spectrum of dihydropycnanthine with that of vindolinine it is possible to relate the compound to 2,7-dihydropleiocarpamine and 19 -decarbomethody-14',15'-dihydrovindolinine units with these joined at C-12 in the latter and *via* a methylene to N-1. Hence the structure has been deduced.

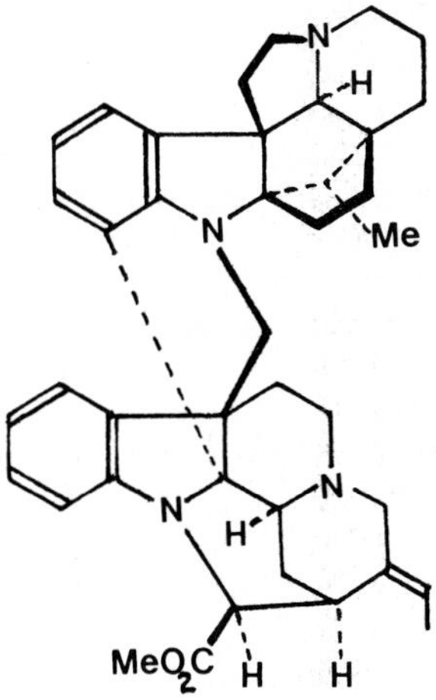

14' - 15' - Dihydropycnanthine

A number of compounds have been isolated containing the opened sarpagine unit (related to vobasine) joined to representatives of the other major skeletal types. Accedinisine and accedinine have been obtained from the root bark of *Tabernaemontana accedens* (H. Achenbach and E. Schaller, Ber., 1976, 109, 3527). The structures follow from the similarity of the spectral properties to those of the monomeric units. They may be synthesised by condensation of vobasinol with the corresponding monomeric sarpagine alkaloid.

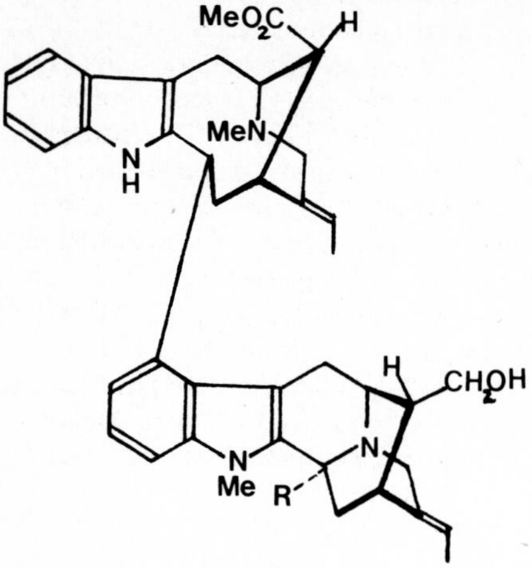

Accendinisine R = H

Accendinine R = OH

Capuronetta elegans contains a number of alkaloids having the vobasine unit joined to a dihydrocleavamine unit or to the related *pseudo*-Aspidosperma skeleton. The major alkaloid, capuvosine, is hydrolysed by hydrochloric acid to identifiable monomers; the structure has been confirmed by partial synthesis from vobasinol and capuronine acetate in methanolic hydrogen chloride to give the acetate of the alkaloid (I. Chardon-Loriaux and H.P. Husson, Tetrahedron Letters, 1975, 1845).

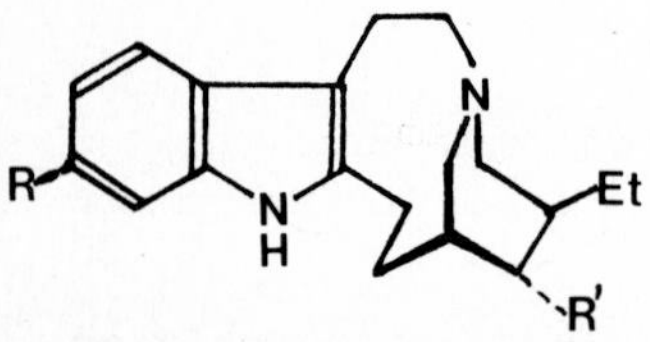

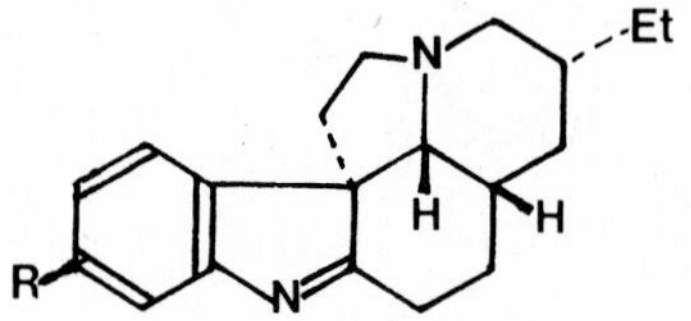

Capuvosidine

Capuvosine R^1 = OH

Deoxycapuvosine R^1 = H

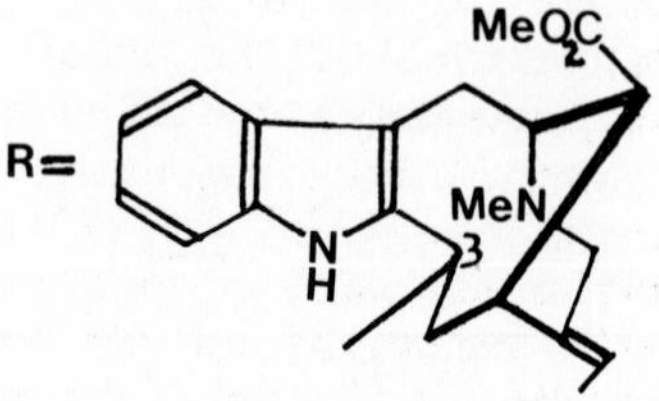

A minor alkaloid from the same source, capuvosidine, displays a combination of the indole and the indolenine chromophore. Reduction with sodium tetrahydroborate gives an indoline which shows the fragmentation pattern of a *pseudo*-aspidospermidine derivative in its mass spectrum (*idem.*, Phytochem, 1978, 17, 1605). The structure follows from a comparison of the nmr spectra of capuvosine and capuvosidine. Reduction of cupuvosidine with sodium tetrahydroborate in methanol leads to deoxycapuvosine (*idem.*, J. Indian Chem.Soc., 1978, 55, 1099). The number of alkaloids having the dihydrovobasine (dregamine) unit joined to the "*Iboga*" unit is considerable. Particularly interesting are the tabernaelegantinines C and D (from *T. elegans*) which have the unusual feature of a nitrile group attached to the isovoacangine components. (B. Danielli *et al.*,J.Chem. Soc. Perkin Trans. 1, 1980,601). This may be removed by reduction with sodium tetrahydroborate in the present of cobalt (II) salts, a feature of 2-amino-nitriles. Presumably cyanide anion attacks some intermediate immonium species to give these compounds in the plant. The products obtained by loss of cyanide are their congeners, tabernaelegantines C and D.

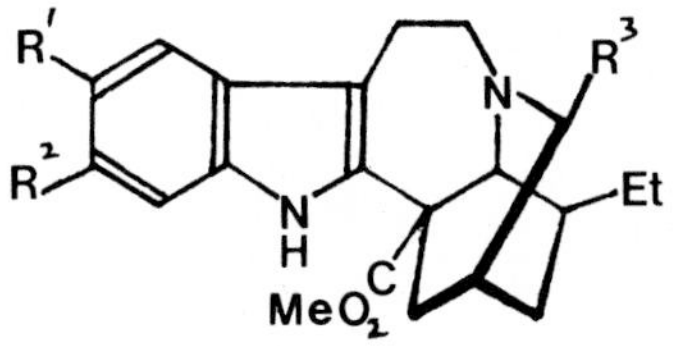

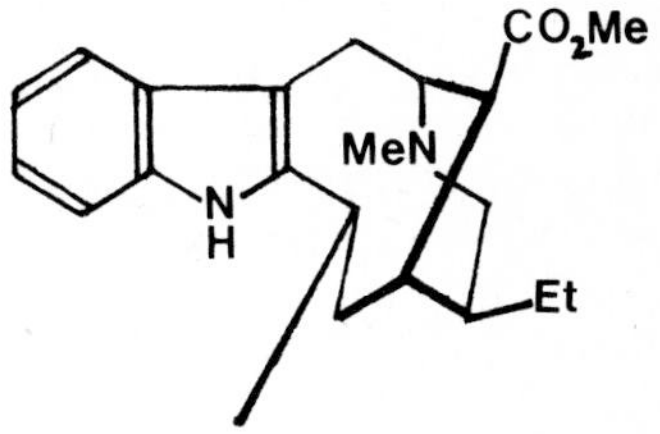

Tabernaelegantinine C
R^1 = MeO
R^2 = 3=dregamyl
R^3 = CN

Tabernaelegantine C
R^1 = MeO
R^2 = 3-dregamyl
R^3 = H

Tabernaelegantinine D
R^1 = 3-dregamyl
R^2 = OMe
R^3 = CN

Tabernaelegantine D
R^1 = 3-dregamyl
R^2 = OMe R^3 = H

3 - Dregamyl

A molecule having isovoacangine and hydrocanthine components is bonafousine from *Bonafousia tetrastachya*. This is an example of an alkaloid having two recognisable alkaloid halves, one derived from a monoterpenoid indole alkaloid and the other a simple indole. The structure has been determined by X-ray methods (M. Damak *et al.*, Chem. Comm., 1976, 510).

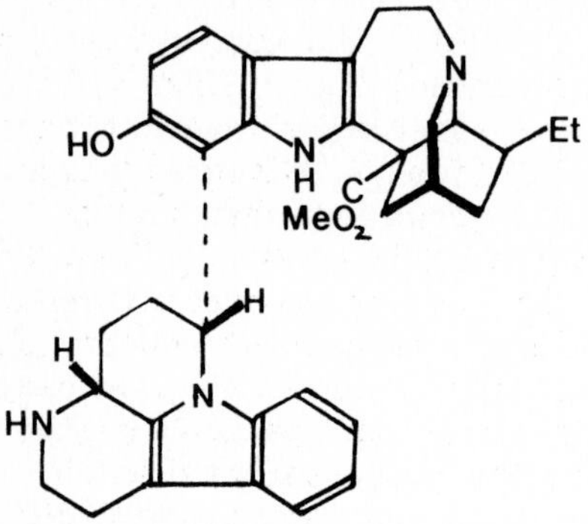

HO, N H, N, Et, MeO_2C, H, H, HN, N

Bonafousine

Similarly cimiciphytine and norcimiciphytine, two minor lactonic alkaloids from *Haplophyta cimicidum*, contain the canthine type unit coupled to an Aspidosperma unit (M.P. Cava *et al.*, Heterocycles, 1978, 9, 1009). Related to these is haplophytine, from the same source.

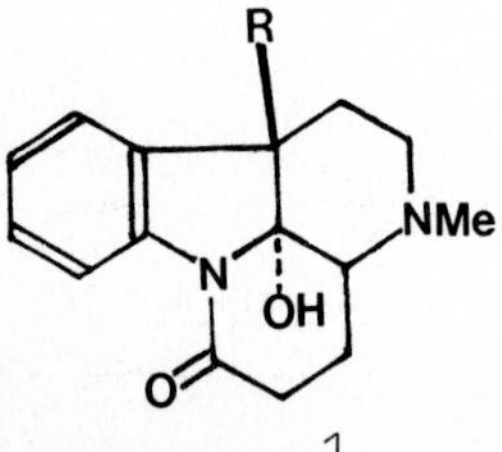

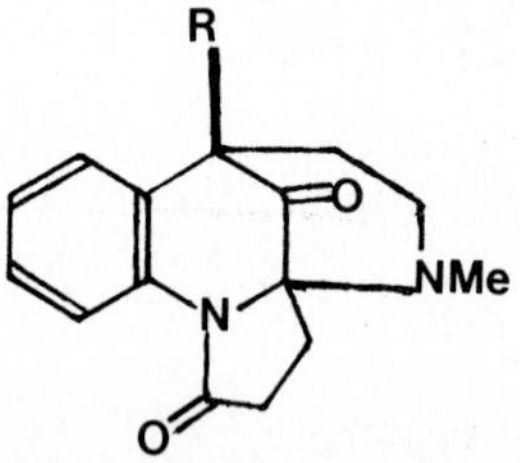

R, NMe, N, OH, O; R, O, NMe, N, O

Cimiciphytine R^1 = Me

Norcimiciphytine R^1 = H

Haplophytine R^1 = H

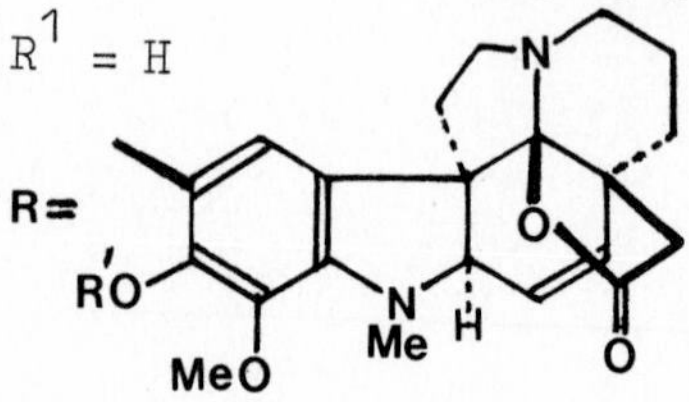

R=, R'O, MeO, N, Me, H, N, O, O

A new dimeric Strychnos type alkaloid is sungucine, from the roots of *S. icaja*; the structure has been determined by X-ray methods (Angenot *et al.*, Tetrahedron Letters, 1979,4227). It is a dimer of deoxyisostrychnine and shows a novel attachment between the two units.

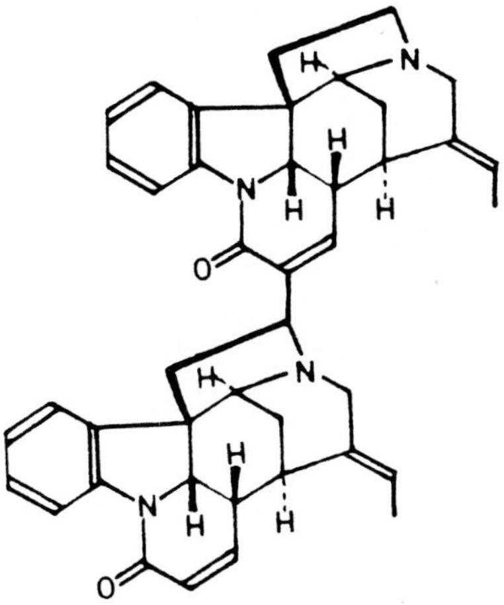

Sungucine

Chapter 10

FIVE-MEMBERED MONOHETEROCYCLIC COMPOUNDS AMARYLLIDACEAE ALKALOIDS

M. SAINSBURY

1. *Introduction*

The *Amaryllidacea* group of alkaloids continues to attract the attention of organic chemists, and since the main work was prepared in 1975 a number of important review articles have been published. The Alkaloids (Specialist Periodical Report, Royal Society of Chemistry) provides an annual survey of development in the study of biosynthesis, synthesis and the chemistry of alkaloids in general and includes, when appropriate, sections devoted to the *Amaryllidacea* bases (see Vol. 9, 1980 and previous volumes). In addition W. Döpke (Heterocycles, 1977, 6:551) has reviewed progress relating to the alkaloids bearing a lactone ring, and bases belonging to the crinane and galanthane series have been discussed by C. Fuganti (The Alkaloids, Vol. 15, ed. R.H.F. Manske, Academic Press, New York, p.83).

General methods of alkaloid synthesis, including some examples from the *Amaryllidacea* group, have been reviewed by R.V. Stevens (Acc. chem. Res., 1977, 10:193; Total Synth. Nat. Prod., 1977, 3:439) and the synthesis of crinane is described in an article exemplifying the application of anodic oxidative coupling reactions to the construction of natural products (S. Tobinaga, Bioorg. Chem., 1975, 4:110).

Aspects of the pharmacology of some of the alkaloids are considered by W.D. Wiezorik *et al.* (Pharmazie, 1975, 30:618) and by G.V. Selezhinskii (Khim. Zhizn., 1977, 3:50; Chem. Abs., 1975, 86:152586).

2. *Biosynthesis*

Since the biosynthetic routes to the major structural types of alkaloids within the *Amaryllidacea* family are known in broad outline, most work in this area in the period 1974 - 1980 has centred on resolving points of fine detail. Surprisingly a number of conflicting results have been presented which appear to indicate stereochemical variation in the way different plants utilise precursors *en route* to a common alkaloid. Thus it has been shown that *Rhodophiala bifida* converts norbelladine (1) into both haemanthamine (4) and montanine (5) with loss of the 2-pro-*R* hydrogen atom (W.C. Wildman and B. Olesen, Chem. Comm., 1976, 551), whereas C. Fuganti, D. Ghiringhelli and P. Grasselli (*ibid.*, 1972, 1152; 1973, 430) have observed that in *Haemanthus coccineus* montanine is derived from the same precursor but with elimination of the 2-pro-*S* hydrogen atom.

In *R. bifida* vittatine (2) is a precursor for both haemanthamine and montanine (A.I. Feinstein and W.C. Wildman, J. Org. Chem., 1976, 41:2447), although the incorporation is greater into the former alkaloid. This result is taken to mean that 11-hydroxyvittatine (3) is formed first, and converted directly into haemanthamine by 3-*O*-methylation. In the case of montanine a rearrangement of 11-hydroxyvittatine is necessary before 2-*O*-methylation can occur - hence the lower efficiency of precursor utilisation. It is noteworthy that in *R. bifida* there is no equilibration of the enantiomeric (-)- and (+)-crinine ring-systems.

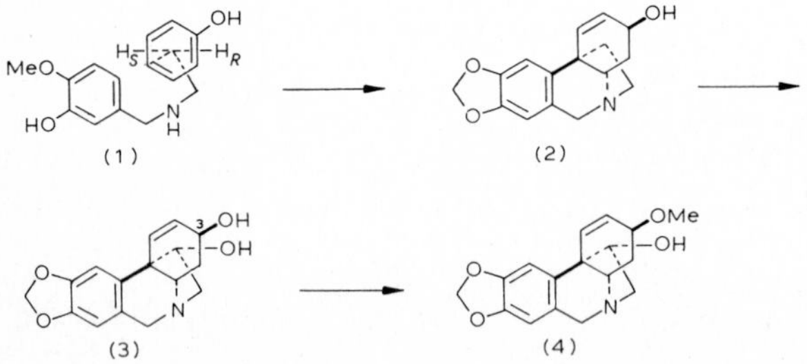

MeO HO OH H_S H_R N H

(1)

O O OH N

(2)

O O 3 OH OH N

(3)

O O OMe OH N

(4)

(3)

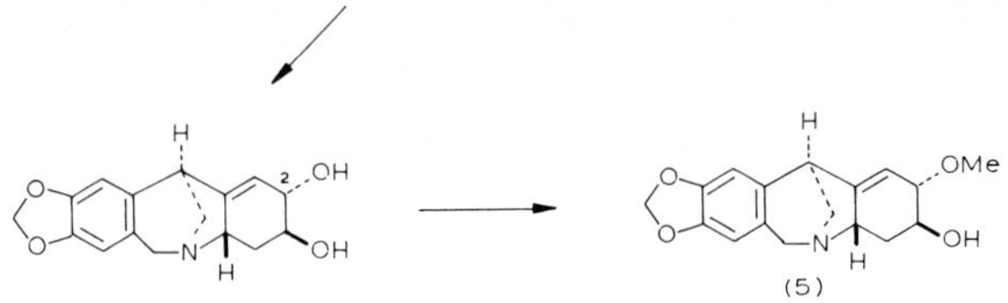

(5)

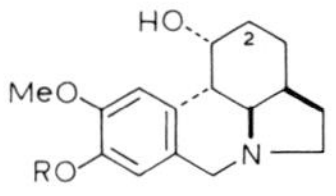

(6; R = Me)
(8; R = H)

(7; R = OMe, R' = Me)
(9; R,R = OCH_2O, R' = H)

A similar controversy exists in the lycorine series: the conversion of pluviine (6) into galanthine (7) in King Alfred daffodils (*Narcissus pseudonarcissus* L.) takes place with hydroxylation at C-2 and inversion of configuration (R.D. Harken, C.P. Christensen and W.C. Wildman, *ibid.*, 1976, 41: 2450). However, C-2 hydroxylation of nor-pluviine (8) leading on to lycorine (9) proceeds with retention of configuration in one plant (Fuganti and M. Mazza, Chem. Comm., 1972, 936) but with inversion in another (I.T. Bruce and G.W. Kirby, *ibid.*, 1968, 207; Wildman and N.E. Heimer, J. Amer. chem. Soc., 1967, 89:5265).

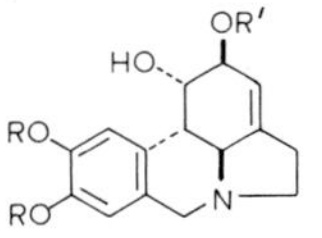

The conversion of *O*-methylnorbelladine (10) into narcissidine (11) by *Sempre avanti* daffodils involves loss of the C-4 pro-*S* hydrogen atom from galanthine (7) acting as an intermediate (Fuganti, Ghiringhelli and Grasselli, Chem. Comm., 1976, 350).

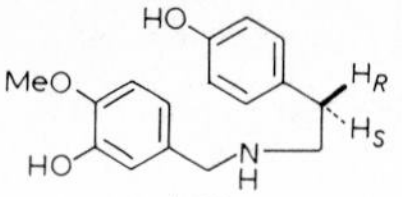

(10)

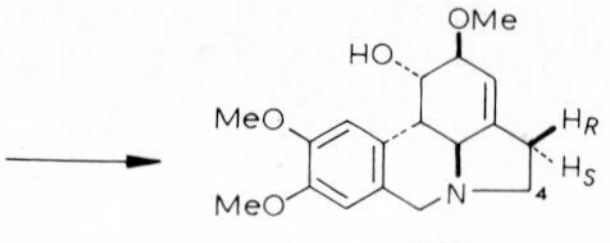

(7)

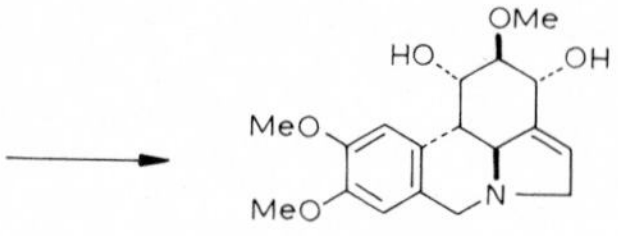

(11)

3. *New Alkaloids and Plant Sources*

(a) *Lycorine and pretazettine*

The co-occurance of lycorine and pretazettine (12) is often reported, see table 1, and if the view is accepted that tazettine (13) is simply an artefact of pretazettine formed during fractionation then this situation is even more common. Thus in the period 1976 - 1980 the authors of at least 10 papers have noted the coexistence of lycorine and pretazettine or lycorine and tazettine in plant extracts.

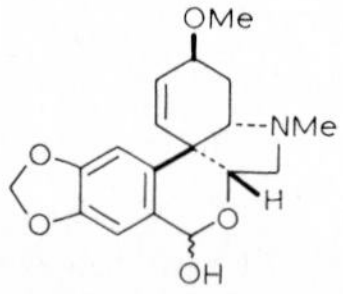

(12)

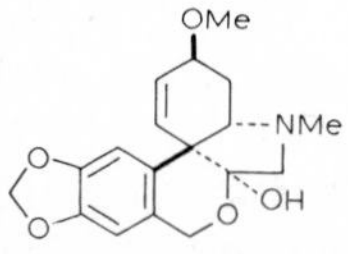

(13)

TABLE 1

Co-occurence of lycorine and pretazettine (tazettine)

Plant	Alkaloids present	References
Narcissus tazetta L.	Lycorine, pseudolycorine, homolycorine, pretazettine.	1
Zephyrathes carinata Herb.	Lycorine, pretazettine, galanthine, haemanthamine, carinatine*.	2
Hymenocallis arenicola Northrop (*H. caribea* Baker) (syn. *Lirio sanjuanera*)	Lycorine, tazettine, galanthamine, haemanthamine, haemanthidine, havanine*, varadine*, zaidine*, caribine*.	3,4,5,6
Panacratium maritimum	Lycorine, tazettine.	7
Ungernia vvedenskysi S. Khamidkh	Lycorine, tazettine, methoxytazettine*, ungminorine, ungminoridine*, hippeastrine, galanthamine, narwedine, pancratine, hordenine.	8,9
Ungernia spiralis	Lycorine, tazettine, dihydroepimacronine*, *N*-demethylepimacronine (ungspiroline)*.	10
Hippeastrum vittatum L. Her.	Lycorine, tazettine, vittatine, hippacine, hippadine*, hippagine*, hippafine*.	11
Lycoris sanguinea Maxim. var. *kiushiana* Makino	Lycorine, pretazettine, galanthamine, sanguinine*.	12

(*Continued on p.156*)

Plant	Alkaloids present	References
Lycoris sanguinea Maxim.	Lycorine, tazettine, haemanthamine, lycoramine, haemanthidine galanthamine, lycoricidinol (narciclasine), arolycorcidine, narciprimine (arolycoricidinol), 4,5-etheno-8,9-methylenedioxy-6-phenanthridone*	13
Lycoris radiata Herb.	Lycorine, pretazettine, homolycorine, demethylhomolycorine, *O*-demethyllycoramine*, lycorenine, hippeastrine.	14

* denotes new compounds not listed in the main work.

References

1. E. Furusawa *et al.*, Chem. pharm. Bull. Japan, 1976, 24:336.
2. S. Kobayashi *et al.*, *ibid.*, 1977, 25:2244.
3. W. Döpke and Z. Trimiño, Z. chem., 1977, 17:101.
4. Döpke, E. Sewerin and Z. Trimiño, *ibid.*, 1979, 19:215.
5. Döpke and Z. Trimiño, *ibid.*, 1979, 19:377.
6. Döpke, E. Sewerin and Z. Trimiño, *ibid.*, 1980, 20:26.
7. A. Amico, S. Bruno and V. Bonvino, Ann. Fac. Agrar. Univ. Bari, 1972, 25:129; Chem. Abs., 1975, 82:54130D.
8. Kh.A. Kadyrov and S.A. Khamidkhodzhaev, Khim. prir. Soedin., 1979, 418.
9. Kadyrov, A. Abdusamatov and S.Yu. Yunusov, *ibid.*, 1979, 585.
10. Kadyrov and Abdusamatov, Khim. prir. Soedin., 1977, 426; Kadyrov, Abdusamatov and Yunusov, *ibid.*, 719.
11. A.A. Ali and M.K. Mesbah, Planta Med., 1975, 28:336.
12. Kobayashi *et al.*, Chem. pharm. Bull. Japan, 1976, 24:1537.
13. S. Takagi and M. Yamaki, Yakugaku Zasshi, 1974, 94:617.
14. Kobayashi *et al.*, Chem. pharm. Bull. Japan, 1980, 28:3433.

Lycorine is found in the bulbs of *Hippeastrum añañuca* Phil. together with homolycorine (14) 17-epihomolycorine (15), m.p. 160–163°, $[\alpha]_D$ + 34.5 (*c* 2.4, $CHCl_3$), maritidine (16) and hippeastidine (17), m.p. 196°, $[\alpha]_D$ 6.94 (*c* 2.3, $CHCl_3$) (P. Pacheco *et al.*, Rev. Latinoam. Quim., 1978, 9:28; Chem. Abs., 1978, 89:103746).

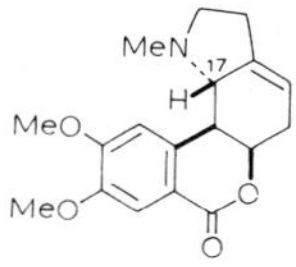

(14)

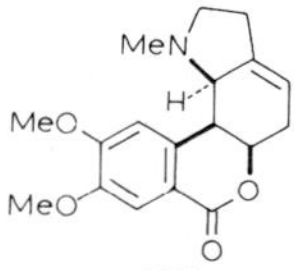

(15)

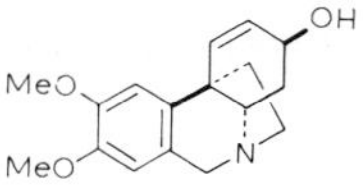

(16)

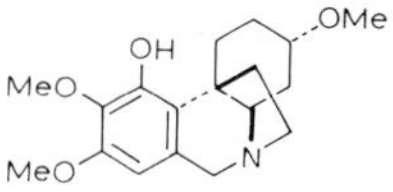

(17)

The structure of epihomolycorine has been confirmed by X-ray crystallography (E.M. Gopalakrishna and W.H. Watson, Cryst. struct. Comm., 1978, 7:41).

Lycorine is also present in the rhizomes of *Curculigo orchiodes* (R.V. Rao *et al.*, Ind. J. pharm. Sci., 1978, 40:104) and its extraction from *Ungernia sewerzowi* has been studied and optimised (T. Sadikov, I.N. Zatorskaya and T.T. Shakirov, Uzb. khim. Zh., 1974, 18:74; Chem. Abs., 1975, 82:95300; Sadikov *et al.*, D8DEP2, 1974, VINITI, 15, deposited document).

(b) *4,5-Etheno-8,9-methylenedioxy-6-phenanthridone*

The bulbs of *Lycoris sanguinea* Maxim afford a number of conventional *Amaryllidacea* alkaloids (see refs. 11 and 12, table 1) as well as some bases in which two benzenoid rings are present. Among the latter is alkaloid N-3, m.p. 235°, $C_{16}H_9O_3N$, which has been shown (ref. 11) to be 4,5-etheno-8,9-methylenedioxy-6-phenanthridone (18).

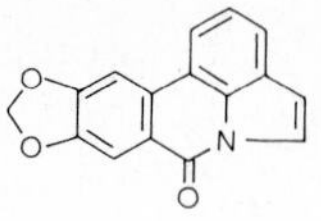

(18)

Such a structure can be formally derived from lycorine (9) by dehydration and oxidation, but nothing is known of its true biosynthesis.

(c) *Carinatine and goleptine*

Carinatine $C_{17}H_{21}NO_4$ is a new phenolic alkaloid isolated from the bulbs of *Zephyranthes carinata* (see ref. 2, Table 1). Carinatine has $[\alpha]_D^{27}$ - 68.97° (*c* 0.696, $CHCl_3$), m.p. of picrate 195-197° (decomp.). Mass spectrometry indicates that this base is of the lycorine type and comparison of its ultraviolet spectrum and optical rotatory dispersion curve with those of galanthine (7), which co-occurs with carinatine in *Z. carinata*, leads to the conclusion that the new alkaloid is *O*-demethylgalanthine (19, R=H; R'=Me).

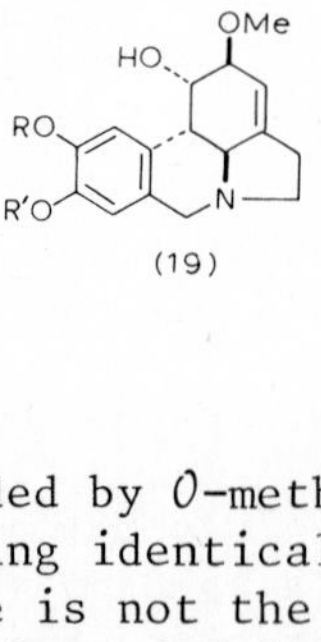

(19)

Confirmation is provided by *O*-methylation using diazomethane, the product being identical with natural galanthine. Interestingly carinatine is not the same as goleptine $C_{17}H_{21}NO_4$, m.p. 141°, $[\alpha]_D$ - 99° ($CHCl_3$), picrate m.p. 174°, previously isolated from the Narcissus hybrid 'Golden Sceptre' by Döpke (Naturwiss., 1963, 50:645), and it seems highly probable that goleptine has the isomeric structure (19, R=Me; R'=H).

(d) *Alkaloids from plants of the genus Crinum*

The alkaloid content of four West African species of the genus *Crinum* has been determined by J.W. Powell and D.A.H. Taylor (J. West Afr. sci. Assoc., 1967, 12:50). *C. glaucum* was shown to metabolise lycorine, ambelline (20) and two unidentified bases named criglaucine and criglaucidine, whereas *C. jagus* affords lycorine, crinamine (21) and pseudolycorine (22).

Subsequently D.S. Millington, D.E. Games and A.H. Jackson (Proc. Int. Symp. Gas Chromat. Mass Spectrom., A. Frigerio ed., Tamburini, Milan, 1972, 275) showed that *C. glaucum* also gives rise to crinamine, criwelline (23), criglaucine and four additional alkaloids of unknown structure. The mass spectrum of criglaucine is similar to that of deacetylbowdensine (24) and the spectrum of one of the unknown bases is similar to that of ambelline.

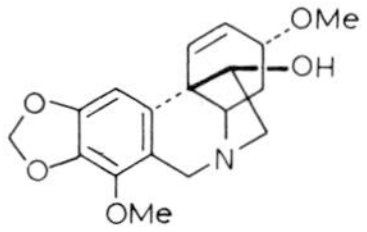

(20) 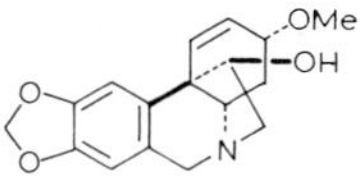 (21) 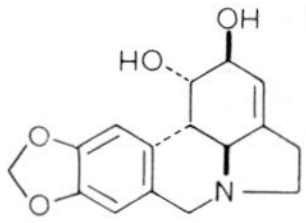 (22)

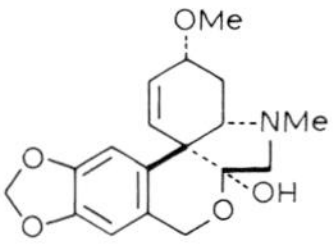

(23) 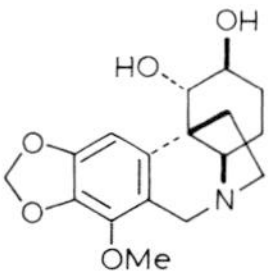 (24)

J.W. Powell and D.A.H. Taylor (*loc. cit.*) also examined *C. ornatum* and *C. natans* and reported that they were devoid of alkaloids, however, O.S. Onyiriuka and A.H. Jackson (Israel J. Chem., 1978 17:185) have re-investigated these two plants and have established that alkaloids are present, at least at certain seasons of the year. *C. ornatum*, for example, produces at least seven bases including lycorine and crinamine. Three of the remaining compounds have been named and their molecular formulae determined, they are ornamine ($C_{18}H_{21}NO_3$), m.p. 213-214°, ornazamine ($C_{18}H_{21}NO_4$), m.p. 218-220°, and ornazidine ($C_{16}H_{19}NO_3$), m.p. 217°. Ornazidine is tentatively assigned structure (25) and the other alkaloids are probably also of this type. *C. natans* elaborates crinatine ($C_{18}H_{21}NO_5$), m.p. 262°.

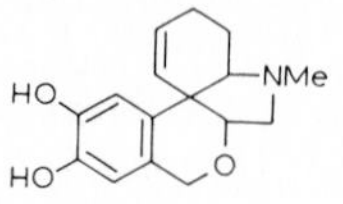

HO HO NMe O

(25)

Hamayne, m.p. 79-80° $[\alpha]_D^{12}$ + 43 (c 0.1, EtOH) is a new alkaloid which co-occurs with crinamine in *C. asiaticum* var. *japonicum*, together with *N*-desmethylgalanthamine (26) (M. Ochi, H. Otsuki and K. Nagao, Bull. chem. Soc. Japan, 1976, 49:3363). From the ^{1}H nmr spectrum of *O,O*-diacetylhamayne, and the fact that the same apohaemanthamine (27) is formed by acid treatment of hamayne and crinamine, the Japanese workers propose that the former alkaloid is *O*-desmethylcrinamine (28).

MeO O OH N H

(26)

O O O N

(27)

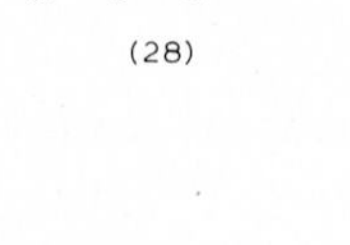

O O OH OH N

(28)

(e) *Havanine, varadine, zaidine and caribine*

Havanine, $C_{18}H_{25}NO_5$, m.p. 135-137° is an alkaloid of the crinine series and has been allocated structure (29) (ref. 5, table 1). It co-occurs with varadine, zaidine and caribine in the plant *Hymenocallis arenicola* which is indigenous to Florida, Cuba and the West Indies.

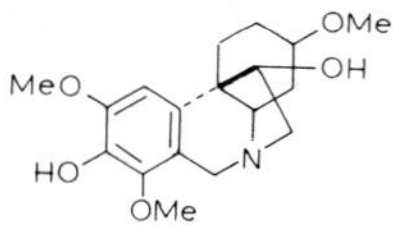

(29)

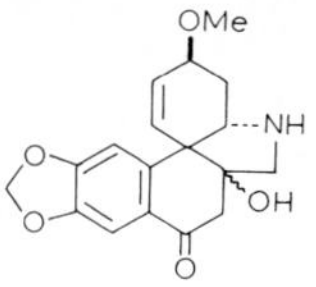

(30)

Varadine (ref. 4, table 1), $C_{18}H_{19}NO_5$, m.p. 95°, shows M-15 and M-31 peaks in the mass spectrum, but the major fragmentation gives rise to an ion at *m/e* 245. This and the data from other spectral studies are taken to favour structure (30) for the alkaloid. The ion (32) *m/e* 245 is assumed to arise by a retro Diels Alder cleavage of the species (31), itself derived by a hydrogen transfer reaction within the molecular ion.

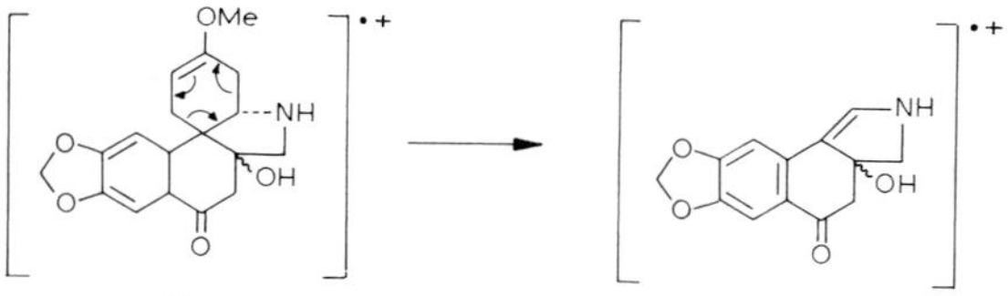

(31)

(32)

At the moment insufficient evidence is provided to confirm this structural proposal, particularly the stereochemistry, and it is of interest to note that the representation (30) lacks the ring B oxygen atom common to alkaloids of the pretazettine type. Instead varadine is represented as an octahydro-1*H*-naphtho [2,1-*c*] indole derivative.

Zaidine, m.p. 192-193°, $[\alpha]_D$ - 4° (*c* 0.2, $CHCl_3$) is a typical lycorine-type alkaloid. Its basic structure (33) has been determined by mass spectrometry, but stereochemical details are as yet unknown (ref. 3, table 1).

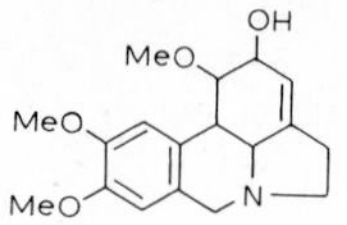

(33)

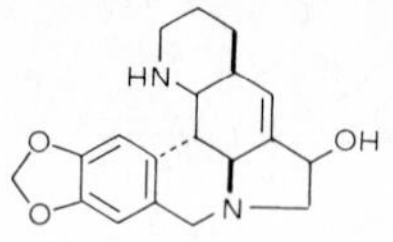

(34)

Not so straightforward is caribine (ref. 6, table 1). This alkaloid $C_{19}H_{22}N_2O_3$, m.p. 200-204°, is most unusual since, although its spectral characteristics point to a lycorine-type structure, it contains an additional piperidine ring system fused on to ring D. Structure (34) is postulated for this base, but it must be said that the evidence presented for it, particularly in relation to the stereochemistry, is limited.

(f) Ungvedine and ungspiroline

Ungernia vvedenskyi S. Khamidh is a rich source of alkaloids (see refs. 8 and 9, table 1). A new alkaloid from this plant is ungvedine ($C_{19}H_{25}NO_5$), m.p. 148-150°, $[\alpha]_D$ + 12.5 (c = 0.5, $CHCl_3$) which has been shown to be *O*-methyltazettine (35). In view of the fact that tazettine is generally regarded as an artefact of pretazettine this structural allocation is rather interesting.

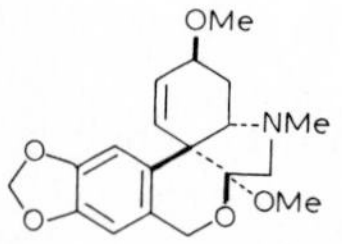

(35)

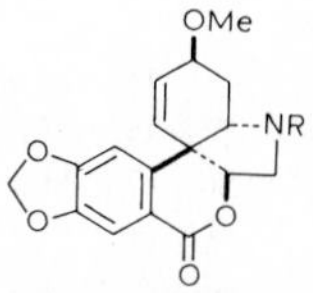

(36, R = H)
(37, R = Me)

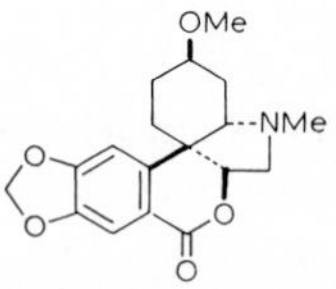

(38)

Ungspiroline $C_{17}H_{17}NO_5$, m.p. 148-149°, is present in *U. spiralis* (Kh.A. Kadyrov, A. Abdusamatov and S.Yu. Yunusov, Khim. prir. Soedin., 1977, 719). It is formulated as *N*-demethylepimacronine (36) since on methylation it affords epimacronine (37).

Dihydroepimacronine (38), m.p. 98-99°C, $[\alpha]_D^{20}$ + 10.7° (*c* 0.65, $CHCl_3$) also occurs in the plant (*idem.*, *ibid.*, 1976, 826; Kadyrov and Abdusamatov, *ibid.*, 1977, 426), as well as two other bases $C_{17}H_{19}NO_5$, m.p. 148-149°, $[\alpha]_D^{20}$ + 105° (*c* 0.6, $CHCl_3$) and $C_{17}H_{21}NO_5$, m.p. 142-143°, $[\alpha]_D^{20}$ + 11° (*c* 0.45, $CHCl_3$).

(g) *Hippadine, hippafine and hippagine*

Specimens of *Hippeastrum vittatum* L. Her. cultivated in the Assiut and Cairo regions of Egypt contain at least thirteen bases (ref. 11, table 1). In addition to alkaloids of known structure, four others have been partly described. These are hippadine, $C_{16}H_{17}NO_4$, picrate, m.p. 155-157° (decomp.), hippafine, picrate 193° (decomp.), hippagine $C_{16}H_{17}NO_4$, m.p. 261-263°, and alkaloid I which is a phenol and forms a picrate, m.p. 187-189° (decomp.).

H. bulbispermum Milne contains crinine (crinidine) (39), vittatine, crinamine (21), powelline (40), lycorine, hippacine (Döpke, Pharmazie, 1966, 21:323) and alkaloid BII, m.p. 130-132° (A.M. El-Moghazi and A.A. Ali, Planta Med., 1976, 30: 369).

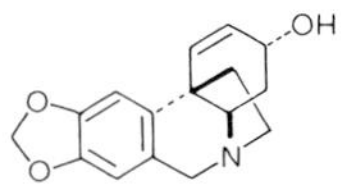

(39) 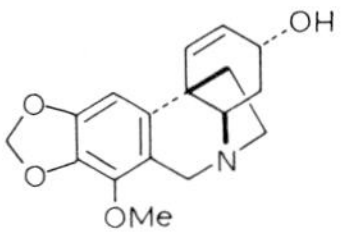 (40)

(h) *Galanthamine and its* N- *and* O-*demethyl derivatives*

The isolation and distribution of galanthamine (41) in plants of the *Lencojum* genus has been studied by several groups (S. Kohlmuenzer and E. Cyunel, Postep. Dziedzinie Leku. Rosl. Pr. ref. Dosw. Wyglossone Symp., 1970, 109; Chem. Abs., 1973, 78:55297t; Zh. Stefanov, P. Savchev and I. Mitkov, Farmatsiya (Sofia), 1974, 24:16). It seems that the alkaloid occurs in *L. aestirum* and *L. vernum*, but not in *L. vernum* var. *carpaticum*.

Ungernia species also contain galanthamine (S.A. Khamidkhodzhaev and A. Abdusamatov, Uzb. Biol. Zh., 1978, 36; Chem. Abs., 1978, 88:148987), as do a wide variety of plants from other genera within the *Amaryllidacea* family (O.A. Cherkasov, Khim.-Farm. Zh., 1977, 11:84; Chem. Abs., 1977, 87:81252; G.M. Gorbunova, A.V. Patudin and V.D. Gorbunov, Khim. prir. Soedin., 1978, 420; Chem. Abs., 197 , 89:126172). Galanthamine, lycorine and *O*-demethylgalanthamine (42) occur in *Lycoris sanguinea* Maxim. var. *kiushiana* Makino (S. Kobayashi *et al.*, Chem. pharm. Bull. Japan, 1976, 24:1537).

This last compound, m.p. 210.5-213° (decomp.), $[\alpha]_D^{27}$ - 133° (*c* = 0.23, EtOH) also has the trivial name sanguinine. Its structure has been confirmed by comparison of its spectral data with those of galanthamine and also by conversion into galanthamine by methylation.

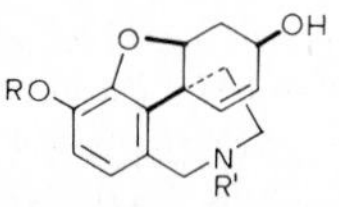

(41; R = R' = Me)
(42; R = OH, R' = Me)
(43; R = Me, R' = H)

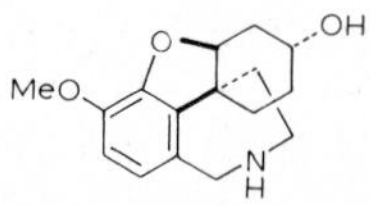

(44)

(-)-*N*-Demethylgalanthamine (43), m.p. 152.5-153°, $[\alpha]_D^{22}$ - 90.7° (*c* = 0.717, EtOH), has been isolated as a single compound from the bulbs of *Crinum asiaticum* var. *japanicum* Baker (Kobayashi *et al.*, *ibid.*, 1976, 24:2553). Previously this compound has only been found together with (+)-*N*-demethyldihydrogalanthamine (44) in the form of a quasi-racemate known as narcissamine (S.M. Laito and H.M. Fales, J. Amer. chem. Soc., 1964, 86:4434).

(*i*) *Clivatine and clivacetine*

Clivatine $C_{21}H_{25}NO_7$, m.p. 166-169°, picrate m.p. 150° $[\alpha]_D^{25}$+ 52° ($CHCl_3$) has been obtained from the plant *Clivia miniata* Regel (B. Mehlis Dissert. Humboldt Univ. Berlin, 1962; Döpke, Heterocycles, 1977, 6:551), and allocated structure (45). The same alkaloid, together with lycorine, clivonine (46), clivimine (50) and a new alkaloid clivacetine, $C_{21}H_{25}NO_7$, m.p. 152-155°, $[\alpha]_D^{24}$ + 53.8° (*c* = 0.67 $CHCl_3$), have also been isolated from specimens of C. *miniata* growing in Japan by Kobayashi *et al.* (Chem. pharm. Bull. Japan, 1980, 28:1827).

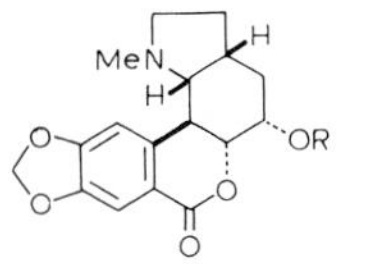

(45; R = $COCH_2CH(OH)Me$)
(46; R = H)
(47; R = Ac)
(48; R = $COCH_2COMe$)
(49; R = $COCH_2CH(OAc)Me$)

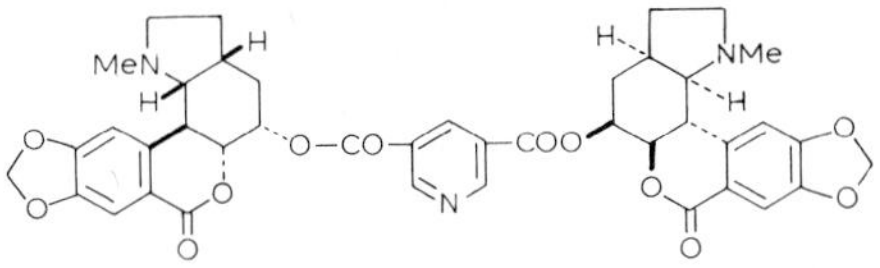

(50)

Clivacetine exhibits the characteristic fragment ion peaks of clivonine - type alkaloids in the mass spectrum, these are: *m/e* 83 (C_5H_9N, 100%), 96 ($C_6H_{10}N$, 63%), 82 (C_5H_8N, 23%), 126 ($C_7H_{12}NO$, 1.5%) (Döpke *et al.*, Tetrahedron Letters, 1967, 45; H.K. Schnoes *et al.*, Tetrahedron, 1968, 24:2825) and also shows a fragment *m/e* 84 due to acetylketene ($C_4H_4O_2$). Its ^1H nmr spectrum is very similar to that of *O*-acetylclivonine (47) except that a methylene singlet is present at δ3.49 and a methyl singlet occurs at δ2.28. The chemical shifts of these protons and the fact that the methylene protons exchange with deuterium, points to the unit $-OCOCH_2COMe$ and thus clivacetine is regarded as *O*-acetoacetylclivonine (48). *O*-Acetylclivatine (49) is also obtained by the reduction of clivacetine with sodium tetrahydridoborate, followed by *O*-acetylation of the resultant product.

The structure of clivacetine indicates the possibility that it is the natural precursor of the unique alkaloid clivimine and this speculation is supported by the fact that this conversion has been achieved *in vitro* (see page 178).

The leaves of C. *miniata* are also a source of hippeastrine (51) (A. Abdusamatov, S.A. Khamidkhodzhaev and S.Yu. Yunusov, Khim. prir. Soedin., 1975, 11:273).

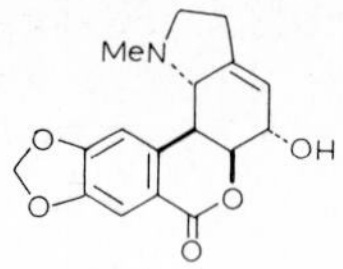

(51)

4. *Crystal Structure Determinations*

(a) Maritidine. The absolute configuration of maritidine shown in formula (16) has been confirmed by X-ray crystallography (V. Zabel and W.H. Watson, Cryst. Struct. Comm., 1979, 8:371).

(b) Norgalanthamine. (*N*-Demethylgalanthamine). R. Rogues and J. Lapasset (Acta Crystallogr., 1976, B32:579; 3358) have shown that norgalanthamine has structure (43).

(c) Cocculine and cocculidine. Cocculine and cocculidine, first isolated in 1950 from *Cocculus laurifolius* D.C. by S.Yu. Yunusov (Zhur. obshchei Khim., 1950, 20:368), have been subjected to X-ray diffraction analysis and the structures proposed originally, (52, R=H) and 52, R=Me) respectively, are shown to be correct (R. Razakov *et al.*, Chem. Comm., 1974, 150; S.M. Nasirov *et al.*, Khim. prir. Soedin., 1975, 11:395).

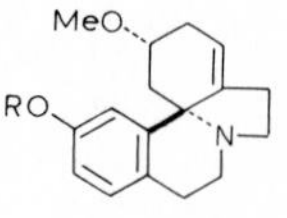

(52)

(d) Lycorine chlorohydrin. Lycorine was first converted into its 'so called' *cis*-chlorohydrin by K. Takeda, K. Kotera and S. Mizukami (J. Amer. chem. Soc., 1958, 80:2562), this compound has now been subjected to single crystal X-ray analysis and has been shown to have a *trans*-configuration (53) (J. Toda *et al.*, Tetrahedron Letters, 1980, 369).

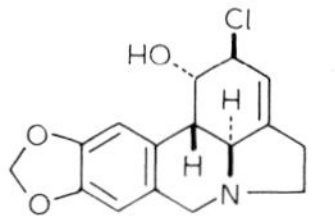

(53)

(e) Lycorine (9). X-Ray diffraction analysis (E.M. Gopalakrishna *et al.*, Cryst. struct. Commum., 1976, 5:795) shows that the B and C rings of lycorine are *trans*-fused, with ring B in a half-chair conformation. Ring C has a conformation intermediate between half-chair and diplanar with the two oxygen atoms occupying what would normally be axial sites. Both ring D and the dioxolane system are in evelope conformations, although the latter is flattened. The hydrogen atom at position C-11 and the nitrogen lone pair electrons bear a *trans*-relationship. Interestingly, M. Shiro, T. Sato and H. Koyama (J. chem. Soc. (B), 1968, 1544) report that H-11_{γ} and the nitrogen proton in dihydrolycorine hydrobromide have *cis*-geometry. These earlier workers also record that in this derivative ring B has a distorted boat conformation.

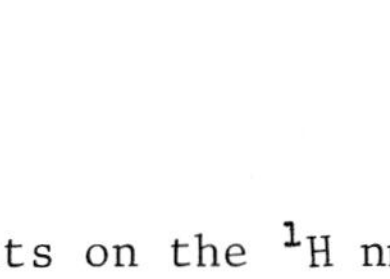

(9)

5. *Spectroscopy*

The effect of solvents on the ^{1}H nmr parameters of alkaloids of the *Amaryllidacea* group has been examined by Russian workers

(K.L. Seitanidi and M.R. Yagndaev, Khim. prir. Soedin., 1976, 500).

6. *Photochemical Reactions*

The irradiation of crinamine (21≡54) in methanol solution with a high pressure mercury lamp (200W) yields a compound, m.p. 189-190°, which has been named photocrinamine (Y. Tsuda *et al.*, Tetrahedron Letters, 1978, 1199). A single crystal X-ray of this product demonstrates that it has structure (55) and thus forms by an intramolecular [$\pi 2a + \sigma 2a$] addition, followed by photo-reduction of the intermediate aryl-conjugated cyclopropane. If this mechanistic interpretation is correct then the latter reaction may be the first of its type to be reported.

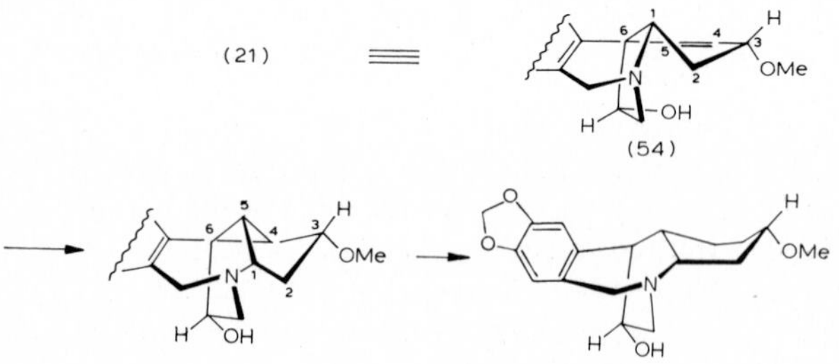

(21) ≡ (54)

(55)

7. *Synthesis*

(a) *The synthesis of lycorine and related structures*

Lycorine, the most important alkaloid within the group, has finally been synthesised by a number of research groups.

One approach (Y. Tsuda *et al.*, Chem. Comm., 1975, 933; Yuki Gosei. Kagaku Kyokai Shi., 1976, 34:625; Chem. Abs., 1975, 86: 72929h; J. chem. Soc. Perkin 1, 1979, 1358) utilizes the racemic urethane-ester (56) previously made by H. Irie *et al.* (Chem. Comm., 1973, 302). Cyclisation of this compound with tin (IV) chloride affords the lactam (57) which is converted by a series of standard procedures into the chloro derivative (58). This product with Meerwein's reagent, triethyloxonium tetrafluoroborate and triethylamine, gives the tetracyclic lactam (59). The latter with 3-chloroperbenzoic acid forms the epoxide (60), and reaction with diphenylselenide and then with sodium tetrahydridoborate yields the hydroxyselenide (61). When treated with sodium periodate the hydroxyselenide affords the hydroxyalkene (62). This compound is protected by acetylation with acetic anhydride before epoxidation with 3-chloroperbenzoic acid. The β-epoxide product (63) treated in turn with diphenylselenide, sodium tetrahydridoborate and sodium periodate gives the amide (64) which as the di-*O*-acetate derivative is converted into (±)-lycorine by reduction with lithium tetrahydridoaluminate.

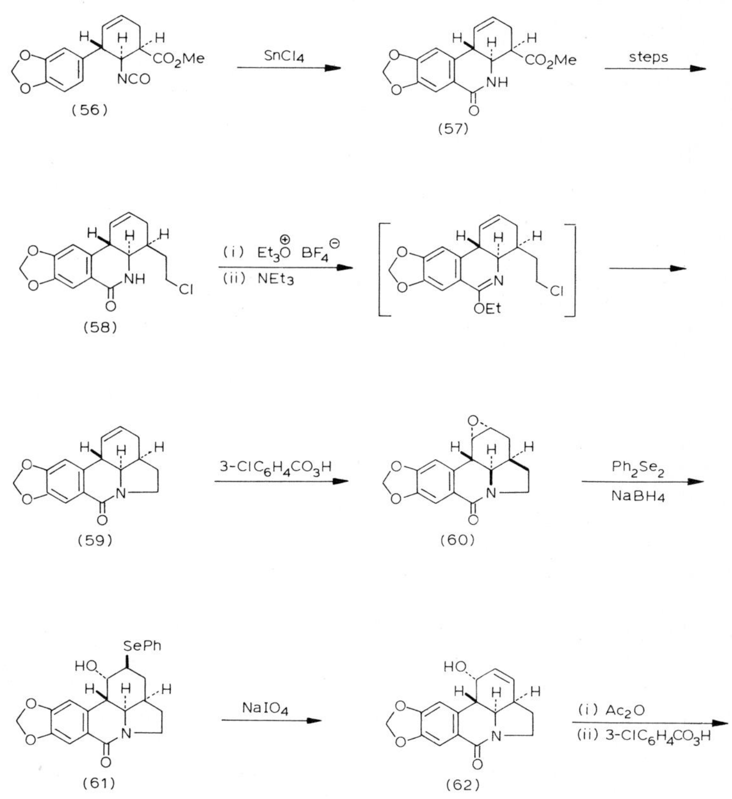

(56) $\xrightarrow{SnCl_4}$ (57) $\xrightarrow{steps}$

(58) $\xrightarrow[(ii)\ NEt_3]{(i)\ Et_3O^{\oplus}\ BF_4^{\ominus}}$ [ ] ⟶

(59) $\xrightarrow{3\text{-}ClC_6H_4CO_3H}$ (60) $\xrightarrow[NaBH_4]{Ph_2Se_2}$

(61) $\xrightarrow{NaIO_4}$ (62) $\xrightarrow[(ii)\ 3\text{-}ClC_6H_4CO_3H]{(i)\ Ac_2O}$

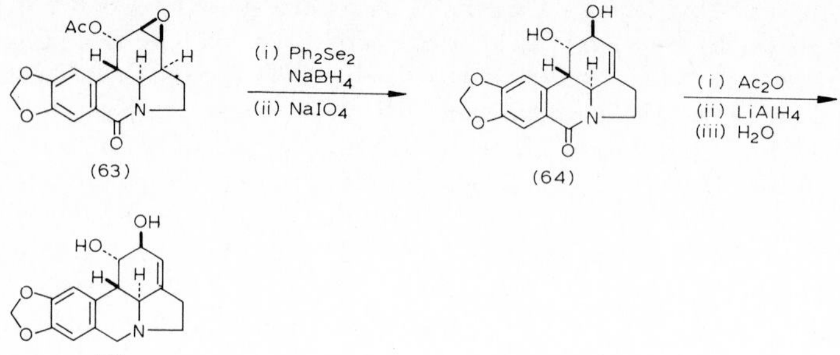

(63) → (i) Ph_2Se_2, $NaBH_4$ (ii) $NaIO_4$ → (64) → (i) Ac_2O (ii) $LiAlH_4$ (iii) H_2O → (9)

The oxirane (60) can be cleaved to yield a mixture of stereoisomeric diols, the major component of which is the *trans*-form (65). This as the di-*O*-acetate can be reduced with lithium tetrahydrioaluminate to give (±)-zephyranthine (66).

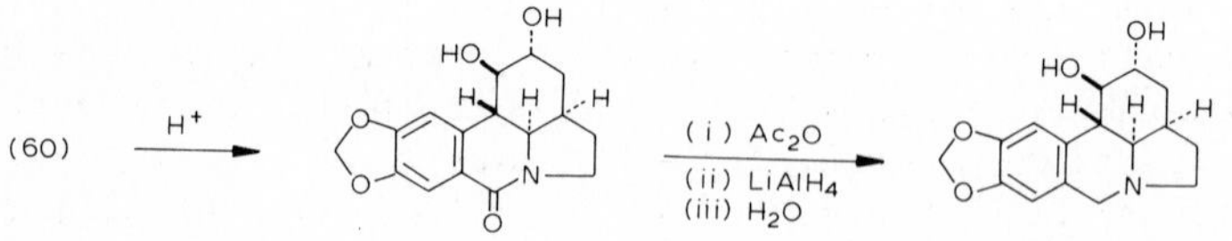

(60) → H^+ → (65) → (i) Ac_2O (ii) $LiAlH_4$ (iii) H_2O → (66)

Tsuda *et al.* (Heterocycles, 1976, 5:163) also record another synthesis of (±)-zephyranthine from the lactam (59). Here the lactam is treated with osmium tetraoxide in pyridine, followed by sodium hydrogen sulphite in pyridine and the mixture of diols formed converted by acetic anhydride in pyridine into the corresponding diacetates (67) and (68). These are separated and the first compound, identical with a degradation product of zephyranthine, reacts with lithium tetrahydridoaluminate to yield a racemic modification of the alkaloid itself.

(59) (i) OsO_4 (ii) $NaHSO_3$, pyridine (iii) Ac_2O, pyridine → (67) + (68)

Another modification (T. Sano *et al.*, *ibid.*, 1980, 14:1097) requires the selective reduction of the lactam carbonyl group of the ester (57) using first treatment with Meerwein's reagent and then reaction of the product imidic ester (69) with sodium tetrahydridoborate. This gives a mixture of the amine (70) and borazine (71), which without separation are heated in methanolic potassium carbonate solution to yield the lactam (72). Epoxidation of this product with 3-chloroperbenzoic acid affords the oxirane (73) which on reduction with lithium tetrahydridoaluminate gives (±)-dihydrocaranine (74). The formation of dihydrocaranine confirms that the oxide ring of the oxirane (73) has the α-configuration, and this compound has been used as the starting material for a synthesis of lycorine the steps of which parallel those used previously (Tsuda *et al.*, J. chem. Soc. Perkin 1, *loc. cit.*).

(57) $Et_3\overset{\oplus}{O}\ BF_4^{\ominus}$ → (69) $NaBH_4$ → (70) + (71)

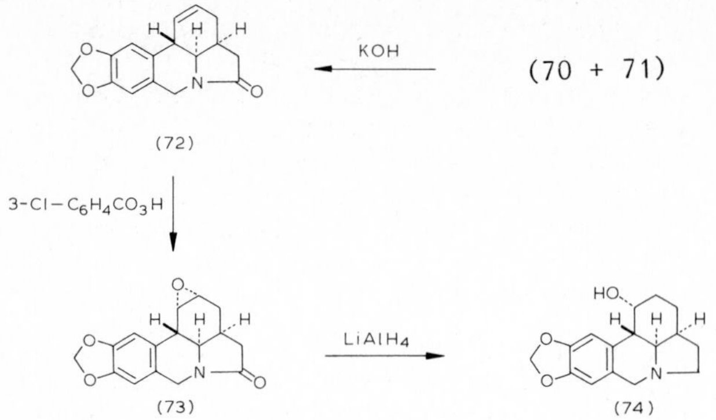

(72) ← KOH — (70 + 71)

(72) → $3\text{-}Cl-C_6H_4CO_3H$ → (73) → $LiAlH_4$ → (74)

K. Torsell (Tetrahedron Letters, 1974, 623) has developed a synthesis of the lactam (79) in six steps, the first of which is a Diels-Alder cycloaddition reaction between the β-nitrostyrene (75) and methyl hexa-3,5-dienoate (76). This gives the adduct (77) of appropriate stereochemistry for direct conversion into the α-dihydrolycorine series. Reductive cyclisation and further reduction affords the pyrrolidine derivative (78) which reacts with ethyl chloroformate and is cyclised to the lactam (79) with phosphoryl chloride.

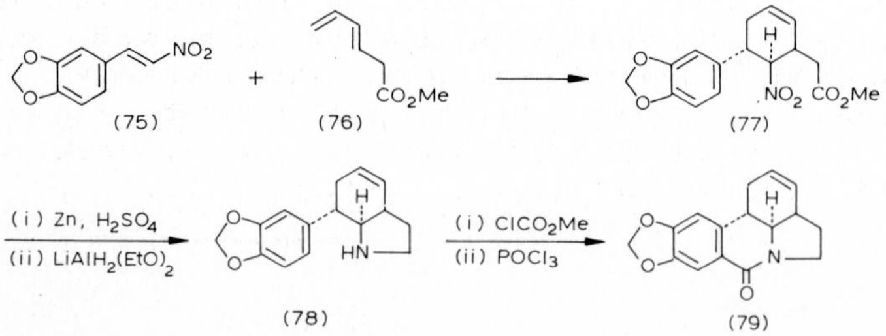

(75) + (76) → (77)

(77) → (i) Zn, H_2SO_4 (ii) $LiAlH_2(EtO)_2$ → (78) → (i) $ClCO_2Me$ (ii) $POCl_3$ → (79)

In order to continue this sequence on to lycorine itself K. Torsell and his colleagues (O. Møller, E.-M. Steinberg and K. Torsell, Acta. chem. Scand., 1978, 32B:98) oxidised the lactam with 3-chloroperbenzoic acid to give the oxirane (80). Oxidative rearrangement of this compound is achieved in three steps affording the enol (81). Finally an acid catalysed allylic rearrangement of the enol in the presence of acetic anhydride yields the *O*-acetate (82). Since this last compound has already been converted into lycorine in the natural series (Tsuda *et al.*, *loc. cit.*), this constitutes a formal synthesis of the alkaloid.

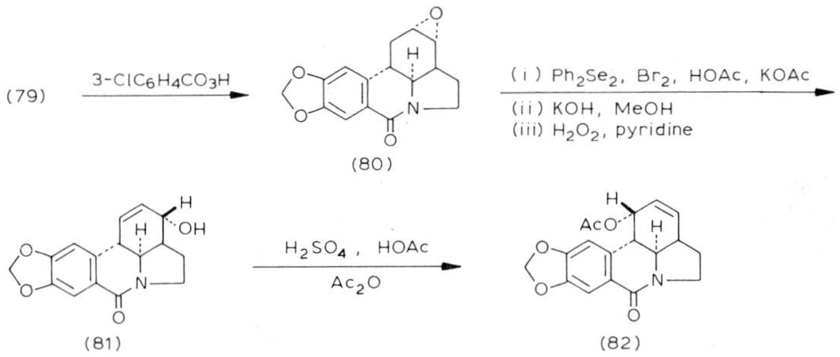

(79) —3-$ClC_6H_4CO_3H$→ (80) —(i) Ph_2Se_2, Br_2, HOAc, KOAc; (ii) KOH, MeOH; (iii) H_2O_2, pyridine→

(81) —H_2SO_4, HOAc; Ac_2O→ (82)

12b-α-Lycorane (γ-lycorane) (85) has been prepared from the enone amide (83), itself synthesised by the photochemical cyclisation of the enamide ketone (84). The reduction of the enone amide is achieved in three steps, two catalytic and one using lithium tetrahydrioaluminate (H. Iida, S. Aoyagi and C. Kibayashi, J. chem. Soc., Perkin 1, 1975, 2502).

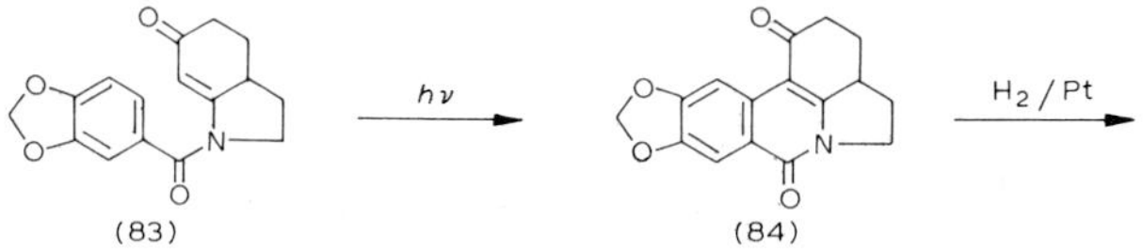

(83) —$h\nu$→ (84) —H_2/Pt→

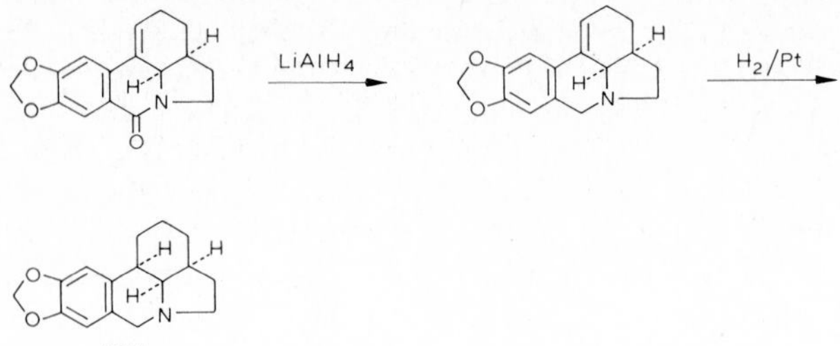

LiAlH$_4$ H$_2$/Pt

(85)

The same group (Iida, Y. Ynasa and Kibayashi, J. Amer. chem. Soc., 1978, 100:3598) have obtained the enaminone (87) by the base catalysed intramolecular cyclisation of the bromide (86). Stereoselective reduction of this compound with lithium tetrahydroaluminate at room-temperature in tetrahydrofuran solution affords (±)-α-dihydrocaranone (88) and some (±)-1-epi-γ-dihydrocaranine (89). The conversion of α-dihydrocaranone into γ-lycorane has already been reported (K. Kotera, Tetrahedron, 1961, 12:248; N. Veda, T. Tokuyama and T. Sakan, Bull. chem. Soc. Japan, 1966, 39:2021).

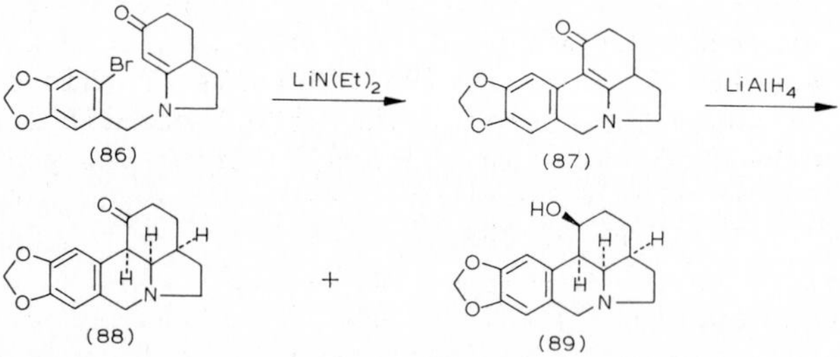

(86) LiN(Et)$_2$ (87) LiAlH$_4$

(88) + (89)

B. Umezawa *et al.* (J. org. Chem., 1977, 42:4272) have synthesised both (±)-α- and (±)-γ-lycoranes by the development of routes previously described (see Chemistry of Carbon Compounds, 2nd ed., Vol. IVB, p. 186). Similarly (±)-α-, (±)-β- and (±)-γ-lycoranes have been constructed from the two stereoisomeric 3-(3,4-methylenedioxyphenyl)-1,2,3,6-tetrahydrophthalic anhydrides (90) and (91) which are obtained by Diels-Alder addition of 1-(3,4-methylenedioxyphenyl) but -3-en-1-ol and fumaric acid (H. Tanadk *et al.*, J. chem. Soc. Perkin 1, 1979, 874).

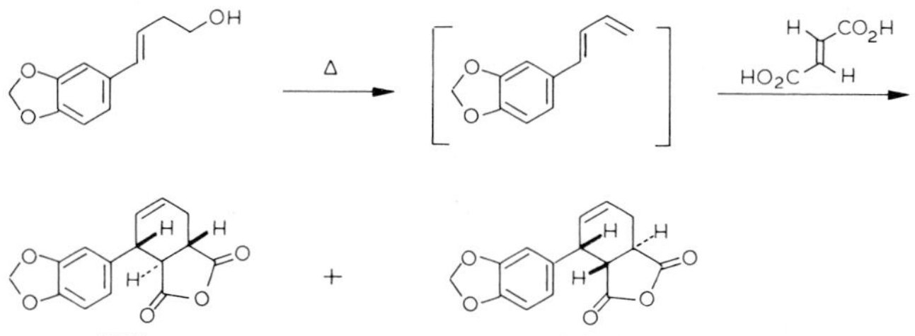

(90) (91)

Umezawa *et al.* (Heterocycles, 1979, 12:1475) have also developed an alternative route to (±)-α-Δ^2-lycoren-7-one (79). The lactam acetal (92) is reduced with lithium tetrahydridoaluminate to give the amino acetal (93). Acid treatment of this product affords the amino ketone (94) which is converted into the tetracyclic keto lactam (95) (details of this conversion are not given in the paper). The keto lactam is then condensed with dimethylamine and the product iminium salt reduced with sodium cyanotrihydridoborate to yield the amine (96). The latter with 3-chloroperbenzoic acid is converted into its *N*-oxide (97) which on pyrolysis at 200° affords the alkene (98). The lactam function in this compound is transposed from C-5 to C-7 by reduction with lithium tetrahydridoaluminate followed by oxidation with manganese (IV) oxide.

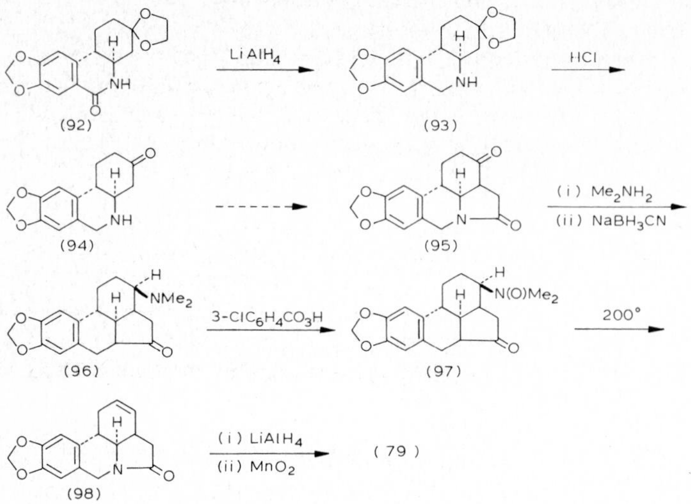

Perhaps the simplest stereospecific route to the skeleton of the lycorine alkaloids devised so far is due to G. Stork and D.J. Morgans Jr. (J. Amer. chem. Soc., 1979, 101:7110). Here the amide alcohol (100), prepared by the reaction of the acid (99) with 3-pyrrolidinol, is oxidised with pyridine-SO_3 complex in dimethylsulphoxide and triethylamine to give the keto amide (101). Wittig olefination, or an Emmons-Horner reaction, [$(EtO)_2P(O)CH_2CO_2Me$] and NaH, affords the β,γ-unsaturated ester (102), which when reduced with lithium tetrahydridoborate and the product treated with 2-nitrophenylselenocyanate and tributylphosphine yields the selenide (103). The latter compound with sodium periodate and sodium bicarbonate gives the triene (104) which cyclises on heating at 140° in chlorobenzene solution, containing 2-t-butyl-4-hydroxy-5-methylphenyl sulphide and *O*,*N*-bis(trimethylsilyl) acetamide, to a single compound (105) which has the correct stereochemistry at the ring B/C junction and the appropriately positioned double bond in ring C for direct conversion into the lycorine system.

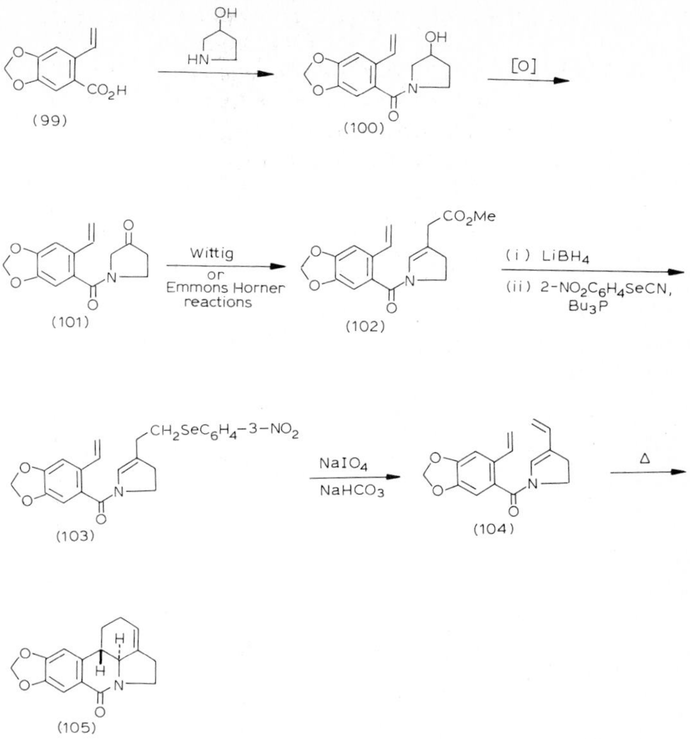

(99) (100) (101) (102) (103) (104) (105)

(b) *Clividine and Clivonine*

(±)-Clividine (113) and (±)-clivonine (114) have been synthesised from c-2-methoxycarbonyl-t-6-(3,4-methylenedioxyphenyl) cyclohex-4-ene-r-carboxylic acid (106) (H. Tanaka *et al.*, J. chem. Soc. Perkin 1, 1979, 535). A Curtis rearrangement of this acid ester, followed by treatment with methanol gives methyl 6-(*N*-methoxycarbonylamino-5-(3,4-methylenedioxyphenyl) cyclohex-3-ene-carboxylate (107), which is converted into its homologue (108) by an Arndt Eistert reaction. Hydrolysis with hydrochloric acid and cyclisation of the product acid with acetic anhydride affords the lactam (109) which, after chloromethylation and acetylation, yields the acetoxy compound (110). Reduction of this product with lithium tetrahydridoaluminate gives the aminoalcohol (111), which is converted into (±)-tetrahydroclividine (112) and (±)-tetrahydroclivonine (113) in a 1:1 ratio by oxidation with osmium tetraoxide. Further oxidation of these compounds in separate reactions using manganese (IV) oxide as reagent yields (+)-clividine and (±)-clivonine respectively.

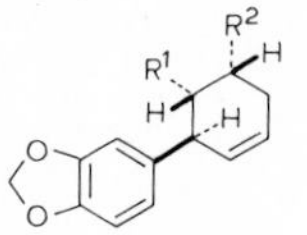

(106; $R^1 = CO_2H$; $R^2 = CO_2Me$)
(107; $R^1 = NHCO_2Me$; $R^2 = CO_2Me$)
(108; $R^1 = NHCO_2Me$; $R^2 = CH_2CO_2Me$)

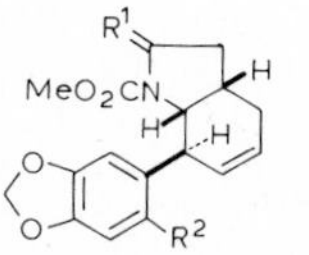

(109; $R^1 = O$; $R^2 = H$)
(110; $R^1 = O$; $R^2 = CH_2OAc$)

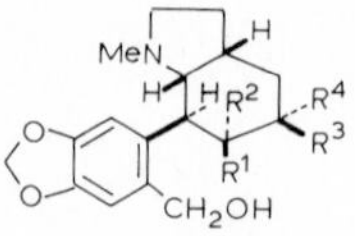

(111; $R^1 = R^3 = OH$; $R^2 = R^4 = H$)
(112; $R^1 = R^3 = H$; $R^2 = R^4 = OH$)

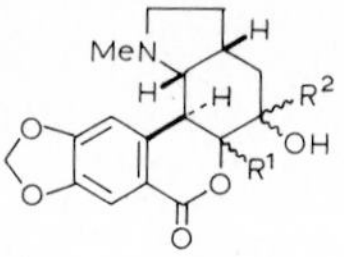

(113; $R^1 = R^2 = \beta$-H)
(114; $R^1 = R^2 = \alpha$-H)

(c) *Clivimine*

Kobayashi *et al.* (Heterocycles, 1980, 14:751) have achieved a partial synthesis of clivimine (50) from clivonine (114). First clivonine is treated with diketone in the presence of triethylamine to give clivacetine (48), which is then subjected to a Hantzsch pyridine synthesis with formaldehyde and ammonia thus yielding dihydroclivimine (115). Dehydrogenation of this last compound to clivimine is achieved by the action of nitrous acid, thus completing what is assumed to be a biogenetic-type conversion of clivacetine into clivimine (see page 165).

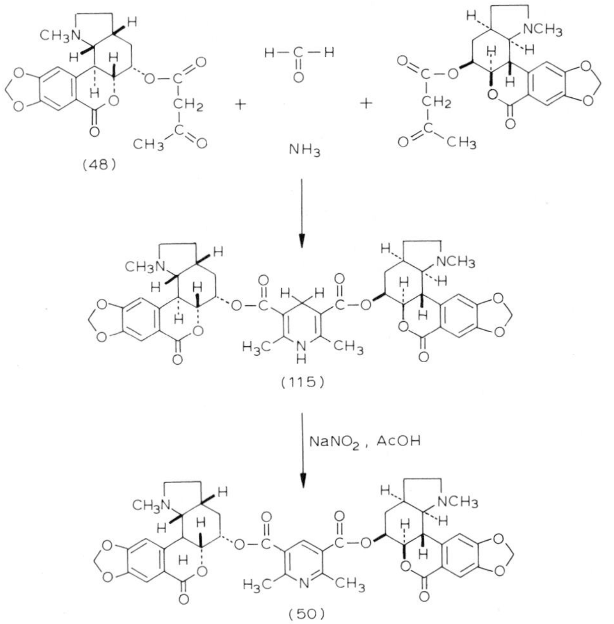
CH3N
H
NCH3
H—C—H
O
NH3
(48)
(115)
NaNO2, AcOH
(50)
H3C
N
CH3

(d) *Maritidine, (+)- and (-)-galanthamine*

The synthesis of (+)-maritidine (16) from L-tyrosine has been described by S. Yamada, K. Tomioka and K. Koga (Tetrahedron Letters, 1976, 57) : L-tyrosine is converted into the *N*-trifluoroacetylnorbelladine derivative (116) and then oxidatively coupled with thallium (III) trifluoroacetate to the quinone methide (117). The reaction of this product with ammonia affords the diamide (118), which with aqueous sodium hydroxide undergoes selective hydrolysis and ring-closure affording only one diastereoisomer (119). The latter with phosphoryl chloride leads to the keto-cyanide (120), which with sodium tetrahydridoborate gives the corresponding alcohol (121) through stereoselective reduction.

The cyanide group is next removed by reaction with sodium in liquid ammonia to yield (+)-epimaritidine (122). Final epimerisation is achieved by heating epimaritidine in 10% aqueous hydrochloric acid at reflux.

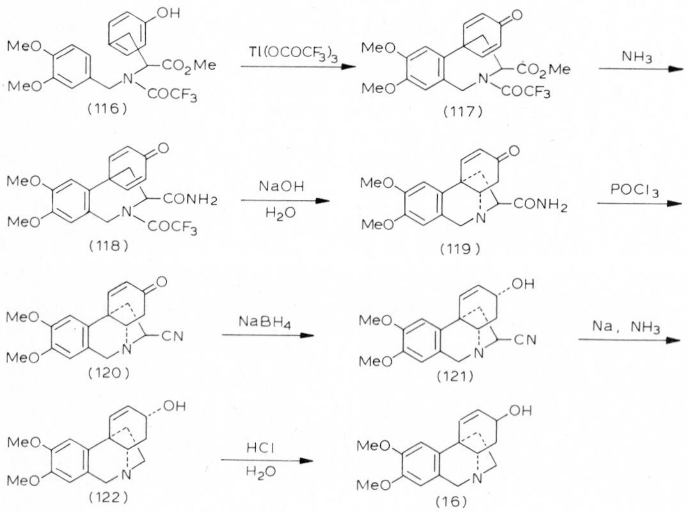

The same authors have published some variations on this theme (Chem. pharm. Bull. Japan, 1977, 25:2681) and also describe the synthesis of (+)- and (-)-galanthamines (K. Shimizu *et al.*, Heterocycles, 1977, 8:277; Chem. pharm. Bull. Japan, 1978, 26:3765) along a biomimetic route. Here the (-)-norelladine derivative (123) is oxidised with manganic trisacetylacetonate affording the phenol (124). Reaction of the phenol with diethyl phosphorochloridate in the presence of triethylamine gives (+)-enone phosphate (125) plus a small amount of its diastereomer (126). Sodium tetrahydridoborate reduction of the phosphate (125) produces a mixture of the isomeric alcohols (127) and (128) in a ratio of 4:1. The major product

(127) is then *N*-methylated and converted into the (+)-amide (129) by reaction with ammonia. Acetylation of the (+)-10-diethylphosphoroxygalanthamine (131) which when treated with sodium and liquid ammonia affords (+)-galanthamine (132). The *N*-trifluoroacetyl derivative of the alcohol (127), when treated with lithium diisopropylamide in tetrahydrofuran containing tetramethylethylenediamine and hexamethylphosphoramide, followed by oxidation of the product with pyridinium chlorochromate, gives the (-)-enone phosphate (133) which serves as starting material for a synthesis of (-)-galanthamine (41) utilising the same reagents as described for the conversion of its antipode (125) into (+)-galanthamine. It is noteworthy that this enantiomeric transformation depends upon the fact that the narwedine-type enone skeleton is unstable and readily undergoes an inversion of stereochemistry (see D.H.R. Barton and G.W. Kirby, J. chem. Soc., 1962, 806).

(123) → Mn(III) acetylacetonate → (124)

→ $(EtO)_2POCl$, Et_3N → (125) + (126)

(125) → $NaBH_4$ → (127) + (128)

(127) → (i) HCHO, HCO_2H; (ii) NH_3 → (129) → (i) Ac_2O, pyridine; (ii) $POCl_3$ →

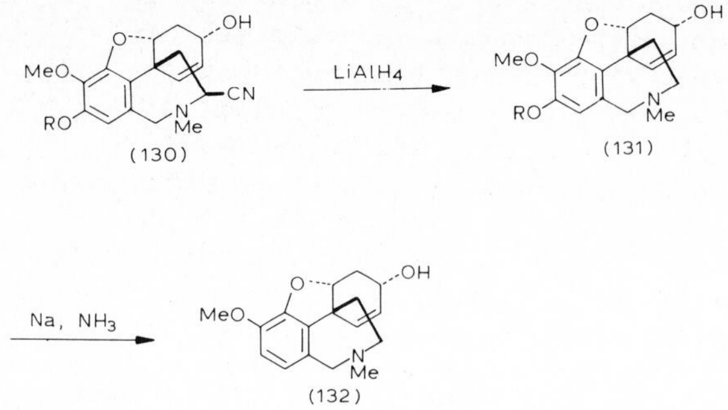

MeO
RO
OH
CN
N
Me
(130)
LiAlH4
(131)
Na, NH3
(132)

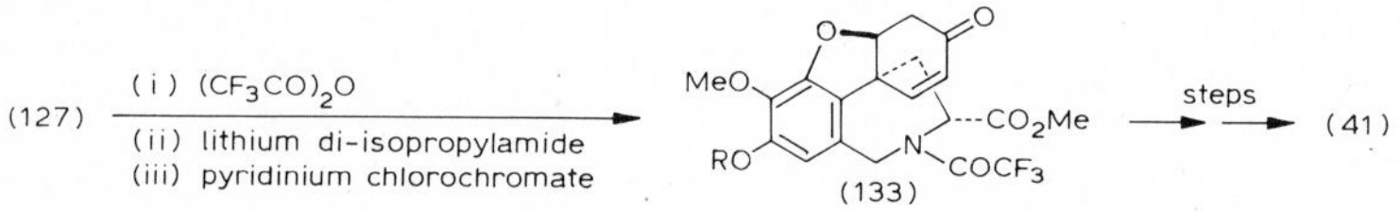

(127)
(i) (CF3CO)2O
(ii) lithium di-isopropylamide
(iii) pyridinium chlorochromate
MeO
RO
O
CO2Me
COCF3
(133)
steps
(41)

R = $(EtO)_2PO$

(e) (±)-*Lycoramine (dihydrogalanthamine)*

A.G. Schultz, Y.K. Yee and M.H. Berger (J. Amer. chem. Soc., 1977, 99:8065) have described a multistage of (±)-lycoramine (143). The enone (134) is prepared and photolysed in benzene/ methanol to give the *cis*-fused dihydrofuran (136). Mechanically this reaction is considered to involve a conrotatory mode of ring-closure and the intermediacy of a carbonyl-ylide (135) (or an equivalent species) from which protonation - deprotonation in methanol leads to the dihydrofuran. The azepine ring is then constructed in the following manner: first the carbonyl group in the dihydrofuran is protected by acetalization and the product reduced with lithium tetrahydridoaluminate to give the amino-alcohol (137). Finally ring-closure and de-acetalization are achieved by the action of thionyl chloride in triethylamine dichloromethane solution, followed by treatment of the product with aqueous sulphuric acid.

Carbonyl transposition within the tetracyclic ketone (138) is carried out by treatment of the lithium enolate derivative with phenyl phenylthiosulphonate thus yielding the thio acetal (139). Reduction with lithium tetrahydridoaluminate then gives the alcohol (140) which is converted into the mesyl ester (141) by the action of methane sulphonyl chloride. Thioacetal hydrolysis with mercury (II) chloride and mercury (II) oxide gives a ketone (142) which on treatment first with chromium (II) chloride in aqueous acetone and then stereospecific reduction with lithium tetrahydridoaluminate affords (±)-lycoramine (143).

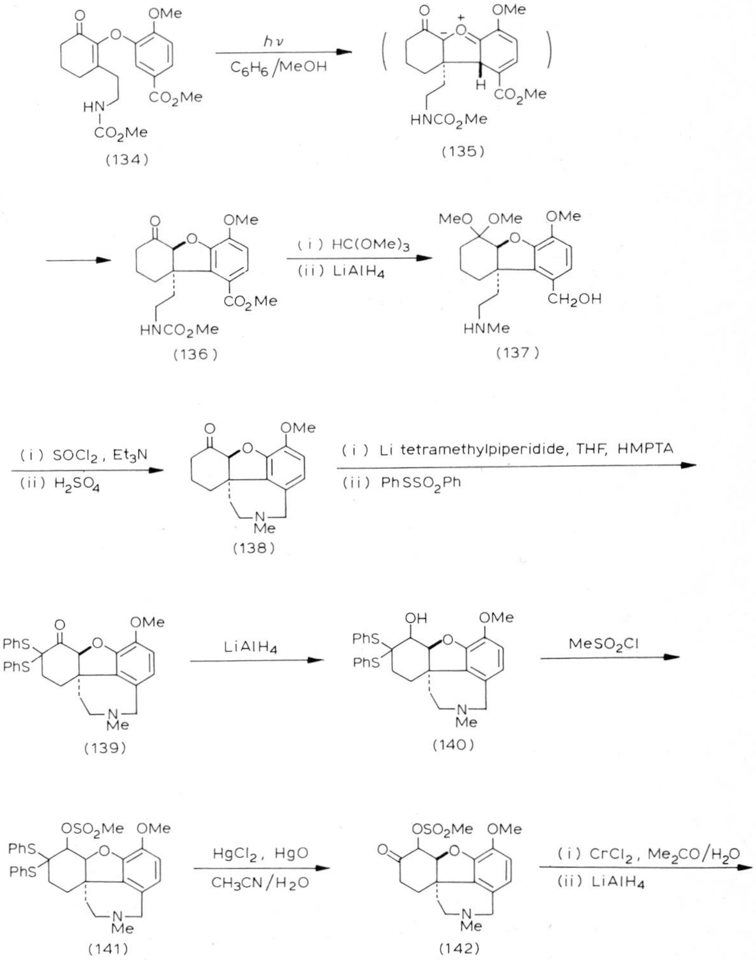

(134) $\xrightarrow[C_6H_6/MeOH]{h\nu}$ (135) $\rightarrow$ (136) $\xrightarrow[(ii)\ LiAlH_4]{(i)\ HC(OMe)_3}$ (137)

$\xrightarrow[(ii)\ H_2SO_4]{(i)\ SOCl_2,\ Et_3N}$ (138) $\xrightarrow[(ii)\ PhSSO_2Ph]{(i)\ Li\ tetramethylpiperidide,\ THF,\ HMPTA}$

(139) $\xrightarrow{LiAlH_4}$ (140) $\xrightarrow{MeSO_2Cl}$

(141) $\xrightarrow[CH_3CN/H_2O]{HgCl_2,\ HgO}$ (142) $\xrightarrow[(ii)\ LiAlH_4]{(i)\ CrCl_2,\ Me_2CO/H_2O}$

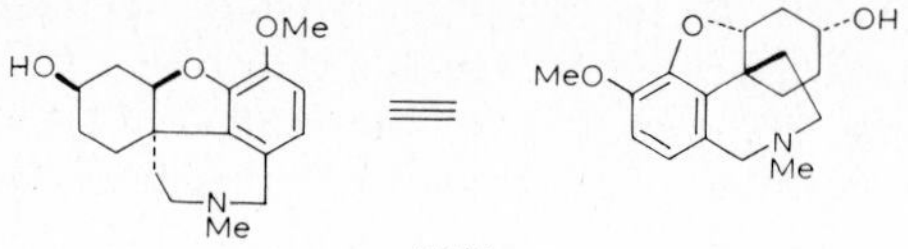

(143)

(f) *Apogalanthamine analogues*

Interest in dibenzo[c,e]azocines as potential α-adrenergic blocking agents has stimulated the study of synthetic approaches to apogalanthamine analogues (S. Kobayashi *et al.*, Chem. pharm. Bull. Japan, 1977, 25, 3312; 1978 26:635; T. Kametani *et al.*, Yakugaku Zasshi, 1977, 97:1353; Chem. Abs., 1978, 88: 152394). In the publication by Kobayashi, M. Kihara and H. Matsumoto (Yakugaku Zasshi, 1978, 98:863; Chem. Abs., 1978, 89: 197764) aryl-aryl bond formation is reported through ultraviolet irradiation of mono-iodoarylalkylamines (e.g. 144).

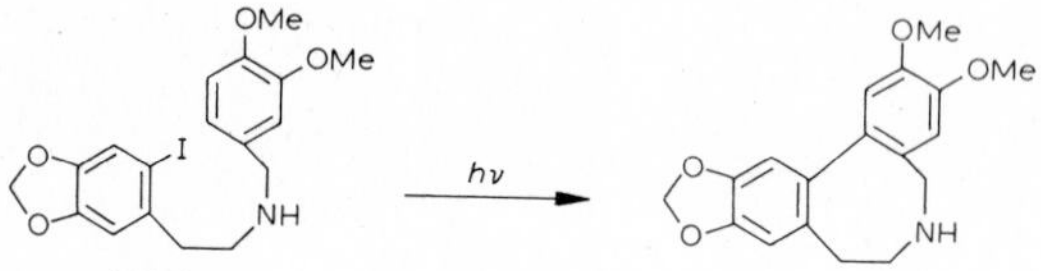

(144)

(g) *Tetrahydrometinoxocrinine and crinine*

Tetrahydrometinoxocrinine (155), an important degradation product of crinine (39), has been synthesised previously by W.C. Wildman (J. Amer. chem. Soc., 1958, 80:2567) and by S. Uyeo *et al.* (Chem. pharm. Bull. Japan, 1965, 13:427). A new isomer-free construction is described by I.H. Sańchez and M.T. Mendoza (Tetrahedron Letters, 1980, 3651). Piperonyl nitrile is reacted with ethyl acrylate and Triton B in dry acetonitrile solution to give the pimelate ester (145). This compound is then cyclised under Dieckman conditions [NaH/dimethoxyethane (DME)] to the enol ester (146) which is hydrolysed and decarboxylated to the hexanone (147). Reduction of this ketone with diisobutylaluminium hydride (DIBAL-H) in boiling benzene gives the syn-hydroxyaldehyde (148) in a stereospecific high yielding reaction. The *O*-acetate derivative (149) is then condensed with diethyl cyanomethylphosphonate and sodium hydride to yield the (*E*)-acrylonitrile (150) which on hydrogenation affords the saturated nitrile (151). Hydrolysis of this compound gives the hydroxyacid (152).

Reacetylation of this product, followed by a Curtius rearrangement yields the isocyanate (153) which is ring-closed to the acetoxylactam (154) by its reaction with polyphosphoric acid.

Since the acetoxylactam has already been converted into tetrahydrometinoxocrinine (S. Uyeo *loc. cit.*) and thence into crinine itself (S. Minami *et al.*, *ibid.*, 1965, 13:1084), this synthesis represents an isomer-free synthesis of the alkaloid.

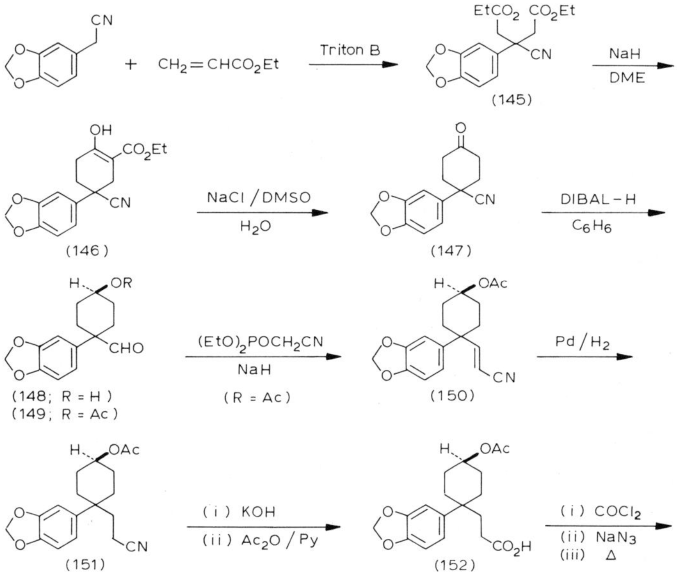

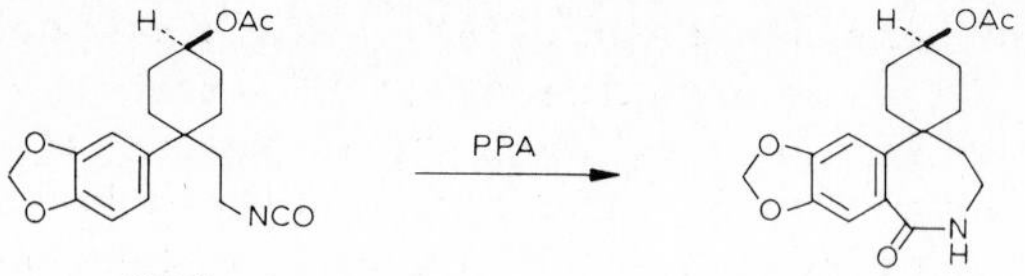

(153) —PPA→ (154)

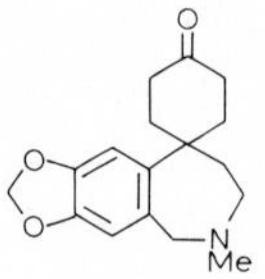

(155)

(h) (±)-*Crinamine*, (±)-*6-hydroxycrinamine*, (±)-*criwelline and* (±)-*macronine*

Racemic modifications of crinamine (163), 6-hydroxycrinamine (160), criwelline (161) and macronine (162) have all been obtained from the same starting material - the lactam (156) (K. Isobe, J. Taga and Y. Tsuda, Tetrahedron Letters, 1976, 2331). This useful substrate, available through the development of earlier work (Tsuda and Isobe, Chem. Comm., 1971, 1555), is converted stereospecifically into the allylic ether (157) by methoxyselenation and elimination of phenylselenoxide. The carbonyl group of the lactam ring is then reduced in two steps, first by treatment with triethyloxonium tetrafluoroborate and then reduction of the product imino ether with sodium tetrahydridoborate and tin (IV) chloride. The product amine (158) is *N*-formylated and the derivative cyclised with phosphoryl chloride to the carbinolamine (159). This compound on hydrolysis affords (±)-6-hydroxycrinamine (160); methylation and base-catalysed rearrangement gives (±)-criwelline (161) and oxidation, rearrangement and *N*-methylation affords (±)-macronine (162).

(±)-Crinamine (163) is obtained from the lactam (157) by reduction with lithium tetrahydridoaluminate and cyclisation of the corresponding amine with formaldehyde.

(156) → (i) $(PhSe)_2$, $CH_3CONHBr$, MeOH; (ii) H_2O_2 → (157)

(157) → (i) $Et\overset{\oplus}{O}_3\overset{\ominus}{BF_4}$; (ii) $NaBH_4$, $SnCl_4$ → (158)

(157) → (i) $LiAlH_4$; (ii) HCHO → (163)

(158) → HCO_2Ac | pyridine →

→ $POCl_3$ → (159)

(159) → K_2CO_3, MeOH → (160)

(159) → (i) MeI, MeOH; (ii) KOH, H_2O → (161)

(159) → MnO_2 →

→ (i) K_2CO_3; (ii) H^+; (iii) HCHO, $NaBH_4$ → (162)

(i) *Lycoricidine and related compounds*

Interest in the synthesis of lycoricidine (178) and alkaloids of similar structure has been stimulated by reports of their cytotoxicity. (±)-Lycoricidine itself has been prepared by S. Ohta and S. Kimoto (Tetrahedron Letters, 1975, 2279; Chem. pharm. Bull. Japan, 1976, 24:2969; 2977) : the homoallylic alcohol (164) is heated with ethyl acrylate to give a diastereomeric mixture of adducts which are equilibrated and hydrolysed to afford the *trans*-acid (165). This is converted into the corresponding azide which is then rearranged thermally to the isocyanate (166). Cyclisation of the isocyanate using boron trifluoride in diethylether yields the lactam (167). This product after being *N*-acetylated, hydrolysed, and treated with *N*-bromosuccinimide gives the bromolactone (169), which has *cis* geometry at the ring junction. This change in stereochemistry is considered to arise through stereoselective attack by the carboxyl group upon a bromonium ion intermediate (168). When heated in pyridine containing an equimolecular amound of 1.8-diazabicyclo [5,4,0] undecene-7 (DBU) the bromolactone undergoes dehydrobromination and, after treatment with aqueous sodium hydroxide, yields the unsaturated lactam (170).

This compound, protected as its tetrahydropyranyl (THP) ether, is oxidised with 3-chloroperbenzoic acid to give the oxirane (171). Here the β-orientated THP unit is considered to shield one face of the substrate from the approach of the reagent, thereby initiating a stereoselective reaction.

When the oxirane (171) is treated first with diphenyldiselenide and sodium tetrahydridoborate and then with hydrogen peroxide the unsaturated alcohol (172) is obtained, *O*-acetylation and acid catalysed deprotection of the oxygen atom at C-1 gives the unsaturated diol monoacetate (173).

The monoacetate is then treated with osmium tetraoxide and pyridine, and the corresponding osmate decomposed with sodium bisulphite to give the triol (174). This with 2,2-dimethoxypropane, dimethylformamide and a catalytic amount of 4-toluenesulphonic acid gives a single acetonide (175), which on dehydration with thionyl chloride in pyridine gives the (±)-lycoricidine derivative (176).

Conversion of this into a racemic modification of the alkaloid is achieved by treatment, first with toluene-4-sulphonic acid in aqueous methanol-chloroform, then acetic anhydride and finally by gentle hydrolysis of the tetraacetate (177) with ammonia and methanol.

(164)

$CH_2{=}CHCO_2Et$, Δ

CO_2Et + CO_2Et

(i) NaOEt, EtOH (ii) NaOH

(165) CO_2H

(i) $ClCO_2Et$, Et_3N (ii) NaN_3

(166) NCO

(i) Δ (ii) BF_3

(167) NH

(i) Ac_2O (ii) KOH, H_2O

NHAc CO_2H

AcHN CO_2H

NBS

(168) AcHN $Br^{\oplus}$ H

(169) AcHN Br

DBU, pyridine

AcHN

NaOH, H_2O

(170) HO NH

(i) dihydropyran, TsOH (ii) $3\text{-}ClC_6H_4CO_3H$

(171) THPO NH

(i) $(PhSe)_2$, $NaBH_4$ (ii) H_2O_2

(172) OH THPO NH

Ac_2O

OAc THPO NH O TsOH HO OAc NH O (173) OsO_4 pyridine

HO OAc OH OH NH O (174) $Me_2C(OMe)_2$ DMF, TsOH HO OAc NH O (175)

$SOCl_2$ pyridine OAc NH O (176) TsOH OAc OH OH NH O

Ac_2O OAc OAc OAc NH O (177) NH_3, MeOH OH OH OH NH O (178)

Some model experiments *en route* to lycoricidine analogues have been reported by G.E. Keck and S.A. Fleming (Tetrahedron Letters, 1978, 4763). For example, the amide (179) is prepared and reacted first with iodine and silver acetate and then with acetyl chloride to give the iodo-*O*,*O*-diacetate (180), in which the iodine atom is β-orientated. In the presence of dimethoxyethane and sodium acetate the iodine atome is displaced by the oxygen atom of the amide carbonyl group thus generating the oxazine (181). Finally acid hydrolysis and *O*-acetylation of the oxazine affords the triacetate (182).

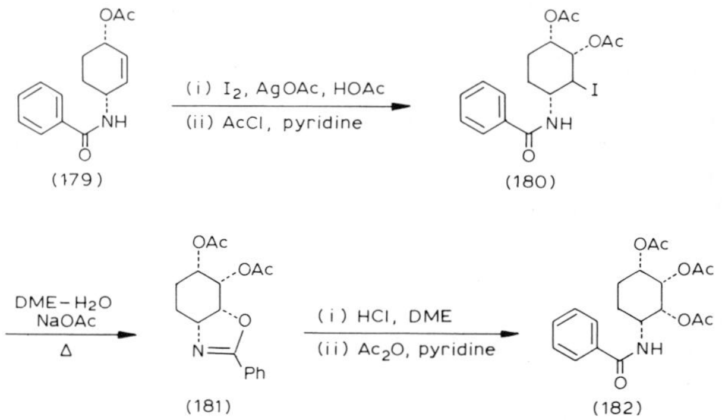
OAc
NH
O
(179)
(i) I2, AgOAc, HOAc
(ii) AcCl, pyridine
OAc
OAc
I
NH
O
(180)
DME-H2O
NaOAc
Δ
OAc
OAc
O
N
Ph
(181)
(i) HCl, DME
(ii) Ac2O, pyridine
OAc
OAc
OAc
NH
O
(182)

(*j*) *Elswesine*

A synthesis of elswesine (185) which is an improvement on that previously reported (H. Irie, S. Uyeo and A. Yoshitake, J. chem. Soc. (C), 1968, 1802) has been described by T. Fushimi *et al.* (Hetereocycles, 1979, 12:1311). β-[4-Acetoxy-1-(3,4-methylenedioxyphenyl)cyclohexyl] propionyl isocyanate (183) when treated first with phosphoryl chloride, then with tin (IV) chloride affords the lactam (184). This is then transformed into elswesine by the original method.

(183)

(i) $POCl_3$
(ii) $SnCl_4$

(184)

(185)

(k) (±)-*Tazettine*

The total synthesis of (±)-tazettine (13) and (±)-6a-epipretazettine (200) has been announced (S. Danishefsky *et al.*, J. Amer. chem. Soc., 1980, 102:2838). This work commences with a seven stage synthesis of the sulphone (187) from 3,4-methylene dioxyphenylacetone (186). A [4 + 2π] cycloaddition reaction between the sulphone and the diene (188) then gives the cyclohexenones (189) and (190), which are not separated, but treated with methylamine to give a 9:1 mixture of the amines (191) and (192). The amine (191) when chromatographed on neutral alumina yields the enedione (193).

The required stereochemistry at C-3 and C-6a in this product is introduced as follows: the enone function is reduced with diisobutylaluminium hydride to yield a 3:1 ratio of alcohols (194) and (195). The α-isomer (194) is converted into the required β-methoxy derivative (196) by the action of mesyl chloride and triethylamine, followed by solvolysis in methanol, while the β-isomer gives the same ether by its reaction with diazomethane in the presence of aluminium (III) chloride. Reduction of the ether with sodium tetrahydridoborate gives a 3:1 mixture of the alcohols (197) and (198); whereas reduction with K-selectride ($KB[CH(Me)CH(Me)_2]_3H$) in tetrahydrofuran produces the same alcohols, but in a ratio of 9:1.

Reaction of the alcohol (197) with triethylformate in the presence of polyphosphoric acid gives 6a-epipretazettine *O*-methylether (199). Acidic hydrolysis then affords (±)-6a-epipretazettine (200) which is reduced with lithium tetrahydrioaluminate and the product (201) converted into the *O*-t-butyldimethylsilyl derivative (202) by treatment with t-butyldimethylsilyl chloride, triethylamine and 4-pyrrolidinopyridine.

Moffat-Pfitzner oxidation (dimethylsulphoxide and *N,N*-dicyclohexylcarbodiimide) of this compound gives the *O*-silyl ether (203) which on desilation with tetra-n-butylammonium fluoride yields (±)-tazettine (13).

(186) — steps → (187) — (188) →

(189; X = α-OMe)
(190; X = β-OMe)

— $MeNH_2$ →

(191; X = α-OMe)
(192; X = β-OMe)

— Al_2O_3 → (193) — DIBAH →

(194; X = α-OH)
(195; X = β-OH)

(194) — (i) $MeSO_2Cl/Et_3N$ (ii) MeOH →

(196) $\xrightarrow{NaBH_4}$ (197; β-OH) (198; α-OH)

(197) $\xrightarrow[PPA]{HC(OMe)_3}$ (199) $\xrightarrow{H^{\oplus}}$

(200) $\xrightarrow{LiAlH_4}$ (201) $\xrightarrow[\text{4-pyrrolidinopyridine}]{{}^{t}Bu(Me)_2SiCl,\ Et_3N}$

(202) $\xrightarrow[H_3PO_4]{DMSO,\ DCC}$ (203) $\xrightarrow{{}^{n}Bu_4\overset{\oplus}{N}\overset{\ominus}{F}}$

(13)

Overall this is a 17-stage synthesis of tazettine, but in addition it allows access to the C-3 epimer criwelline as well as pretazettine and precriwelline (the C-3 epimer of pretazettine).

The transformation of tazettine into pretazettine has been reported by S. Kobayashi, M. Kihara and T. Shingu (Heterocycles, 1979, 12:1547; Chem. pharm. Bull. Japan, 1980, 28:2924) by a method which lends support to the stereochemistry allocated to the latter compound. Lithium tetrahydrioaluminate reduction of tazettine (13) affords a mixture of tazettadiol (205) and 3-epitazettadiol (206), these products indicate that the hydride reaction proceeds *via* the keto-alcohol intermediate (204), although this was not isolated.

Addition of the AlH_4^- nucleophile to the carbonyl function of this intermediate is restricted by the size of the β-orientated phenyl group so that tazettadiol is the major product (63%). The minor product (13.5%) 3-epitazettadiol is formed by attack at the hindered face of the keto-alcohol.

Cyclisation of 3-epitazettadiol with 3% sulphuric acid at 100° gives deoxypretazettine (207), m.p. 112-113°, whereas similar treatment of tazettadiol affords deoxytazettine (208). The ORD and CD curves of pretazettine and deoxypretazettine are similar and show positive Cotton effects centred at λ290 nm. The curves of tazettine and deoxytazettine on the other hand resemble one another, but show a negative Cotton effect at this wavelength. These facts demonstrate that the two pairs of compounds differ in stereochemistry at C-6a. Moreover, the ^{1}H nmr coupling constants $J_{6a-6\alpha}$, and $J_{6a-6\beta}$ in deoxypretazettine are larger than those in deoxytazettine (11 and $8H_z$ *vs* 5 and $3H_z$ respectively), indicating a B/D *trans* configuration in deoxypretazettine and a B/D *cis* stereochemistry in depxytazettine.

Oxidation of 3-epipretazettadiol with manganese (IV) oxide gives three products pretazettine, 3-epimacronine (209), m.p. 125-127°, and tazettine. Similar oxidation of tazettadiol yields 6a-epipretazettine (200) and 6a-epi-3-epimacronine (210), m.p. 105-108°. ^{1}H nmr analysis of these oxidation products shows that pretazettine and 3pepimacronine have the *R*-configuration at C-6a.

(13) $\xrightarrow{LiAlH_4}$ [(204)] ⟶ (205) + (206)

(207; $R_1 = R_2 = H$)
(209; $R_1 = R_2 = O$)

(208; $R_1 = R_2 = H$)
(210; $R_1 = R_2 = O$)

(206) $\xrightarrow{MnO_2}$ ⟶ (12) + (209) + (13)

(205) $\xrightarrow{MnO_2}$ ⟶ (200) + (210)

Vinblastine $R_1 = CH_3$
Vincristine $R_1 = CHO$

Colchicine

Acronycine

Thalisidine

Allamandin

Alatolide

Chapter 11

FIVE-MEMBERED MONOHETEROCYCLIC COMPOUNDS ALKALOIDS (CONTINUED):

TROPANE ALKALOIDS

JACK G. WOOLLEY

Introduction

Plants have been used for medicinal and ceremonial purposes for centuries but it was not until 1803 when Derosne isolated the first alkaloid, narcotine, and Serturner later isolated morphine from opium that the search to obtain single medicinally active agents from plants gathered momentum. In 1833 atropine and hyoscyamine were the first tropane alkaloids to be isolated (L. Geiger and C. Hesse, Ann., 1833, 5, 43; 1833 6, 44; 1833, 7, 269), their names deriving from their plant sources, *Atropa* (deadly nightshade) and *Hyoscyamus* (henbane), both of the Solanaceae family, and it is from the former that the name Tropane originates. The other important medicinal alkaloid, hyoscine, was isolated much later (A. Ladenburg, Ann., 1881, 206, 274) and again independently by E. Schmidt (Arch. Pharm., 1892, 230, 207), under the name scopolamine. Thus, the former name should take precedence although it rarely does.

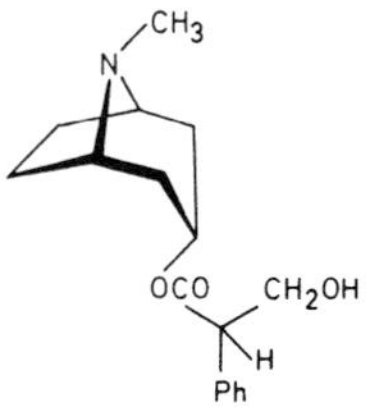

hyoscyamine

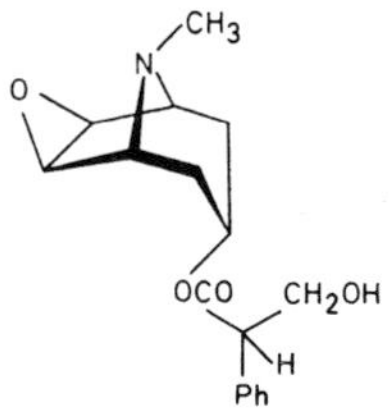

hyoscine

The formidable task of establishing the structures of these and related alkaloids, like cocaine, by purely chemical means has been reported fully elsewhere, (T.A. Henry, The Plant Alkaloids, 4th Edn., 1949, Churchill, London, p 64; H.L. Holmes, in The Alkaloids, ed., R.H.F. Manske and H.L. Holmes, 1950, Academic Press, London, p 272; A.R. Pinder, Chemistry of Carbon Compounds IVC p 1849)

Although these alkaloids have been known and well understood for a long time, there has been renewed interest in them: new synthetic routes to the ring system have been devised; the absolute stereochemistry of many bases has been established; their biosynthesis has attracted much interest (R.B. Herbert, Chemistry of Carbon Compounds IVL p 291); new types of alkaloid have been isolated from plants of the Proteaceae and Erythroxylaceae, and X-ray and spectroscopic studies, particularly nmr, have appeared.

The tropane alkaloids are peculiar to the higher plant families Convolvulaceae, Erythroxylaceae, Euphorbiaceae, Proteaceae, Rhizophoraceae and Solanaceae and reports that atropine is present in the fungus *Amanita* (J.J. Willaman and B.G. Schubert, U.S. Department of Agriculture, Tech. Bull. 1234, 1961) do not appear to have been confirmed. The tropane ring is a fused bicyclic system containing an *N*-methyl-pyrrolidine and a piperidine component in which the numbering proceeds from one or other of the bridgehead carbons. Such alkaloids are now systematically named and tropane itself is 8-methyl-8-azabicyclo[3.2.1] octane. All the alkaloids have, or did have during their biosynthesis, an oxygen function at C(3) and the simplest base tropinone (8-methyl-8-azabicyclo[3.2.1] octan-3-one) is found in the monotypic solanaceous genus *Nicandra* (A. Romeike, Pharmazie, 1965, 20, 738; Naturwiss., 1965, 52, 619).

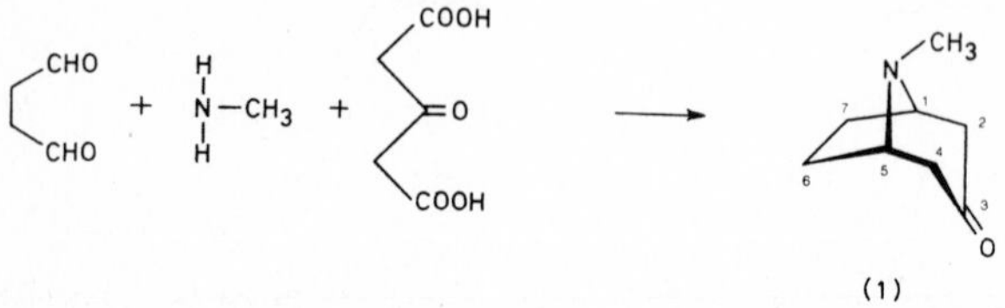

(1)

The classically simple synthesis of tropinone (1) by allowing an aqueous solution of succindialdehyde, methylamine hydrochloride and acetone dicarboxylic acid to stand for several days, reported by R. Robinson in 1917, (J. chem. Soc., 1917, 111, 762), is still the industrial basis for its synthesis. Reduction of the ketone group with sodium borohydride or lithium aluminium hydride yields the isomeric amino alcohols (or alkamines) tropine (2) and ψ-tropine (3), the former having a 3α- the latter and a 3β- hydroxyl group. Both are found in nature but the only natural ester of ψ-tropine is its tigloyl ester, tigloidine (5) a constituent of *Duboisia* (G. Barger, *et al.*, J. chem. Soc., 1937, 1820), *Datura* (W.C. Evans and M. Wellendorf, *ibid.*, 1959, 1406) and *Physalis* (K. Basey and J.G. Woolley, Phytochem., 1973, 12, 2557) which has been used in the treatment of Parkinson's disease.

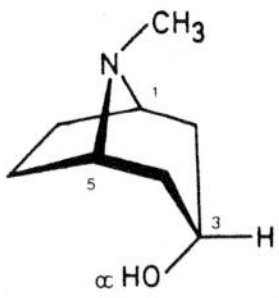

CH_3 N 1 5 3 H α HO

(2)

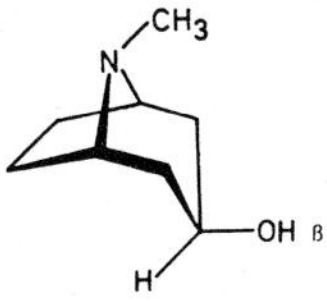

CH_3 N OH β H

(3)

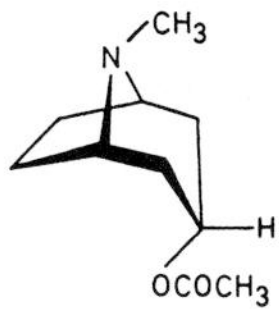

CH_3 N H $OCOCH_3$

(4)

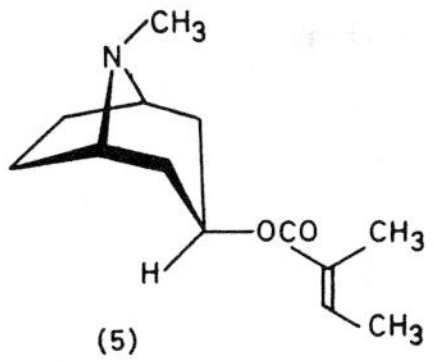

CH_3 N OCO CH_3 H CH_3

(5)

Previously, the only known method of reducing tropinone to tropine (the 3α-ol) was by means of catalytic reduction over Raney nickel, platinum oxide catalyst again giving a mixture, but it has now been shown that di-isobutyl aluminium hydride reduction also gives the 3α-ol almost exclusively (R. Nayori, *et al.*, J. Amer. chem. Soc., 1974, 96, 3336; Y. Hayakawa, *et al.*, *ibid.*, 1978, 100, 1786). The 3α- substituted alkaloids are invariably found esterified with both aromatic and aliphatic acids, the simplest, 3α-acetoxytropane (8-methyl-8-azabicyclo [3.2.1] oct-3α-yl acetate) (4), being a constituent of *Datura sanguinea* (W.C. Evans and V.A. Major, J. chem. Soc., (C) 1966, 1621). In this context it is worth noting that tropic acid, the 3α-esterifying acid in hyoscine and hyoscyamine is unique in that it occurs only in these alkaloids which are themselves found solely in the Solanaceae.

Modern nomenclature is satisfactory but cumbersome. In some cases it can be misleading: thus, bellendine (6) (I.R.C. Bick *et al.*, Austral. J. Chem., 1979, 32, 1827) is 3,10-dimethyl-6, 7,8,9-tetrahydrocyclohepta[b]pyran-5,8-imin-4(5*H*)-one and its literal association with tropane is lost until the structure is examined. The alternative name, 2,3-(2',3'-tropeno)-5-methyl-α-pyrone, may be preferred. To a large extent a compromise has been reached in the literature, the name tropane is still retained, tropine being 1α*H*, 5α*H*-tropan-3α-ol. The designation of the C(1) and C(5) protons as α is, however, purely arbitrary, and in fact models of the system show that the alkaloids display β-orientated protons at these positions.

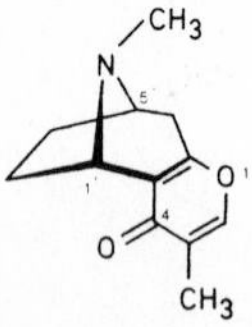

(6) 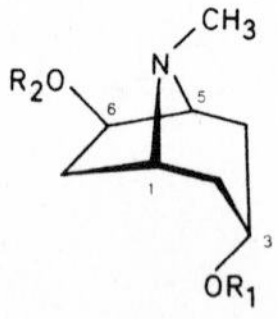 (7) 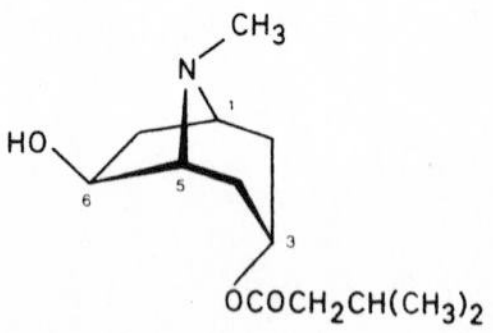 (8)

(a) $R_1 = R_2 = H$

(b) $R_1 = R_2 = CH_3CH{=}C(CH_3)CO$

Two other positions, C(6) and C(7) are found substituted with oxygen, 1α*H*, 5α*H*-tropan-3α,6β-diol (7a) being the nucleus of several alkaloids, e.g. 1α*H*,5α*H*-tropan-3α6β-diol ditiglate (7b) (W.C. Evans and M. Wellendorf, J. chem. Soc., 1958, 1991) and valeroidine [1a*H*,5a*H*-tropan-3α,6β-diol 3-isovalerate (8)] (G. Barger, W.F. Martin and W. Mitchell, *ibid.*, 1937, 1820). The alkamine of the first is *dextro* rotatory, the second *laevo*, however, they are both designated 3α,6β-disubstituted bases since the ring may be counted from either bridgehead carbon and this can lead to confusion. By relating valerine, the *laevo* rotatory form (as in 8), to *S*(-)-methoxy succinic acid Fodor (G. Fodor and F. Soti, J. chem. Soc., 1965, 6830) has shown that it has the 3*S*,6*S* configuration and some authors follow this nomenclature. However, it is clear, even in symmetrical tropine (2), that one bridgehead carbon [shown as C(1)] has the *S* configuration, and the other [shown as C(5)] is *R*. Thus valerine is 1*S*-8-methyl-8-azabicyclo [3.2.1] octan-3α,6β-diol, or more commonly 3*S*, 6*S* -1α*H*,5α*H*-tropan-3α,6β-diol. Resolution of the racemic mixture is easily accomplished by means of the (+)-tartrate salt, the *dextro* isomer crystallizing first (G. Fodor and O. Kovacs, J. chem. Soc., 1953, 2341).

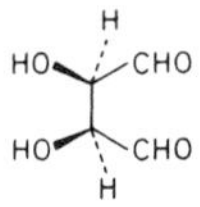

HO CHO HO CHO H H

(9)

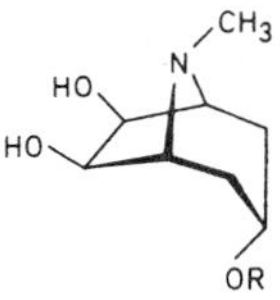

CH_3 N HO HO OR

(10)

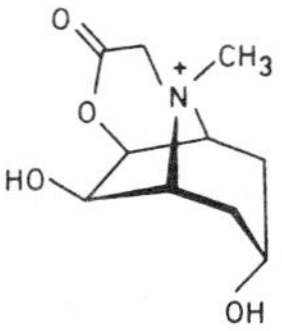

O CH_3 O N HO OH

(11)

(a) R = H

(b) R = $CH_3CH{=}C(CH_3)CO$

Teloidine (10a), first obtained by the hydrolysis of meteloidine (10b) an alkaloid from *Datura meteloides* (F.L. Pyman and W.C. Reynolds, J. chem. Soc., 1908, 93, 2077) is 1α*H*,5α*H*-tropan-3α,6β,7β-triol and it has been synthesised from *meso*-tartaraldehyde (9) by the Robinson method (C. Schöpf and H. Arnold, Ann., 1947, 358, 109). Final proof that the C(6) and C(7) hydroxyl groups are β-orientated has been provided by the fact that teloidine reacts with ethyl iodoacetate to give a quaternary lactone salt (11) (G. Fodor *et al.*, Helv., 1954, 37, 907). This is a useful reaction which has also been used to demonstrate the β-orientation of the C(6) hydroxyl in 1α*H*,5α*H*-tropan-3α,6β-diol (7a).

Scopine (12) which is 6,7β-epoxy-1α*H*,5α*H*-tropan-3α-ol, is mainly found in the alkaloid hyoscine where it is esterified with tropic acid. Normally, alkaline hydrolysis of hyoscine gives oscine (13) (3,6α-epoxy-1α*H*,5α*H*-tropan-6β-ol) and although mild hydrolysis will produce scopine (R. Willstätter and E. Berner, Ber. 1923, 56, 1079) it is unstable. As in the laboratory, in nature hyoscine can only be produced when the 3α-hydroxyl group is protected: thus Romeike (A. Romeike and G. Fodor, Tetrahedron Letters, 1960, 1) has shown that when hyoscyamine is infiltrated into *Datura* plants, it is oxidised to hyoscine.

Other alkaloids in the group are C(2) substituted and the best known example is cocaine (21) which has addictive and local anaesthetic properties. So far, it has only been found in *Erythroxylum coca* and *E. truxillense* but not in other species of coca. On hydrolysis cocaine yields methanol, benzoic acid and ecgonine and since the latter has four centres of asymmetry, C(1), C(2), C(3) and C(5), it has been the focus of much stereochemical investigation. The relative configuration of the C(3) hydroxyl group and the nitrogen is β (or *cis*) because of the ease of N ⟶ O and O ⟶ N migrations observed in *O*-benzoyl *nor*ecgonine (14) and *N*-benzoyl *nor*ecgonine (15) (S.P. Findlay, J. Amer. chem. Soc., 1954, 76, 2855; G. Fodor *et al.*, Nature 1954, 174, 131). The relative positions of the C(2) carboxyl and C(3) hydroxyl groups was shown by similar means: 2-benzamido-1α*H*,5α*H*-tropan-3β-ol (16) prepared by Curtius degradation of *O*-benzoylecgonine easily undergoes N ⟶ O acyl shift to give 2-amino-1α*H*,5α*H*-tropan-3β-yl benzoate (17) (G. Fodor, Nature

1952, 170, 2781).

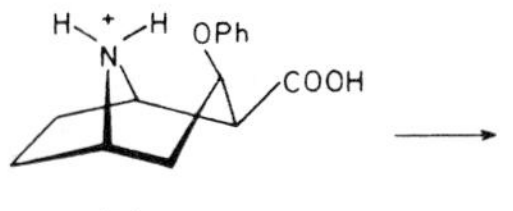

(14) 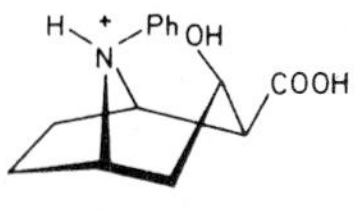 (15)

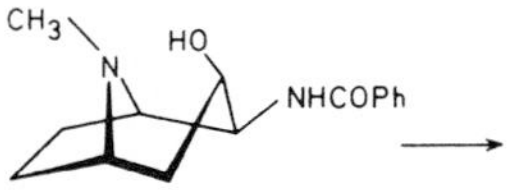

(16) 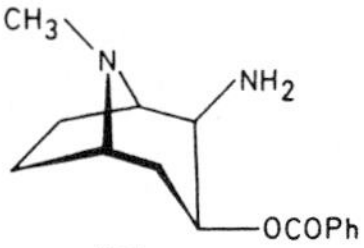 (17)

The absolute configuration of ecgonine, confirmed by X-ray studies (E.J. Gabe and W.H. Barnes, Acta Cryst., 1963, 16, 796), has been established by relating it *via* ecgoninic acid (*N*-methylpyrrolid-2-one-5-acetic acid) (18) to L(+)-pyroglutamic acid (19) and L(+)-glutamic acid (20) (E. Hardegger and H. Ott. Helv., 1955, 38, 312). (-)-Cocaine (21) is, therefore, 1*R*,2*R*,3*S* - 2β-methoxy-carbonyl-1α*H*,5α*H*-tropan-3β-yl benzoate. Other C(2) substituted alkaloids are found particularly in the Proteaceae family.

The free hydroxyl groups of the above alkamines are normally found esterified with a variety of organic acids. With few exceptions these acids are metabolites of phenylalanine and the branched-chain amino acids leucine, valine and isoleucine.

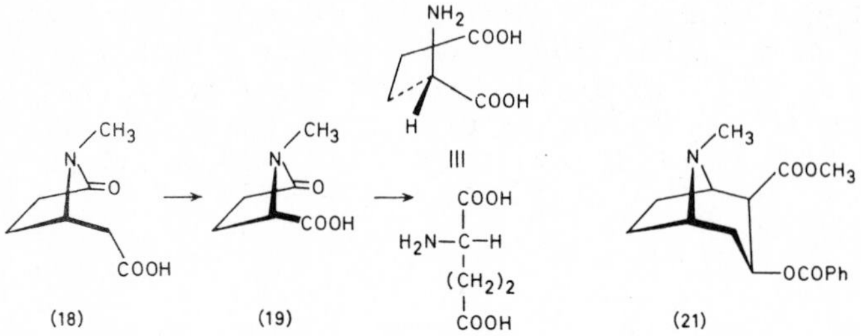

(18) (19) (20) (21)

1. Synthesis

The Robinson route (R. Robinson, J. chem. Soc., 1917, 111, 762; 876) to tropinone is simple, direct and efficient. Acetone, succindialdehyde and methylamine hydrochloride when mixed in aqueous solution for a period of 30 minutes at ambient temperature produce tropinone which is easily characterised as the canary yellow dipiperonylidine derivative. By extending the reaction period to several days and replacing acetone with calcium acetone dicarboxylate, or the dimethyl ester, yields of 50% are achieved. When the reaction is carried out in a buffer at pH 5 it is possible to increase the yield to over 90% (C. Schöpf and G. Lehmann, Ann., 1935, 518, 1). It has been shown that the reaction may be carried out quickly at 55-60° for 90 minutes when an 80% yield may be achieved (N. Elming, Adv. org. Chem., 1960, 2, 110). The success of the method has led to its use for the industrial preparation of tropinone, its adaptation for the synthesis of hydroxy and methoxytropanes, and it is the method of choice when isotopically labelled tropanes are required.

In 1952 two groups independently used furan derivatives as the source of the dialdehyde. 2,5-Diethoxy-2,5-dihydrofuran (22) when allowed to stand with hypobromous acid is converted into the 3-hydroxy-4-bromo derivative which when hydrogenated over Raney nickel during the addition of methanolic potassium hydroxide gives a good yield (60%) of 3-hydroxy-2,5-diethoxytetrahydrofuran (23). Cleavage of the ring with hot dilute acid affords malic dialdehyde. The latter at pH 4 with methylamine hydrochloride and acetone dicarboxylic acid for three days followed by extraction of the liquor with chloroform gives 6β-hydroxy-1α*H*,5α*H*-tropan-3-one (24) in yields varying from 35-55%. (A. Stoll, B. Becker and E. Jucker, Helv., 1952, 35, 1263).

C_2H_5O OC_2H_5 (22) → HO C_2H_5O OC_2H_5 (23) → CH_3 N HO O (24)

In a similar manner, 2,5-dimethoxy-2,5-dihydrofuran gives 3-hydroxy-4-chloro-2,5-dimethoxytetrahydrofuran and when this is refluxed with a suspension of potassium hydroxide in ether it gives the crystalline 3,4-epoxy derivative m.p. 45° in 70% yield. An attempt has been made to prepare 2,3-epoxy succindialdehyde by mild hydrolysis and thence the alkamine scopinone a direct precursor of hyoscine. However, the crystalline product obtained in low but reproducible yield does not resemble scopinone. The synthesis of 6β-hydroxy-1α*H*,5α*H*-tropan-3-one from malic dialdehyde in a somewhat lower yield than that obtained by the Swiss workers (20%) is accomplished by the reaction of malic dialdehyde, prepared by cleaving the ether bridge of 3,4-epoxyfuran with lithium aluminium hydride followed by mild acid hydrolysis, in the usual way. (J.C. Sheehan and B.M. Bloom, 1952, J. Amer. chem. Soc., 24, 3825). The success of the method has been exploited in preparing a wide range of substituted tropanes. Thus, 2,3,5-trimethoxy, 2,3,5-triethoxy, 2,3,5-tri-isopropyloxy tetrahydrofuran, their corresponding dialdehydes and tropane derivatives have been prepared (A. Stoll, A. Lindenmann and E. Jucker, Helv., 1953, 36, 1501), and by replacing methylamine with other substituted amines a considerable number of alternative structures are obtained (A. Stoll, E. Jucker and A. Lindenmann, Helv., 1954, 37, 647; A. Stoll and E. Jucker, 1954, Angew. Chemie, 1954, 66, 376; A. Stoll, E. Jucker and A. Lindenmann, Helv., 1954, 37, 495).

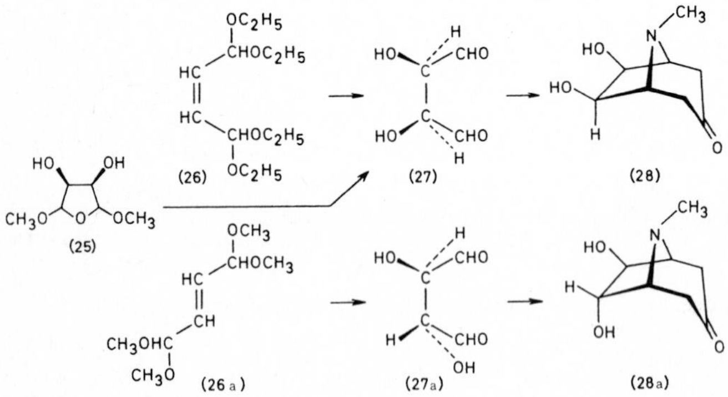

Teloidine (1α*H*,5α*H*-tropan-3α,6β,7β-triol) the alkamine from the alkaloid meteloidine (F.L. Pyman and W.C. Reynolds, J. chem. Soc., 1908, 93, 2077) has been synthesised by C. Schöpf and W. Arnold (Ann., 1947, 358, 109) from *meso* tartaraldehyde (27). The aldehyde prepared by a lengthy route from the tetraethylacetals of acetylene dialdehyde, and ethylene dialdehyde (26) by means of periodic acid gave, over three days at pH 5, a 70% yield of teloidinone (28) m.p. 192°. At other pHs poorer yields of product were obtained. In a later procedure (J.C. Sheehan and B.M. Bloom, J. Amer. chem. Soc., 1952, 74, 3825) 2,5-dimethoxy-2,5-dihydrofuran was oxidized with potassium permanganate in alcohol and after cleavage of the product (25) with acid, the resultant dialdehyde was left with methylamine hydrochloride and acetone dicarboxylic acid at pH 5.2 for 64 hr. This gave a

42% yield of teloidinone (28). The 6,7β-dimethoxy and 6β-hydroxy-7β-methoxy derivatives of teloidine have been prepared from the appropriate tetrahydrofurans (J. Kebrle and P. Karrer, Helv., 1954, 37, 484; K. Zeile and A. Heusner, Ber., 1954, 87, 439). 6α,7β-Dihydroxy-1α*H*,5α*H*-tropan-3-one has been prepared from furan. When furan is treated with methanol and bromine at -35° and the reaction mixture carefully neutralized with ammonia, fumaric dialdehyde tetramethylacetal (26a) is obtained. Oxidation of the latter with permanganate gives (±)-tartaric dialdehyde acetal which is converted into the dialdehyde (27a) and finally into the *trans* 6,7-dihydroxy derivative of teloidinone (28a) (K. Zeile and A. Heusner, Ber., 1957, 90, 1869).

Similar attempts to prepare scopine (32a) (and hyoscine) met with considerable difficulty. The Robinson synthesis with 2,3-epoxysuccindialdehyde failed (C. Schöpf and A. Schmetterling, Angew. Chemie., 1952, 64, 591) as had an earlier attempt to prepare 1α*H*,5α*H*-trop-6-en-3-one from maleic dialdehyde (N.A. Preobrazhenski *et al.*, J. gen. Chem., USSR 1945, 15, 952). However, 1α*H*,5α*H*-trop-6-en-3-one and the 3α-ol (31a) are important synthetic and possibly biosynthetic intermediates (B.T. Cromwell, Biochem. J., 1943, 37, 707; 722) in the formation of hyoscine. Fodor devised a number of methods for their preparation from 1α*H*,5α*H*-tropan-3α,6β-diol (30a) and its derivatives. 6β-Phenylcarbamoyloxy-1α*H*,5α*H*-tropan-3α-ol (29a), (I. Vincze, J. Toth and G. Fodor, J. chem. Soc., 1957, 1349) when converted into the acetyl ester (29b) and pyrolysed at 250°/2 mm gives on recrystallisation from acetone a poor yield of 1α*H*,5α*H*-tropan-6β-ol-3α-yl acetate (30c) m.p. 117°. A better yield (84%) of the latter mono ester can be obtained by partial hydrolysis of 1α*H*,5α*H*-tropan-3α,6β-yl 3,6-acetate (30b). (P. Dobo *et al.*, J. chem. Soc., 1959, 3461), and with toluene-p-sulphonyl chloride it gives a tosyl derivative m.p. 223°. On elimination of tosic acid in collidine at 190° the required 6-ene (31b) is obtained. In an alternative synthesis, 1α*H*,5α*H*-tropan-3α,6β-diol (30a) is dehydrated with phosphoryl chloride at 100° to yield the 3α,6β epoxide, the 'tropene oxide' of Wolfes and Hromatka (Mercks Jahresber., 1934, 47, 45), hydrobromide m.p. 280°. Better yields are attainable with toluene-p-sulphonic anhydride as dehydrating agent (G. Fodor, S. Kiss and J. Rackoczi, Chim. Ind., 1963, 90, 225). The epoxide is cleaved by acetyl bromide in a sealed tube at 120° to give 6β-bromo-1α*H*,5α*H*-tropan-3α-yl acetate (30d) b.p.110-15°/1mm. Dehydrobromination of the latter in collidine containing a few drops of diethylaniline affords 1α*H*,5α*H*-tropan-6-en-3-yl acetate (31b). Attempts to oxidize the 6-ene with

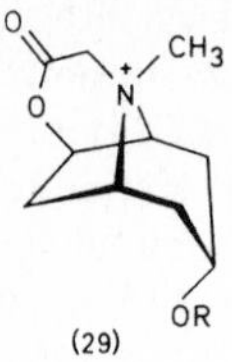

(29) (a) R = H (b) R = CH_3CO

(30) (a) R_1 = H R_2=OH (b) R_1 = CH_3CO R_2=CH_3COO (c) R_1 = CH_3CO R_2=OH (d) R_1 = CH_3CO R_2=Br

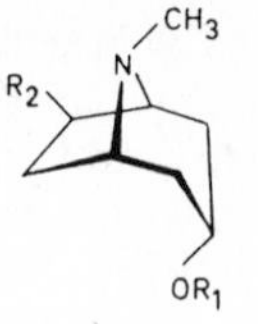

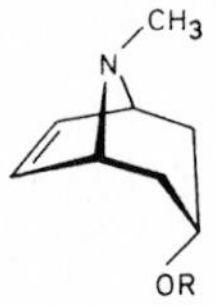

(31) (a) R = H (b) R = CH_3CO

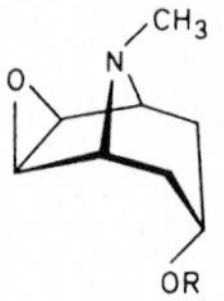

(32) (a) R = H (b) R = CH_3CO

monoperphthalic acid gives appreciable quantities of the *N*-oxide, but both trifluoroperacetic acid in methylene chloride/ acetonitrile at 5° for eight days and formic acid/ hydrogen peroxide at 20° for seven days gives *O*-acetyl scopine (32b), the latter reagent giving a better yield. Even better yields have been achieved using tungstic acid and hydrogen peroxide.

An important route to 1α*H*,5α*H*-trop-6-enes has been reported by H. Hayakawa *et al.*, (J. Amer. chem. Soc., 1978, 100, 1786). Tetrabromoacetone (34), carbomethoxypyrrole (33) and iron ennacarbonyl (1:3:3) in benzene at 50° for 72 hr gives a 70% yield of 2α,4α-dibromo-*N*-carbomethoxy-1α*H*,5α*H*-tropan-6-en-3-one m.p. 155-7° and the corresponding 2α,4β-dibromo product (35) m.p. 114° (2:1 mixture). Iron pentacarbonyl cannot be used as the cyclocoupling reagent except in the presence of uv light, when the ennacarbonyl is produced *in situ*. The dibromo derivatives are easily recognised by their ^{1}H nmr spectra, the former having a two proton doublet (J = 4Hz) at δ 4.8 the latter two proton doublets (J = 2Hz) at δ 4.27 and at δ 5.11 (J = 3.5 Hz) which arise from the two non-equivalent C(2) and C(4) protons. Hydrogenation of the mixture over 10%

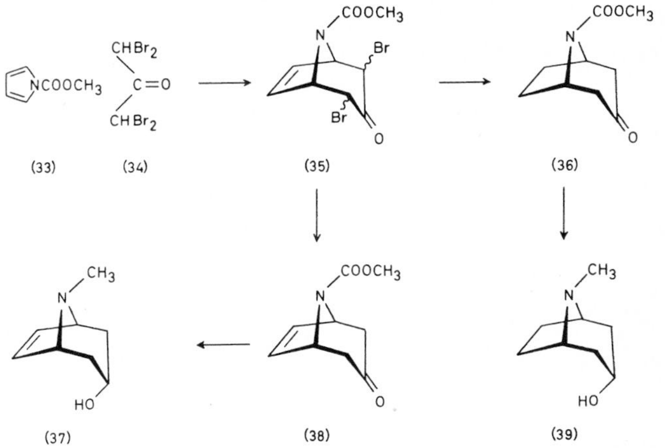

palladium on charcoal gives *N*-carbomethoxy-1α*H*,5α*H*-tropan-3-one (36) m.p. 60-1° and further reduction with di-isobutyl aluminium hydride in THF at -78° gives tropine (39) and ψ-tropine (9:1 mixture) directly in 96% combined yield. Debromination of the mixed dibromotropanes with a Zn/Cu couple in methanol saturated with ammonium chloride gives a quantitative yield of *N*-carbomethoxy-1α*H*,5α*H*-tropan-6-en-3-one (38) m.p. 69-70° which after prolonged reduction with di-isobutyl aluminium hydride produces a 93:7 mixture of 1α*H*,5α*H*-trop-6-en-3α-ol (6-dehydrotropine) (39) and the corresponding 3β-ol (6-dehydro-ψ-tropine). In a second similar synthesis, *N*-carbomethoxypyrrole, tetrabromoacetone Zn/Cu couple (2.1 : 1.2) in dimethoxyethane at -5° affords a 30% yield of the *cis*-dibromoderivative (2α,4α-dibromo) exclusively.

The tosyl derivative of 1α*H*,5α*H*-trop-6-en-3α-ol (37) when heated to 80° with freshly prepared (-)-*O*-acetyltropoyl chloride for 2 hr and then allowed to stand with 6N HCl at room temperature to hydrolyse off the protecting acetyl group results in an excellent yield (88%) of (-)-6-dehydrohyoscyamine (40) m.p. 63-4° $[\alpha]_D^{23}$ - 14°. The latter base and similar 3α-acyl esters are important intermediates since on hydrogenation the saturated tropane nucleus is obtained and as mentioned previously the unsaturated centre facilitates the introduction of an epoxide group (scopine) (41), one hydroxyl group (valerine) (42) or two hydroxyl groups (teloidine) (43). Leete has prepared 6-dehydrohyoscyamine by the action of tungsten hexachloride and n-butyllithium on hyoscine (E. Leete and D.H. Lucast, Tetrahedron Letters, 1976, 3401; K.B. Sharpless *et al.*, J. Amer. chem. Soc., 1972, 94, 6538).

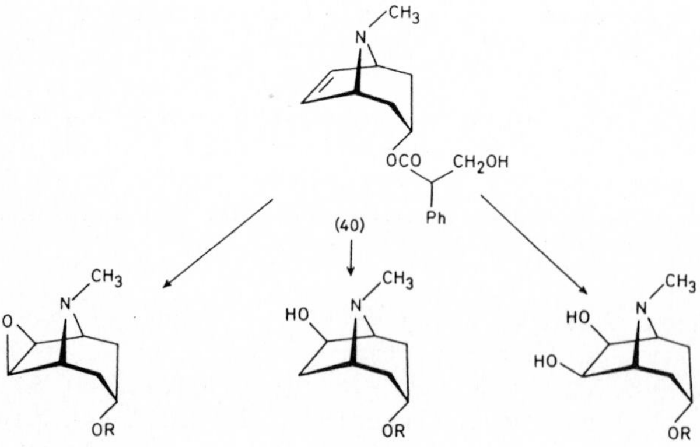

(40)

(41) (42) (43)

Trop-6-enes have also been prepared from αα'-dibromo-ketones and pyrroles (and *N*-methylpyrroles) in acetonitrile with sodium iodide and copper at ambient temperature. (G. Fierz, R. Chidgey and H.M.R. Hoffmann, Angew. Chemie., 1974, 86, 444).

The facile synthesis of cyclohepta-2,6-dione (47) by C(2) and C(7) dibromination of cycloheptanone ethylene ketal followed by elimination of bromine with sodium methoxide (E.W. Garbische, J. org. Chem., 1965, 30, 2109) has allowed the testing of an earlier idea by Robinson that such compounds should react with substituted amines to give tropanes.

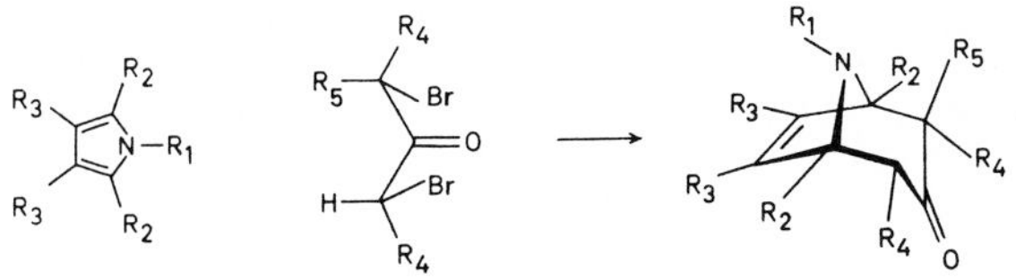

(44) (45) (46)

	(44)			(45)		(46)
	R^1	R^2	R^3	R^4	R^5	% yield
A	H	H	H	Me	H	65
B	Me	H	H	Me	H	89
C	Me	H	H	Me	Me	70
D	H	Me	H	Me	H	66
E	Me	Me	H	Me	H	74
F	H	Me	Me	Me	H	50

The addition of cyclohepta-2,6-dienone to a methanolic solution of methylamine (48) at room temperature leads to an exothermic reaction and after 30 minutes 95% of the dienone (shown by nmr spectroscopy) is consumed. Evaporation of the solvent gives a 64% yield of tropinone (49). In a similar manner, the *N*-benzyl and *N*-ethyl derivatives of tropinone can be prepared in 59 and 56% yields respectively by using the appropriately substituted amines in place of methylamine. Using dimethylamine hydrochloride the tropinone methochlorides can be efficiently prepared directly. (A.T. Bottini and

J. Gal, J. org. Chem., 1971, 36, 1718).

(47) (48) CH_3NH_2 → (49)

A modification of the above synthesis has been reported (T.L. Macdonald and R. Dolan, J. org. Chem., 1979, 44, 4973). Cyclohex-2-enone (50) is converted in 80% yield to 2-[(trimethylsilyl)oxy]-cyclohexa-1,3-diene (51) with di-isopropylamine/n-butyllithium and chlorotrimethylsilane. (G.M. Rubbottom and J.M. Gruber, *ibid.*, 1977, 42, 1051). The ring expansion is effected by the reaction of the dienylsilyl ether in refluxing dimethoxyethane with sodium trichloro-acetate followed by mild hydrolysis with methanolic hydrochloric acid, whereupon the intermediate α,α-dichloro-cyclopropanol (52), which can be isolated (T.L. Macdonald, *ibid.*, 1978, 43, 4241), rearranges to give 2-chloro-cyclo-hepta-2,6-dienone (53) m.p. 8-11°. Addition of excess aqueous methylamine to a methanolic solution of the diene at 0° and after 1 hr, extraction with methylene chloride affords 2β-chloro-1α*H*,5α*H*-tropan-3-one (54) (62%).

(50) → (51) → (52) → (53) $\xrightarrow{CH_3NH_2}$ (54)

It has been shown, by integrating the ^{1}H nmr singlet at δ 4.94 and comparing it with the *N*-methyl signal at δ 2.59 as internal reference, that the 2α-chloro-substituted tropane is formed in 40% yield. Because of the different dihedral angles between the C(1) proton and the C(2) α-proton (2β-chloro derivative) and the C(2) β-proton (2α-chloro derivative) the former gives a doublet at δ 4.7 (J = 4Hz) whereas the latter gives a broad singlet. (*cf.* the similar situation observed with the C(2) protons and the C(3) α and β protons in the spectra of ψ-tropine and tropine (R.J. Bishop, *et al.*, J. chem. Soc., (C), 1966, 74). However, only the 2β-chloro derivative is isolated after work-up. 1α*H*,5α*H*-tropan-3-one is prepared in 70% yield by reducing the latter with tri-n-butyltin hydride and a trace of azoisobutyronitrile in refluxing benzene. This versatile synthetic method has been used to prepare a range of substituted tropanes. For example 3-methylcyclohex-2-enone when subjected to the identical procedure gives 2-chloro-6-methylcyclohepta-2,6-dienone, 2β-chloro-5-methyl-1α*H*,5α*H*-tropan-3-one and 5-methyl-1α*H*,5α*H*-tropan-3-one in an overall 22% yield; the terpene carvone gives an epimeric mixture (yield 18%) of 6α-isopropenyl (75%) and 6β-isopropenyl-5-methyl-1α*H*,5α*H*-tropan-3-ones.

CH_2=CH—CHO

(55)

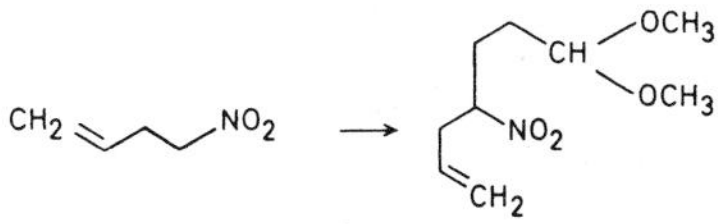

CH_2 NO_2

(56)

CH OCH_3 OCH_3 NO_2 CH_2

(57)

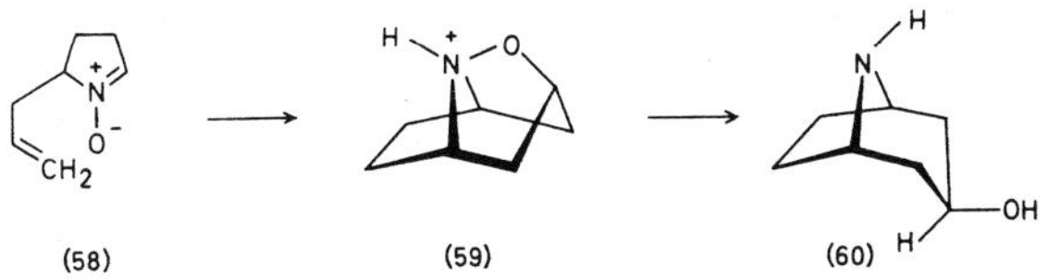

N O CH_2

(58)

H N O

(59)

H N OH H

(60)

The synthesis of tropanes has also been achieved by the intramolecular cycloaddition of nitrones to carbon-carbon double bonds (J.J. Tuffariello, *et al.*, J. Amer. chem. Soc., 1979, <u>101</u>, 2435). Addition of acrolein (55) to a cooled solution of 1-nitro-but-3-ene (56) in sodium methoxide yields the nitroaldehyde which is converted into the dimethylacetal (57) *in situ* by the passage of dry hydrogen chloride. Reduction of the nitro group with powdered zinc and ammonium chloride followed by cleavage of the acetal with acid gives the nitrone, 5-allyl-1-pyrroline 1-oxide (58), in excellent yield. When refluxed in toluene the latter cyclizes to form an isoxazolidine (59) which is effectively a ψ-tropine cyclo-adduct derivative. Reduction with hydrogen over 10% palladium on charcoal gives *nor*-ψ-tropine (60) m.p. 128-30° (66%). Methylation of the isoxazolidine (59) with methyl iodide in ether at room temperature followed by reduction with lithium aluminium hydride in THF, or zinc in acetic acid affords ψ-tropine m.p. 104-5° (in 74 and 51% yields respectively). The significance of the synthesis is that the fixed nature of the isoxazolidine system automatically establishes the 3β-configuration of the hydroxyl group when the N-O bond is cleaved.

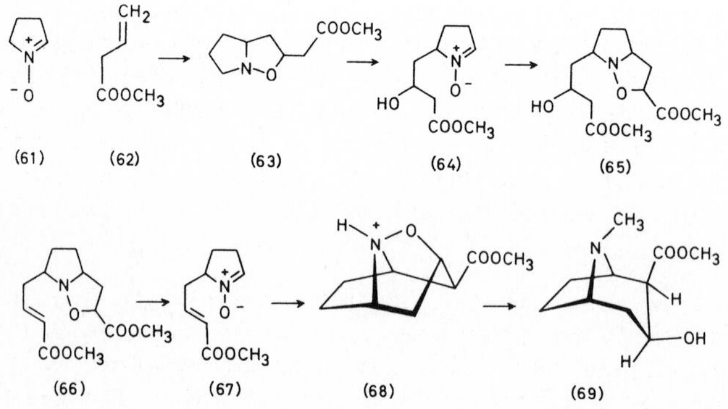

(61) (62) (63) (64) (65)

(66) (67) (68) (69)

In addition the stereochemistry of the cyclization to the nitrone ensures that substituted nitrones with the *trans*-configuration at the olefinic centre give an isoxazolidine substituted in the *exo*-configuration (when cleaved a 2β-substituted tropane). Cocaine has both these features: a 2β-carbomethoxy group and a 3β-benzoyloxy group and its synthesis has been accomplished as follows.

The reaction of 1-pyrroline 1-oxide (61) with methyl but-3-enoate (62) gives an almost quantitative yield of the carbomethoxy pyrrolo isoxazole (63) which, when oxidized with *m*-chloroperbenzoic acid, forms the corresponding δ-carbomethoxy-β-hydroxypropyl nitrone (64). Dehydration of the latter is exceptionally difficult and it is necessary to protect the nitrone group by the reaction of the latter nitrone (64) with methyl acrylate in benzene, whereupon the isoxazole (2-carbomethoxy-7-(δ-carbomethoxy-β-hydroxypropyl) hexahydropyrrolo [1,2-*b*] isoxazole) (65) is reformed. The hydroxy ester can then be dehydrated *via* the methane sulphonate ester, but not the tosyl ester, to yield exclusively the required carbomethoxy isoxazole (66) with a *trans*-configuration about the double bond. Methyl acrylate is eliminated from the molecule by refluxing it in xylene and the resulting nitrone (67) spontaneously rearranges to give the tricycloadduct equivalent of *nor*ecgonine methyl ester (68). Methylation of the latter with methyl iodide followed by cleavage of the N-O bond with zinc dust in 50% acetic acid gives ecgonine methyl ester (69) m.p. 122-6°, which when esterified with benzoyl chloride in benzene containing anhydrous sodium carbonate yields (±)-cocaine m.p. 77-9°. The more direct route to the carbomethoxy nitrone (64) gives unsatisfactory yields, for although the correct carbomethoxy dimethyl acetal can be produced in 94% yield by the reaction of *trans* methyl nitropent-2-enoate with acrolein in sodium methoxide [Cf. synthesis of (57)] , reduction with zinc dust in aqueous ammonium chloride followed by cyclization in boiling toluene gives at best an 11% yield of the required nitrone.

A direct route to the important intermediate 1α*H*,5α*H*-trop-6-en-3-one has been sought by oxidizing tropine (70) with lead tetraacetate and iodine at room temperature (S.Sarel and E. Dykman, Tetrahedron Letters 1976, 3725). The product is a complex mixture of dimers: *N*-(tropan-3'α-yloxycarbonyl)-*nor*tropan-3α-ol (71) m.p. 144-6° (19%); *N*-(tropan-3'α-yloxycarbonyl)-*nor*tropan-3α,6α-oxide (72; R = Me) (11%); *N*'-(acetyl*nor*tropan-3'α-yloxycarbonyl)-*nor*tropan-3α,6α-oxide (72; R = $COCH_3$) and *N*'-formyl-*nor*tropanyloxy carbonyl)-*nor*tropan-3α,6α-oxide (72; R = CHO) (2.5%).

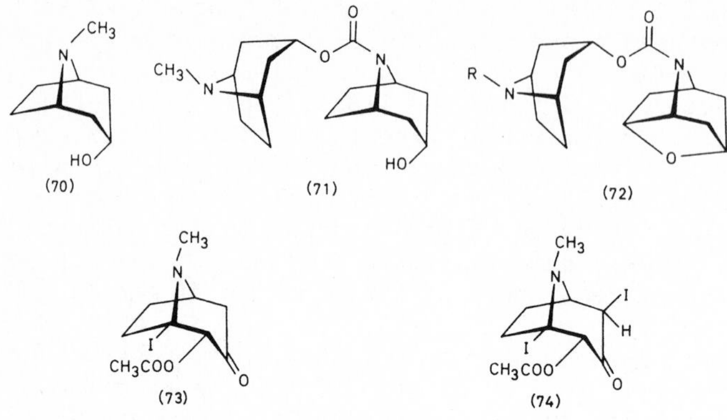

These structures were elucidated by reducing the dimers with lithium aluminium hydride and identifying the monomeric products. In addition 1α,4β-di-iodo-2α-acetoxytropan-3-one (74) m.p. 185°, and 1α-iodo-2α-acetoxytropan-3-one (73) m.p. 137° are formed. The former can be converted into the latter by de-iodination with cold aqueous sodium bisulphite, and into 1α*H*,5α*H*-tropan-2α-ol m.p. 46-48°, (HCl 266°) by removal of the bridgehead iodine by hydrogenolysis over Raney nickel and Wolff Kishner reduction of the 3-carbonyl group.

Atropine (76) has been prepared by the action of nitrous acid on the (±)-phenylalanine ester of tropine (75) (Y. Takenchi *et al.*, Chem. Pharm. Bull., Japan, 1971, 19, 2603). In addition to the aforementioned, littorine (77), 3α-tropanyl-3' hydroxy-3'-phenylpropionate, *apo*atropine and the *cis* and the *trans* cinnamoyl esters of tropine (78) are formed.

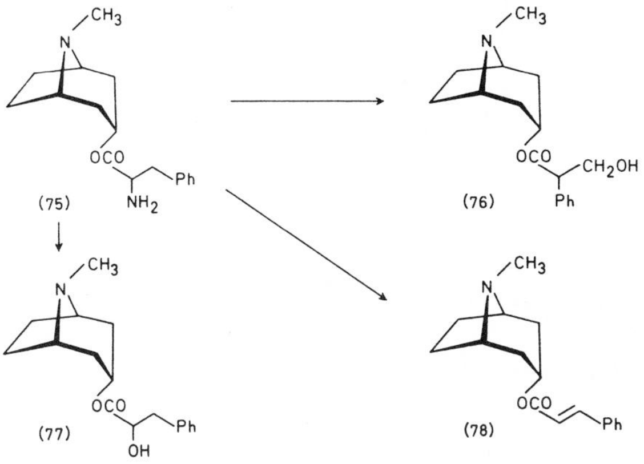

2. *New alkaloids*

(a) Proteaceae

One of the most important events of the 1970's was the discovery of alkaloids in this ancient family of plants which comprises about 1300 sp in 60 genera distributed throughout the southern hemisphere, particularly Australia, Tasmania, New Caledonia and South Africa. A few species are also found in S.E. Asia extending from southern India to Japan and in America. (L. and M. Milne, "Living Plants of the World", Nelson, London 1967, p 78; J. Hutchinson, "The Genera of Flowering Plants", Clarendon, Oxford 1967, vol 2, p 272; J. Hutchinson, "Families of Flowering Plants", Clarendon, Oxford 1959, vol 1, p 217; I.R.C. Bick and H.M. Leow, J. Ind. chem. Soc., 1978, LV, 1103). Alkaloids had been reported in the family as early as 1952 (L.J. Webb, "An Australian Phytochemical Survey", Part II, Bull. 268 CSIRO Melbourne 1952) but the monotypic species *Bellendena montana* which is a small shrub widespread in Tasmania was the first to receive serious attention. From the fresh leaves and stalks a large quantity of methyl (*p*-hydroxybenzoyl)acetate and a trace of the first alkaloid, bellendine (79 ; R = Me, R^1 = H), (0.0013%) were isolated (I.R.C. Bick, J.B. Bremner and J.W. Gillard, Phytochem., 1971, 10, 475). The structure of the pyranotropane 2,3-(2',3'-tropeno)-5-methyl-δ-pyrone was assigned largely on nmr evidence and confirmed by X-ray crystallography (W.D.S Motherwell *et al.*, Chem. Comm., 1971, 133). The base has been synthesised in low yield (I.R.C. Bick, J.B. Bremner and J.W. Gillard, Tetrahedron Letters, 1973, 5099; I.R.C. Bick, J.W. Gillard and H.M. Leow, Austral. J. Chem., 1979, 32, 1827).

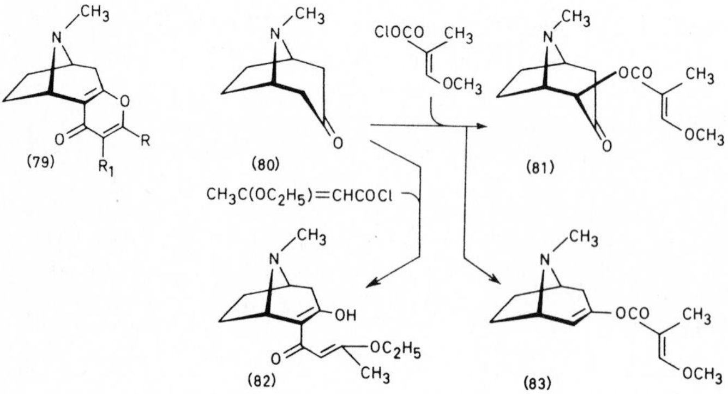

1α*H*,5α*H*-tropan-3-one (80) when refluxed in benzene with sodium hydride for 20 hr and then acylated with 3-methoxy-2-methylprop-2-enoyl chloride yields 2-(3'-methoxy-2'-methylpropanoyl)-1α*H*,5α*H*-tropan-3-one (81) which may be separated as a minor component from starting material and the *O*-acyl derivative (83). Cyclization with 4N acid at 75° gives (±)-bellendine (79; R = Me, R^1 = H). Analysis of plants from an alternative site gave a higher yield (0.12%) of a complex mixture containing at least a dozen alkaloids. Isobellendine (79; R = H, R^1 = Me) is distinguished from bellendine by its nmr spectrum, the olefinic proton signal appearing at δ 6.06

and the C-methyl signal at δ 2.23, compared with signals at δ 7.6 and δ 1.93 for bellendine. The new base is prepared in an analogous way to that used for the latter. Acylation of 1α*H*,5α*H*-tropan-3-one (80) with sodium hydride and 3-ethoxybut-2-enoyl chloride gives a very low yield of the C-acyl product (82) which is cyclized in acid. The 2,3-dihydroderivatives of bellendine and isobellendine have also been isolated, the nmr spectrum of the latter showing two extra proton resonances at δ 4.6 C(2), and 2.78 C(3) when compared with that of the parent compound and the observed couplings indicate a *cis* arrangement of the protons. The structure has been confirmed by synthesis: anhydroecgonine chloride (84) and sodium acetone enolate react vigorously to yield 2-acetoacyltrop-2-ene (85), but the subsequent acid catalysed cyclization proceeds so slowly that only a small yield of dihydroisobellendine (86) is obtained after three months.

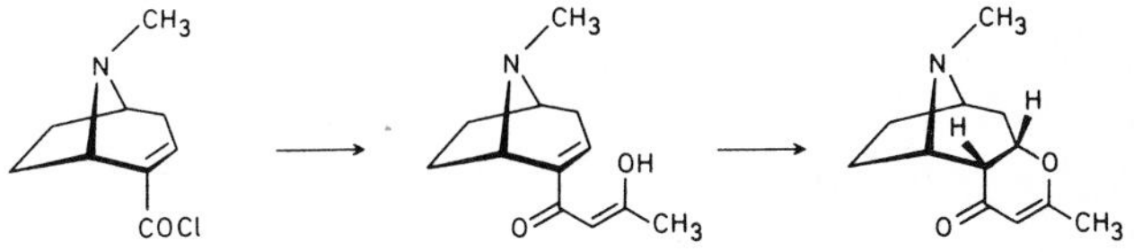

(84) (85) (86)

5,6-Epidihydrobellendine (87; R = H) which differs from 5,6-dihydrobellendine (87; R = H) in the configuration of the C(5) methyl group has also been isolated along with darlingine (79; R = R^1 = CH_3), previously obtained from *Darlingia*, 5,6-dihydrodarlingine(87; R = Me) and the 1α*H*,5α*H*-tropan-3α,6β-diol heterodiesters, 3-acetate 6-isobutyrate (88a) and 3-isobutyrate 3-acetate (88b) and the mono ester, 3-acetate (88c) as racemates. This observation forges a link with the Solanaceae family where (+)-1α*H*,5α*H*-tropan-3α,6β-diol esters are found (*Datura*) and the (-) alkamine (with the 3*S*, 6*S* configuration) is found esterified at the C(3) position with isovaleric acid in the *Duboisia* alkaloid valeroidine.

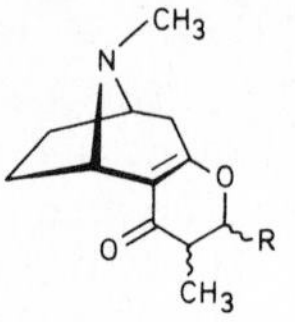

(87) 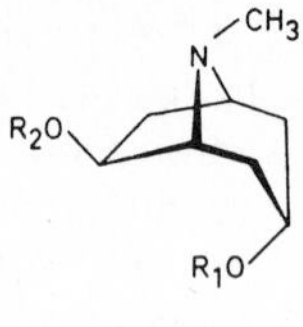 (88)

	R_1	R_2
(a)	$COCH_3$	$COCH(CH_3)_2$
(b)	$COCH(CH_3)_2$	$COCH_3$
(c)	$COCH_3$	H

Two Queensland plants *Darlingia darlingiana* and *D. ferruginea* are new sources of tropane alkaloids. Leaf and stem alkaloids, separated by tlc contained as the major component darlingine (79; R = R^1 = Me), from both plants and from *D. ferruginea* the 2-benzoyl tropane, ferrugine (90) (I.R.C. Bick, J.W. Gillard and M. Woodruff, Chem. and Ind. 1975, 794). The structure of the former base has been established by spectroscopic means and comparison with bellendine. The C(2) location of the benzoyl group of ferrugine follows from the uv, ir and ^{1}H nmr spectra, the C(2) proton appearing as a multiplet at δ 3.25. Because the ^{13}C nmr spectrum shows no δ effect, the C(4) singlet in the proton noise decoupled spectrum being at δ 29.6 ppm, only 0.3 ppm different from unsubstituted tropanes, it follows that the C(2) benzoyl group has the equatorial conformation. A third alkaloid, ferruginine (89) has been obtained from both plants (I.R.C. Bick, J.W. Gillard and H.M. Leow, 1977, 26th IUPAC Symposium (Tokyo) Abstracts, p 1157). Its enantiomer has been synthesized from natural anhydroecgonine acid chloride and it follows that this carbomethoxy substituted base has the 1*S* configuration, i.e. opposite from that of cocaine and it remains to be seen whether other proteaceous alkaloids are the same. A C(2) C(3) disubstituted tropane alkaloid (91) has also been isolated from *D. ferruginea*.

The genus *Knightia* is restricted to New Caledonia (e.g. *K. deplanchei and K. excelsa*). About 12 new alkaloids, 2-benzyltropanes, δ-pyranotropanes, and 1αH,5αH-tropan-3α,6β-diols have been discovered and although there is a general chemical similarity between these and those of the Tasmanian/Australian Proteaceae, the *Knightia* alkaloids display a greater range of structural forms. Because of this complexity the alkaloids are grouped below.

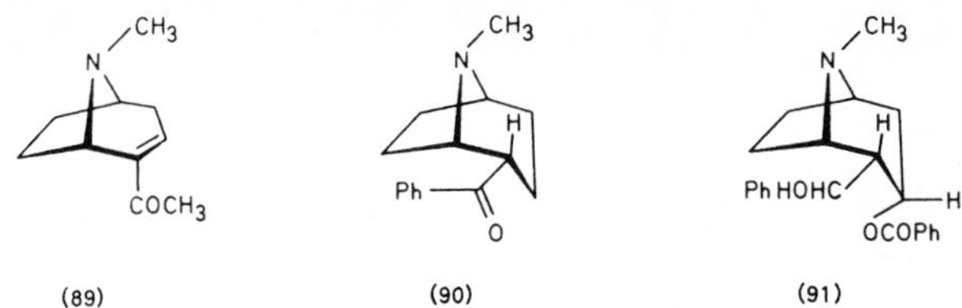

(89) (90) (91)

(b) 2,3-Disubstituted tropanes

Knightia deplanchei leaves contain two new optically inactive alkaloids, (A)(92; R = Ph) and (B) (92; R = Me) which are shown by mass spectroscopy to be C(2) benzyl substituted monoester bases. (C. Kan-Fan and M. Lounasmaa, Acta. Chem. Scand., 1973, 27, 1039). Hydrolysis of both bases gives the same C(3) hydroxy alkamine, but (A) liberates benzoic acid whereas (B) produces acetic acid. ^{1}H nmr double irradiation analysis of the alkamine allows all the tropane ring protons and methylene protons of the benzyl group to be identified and confirms the location of the benzyl group as C(2) and that of the acyl group as C(3). (M. Lounasmaa and G. Massiot, Planta Medica., 1978, 34, 66). The H(2)(3) coupling constant is small (J = 4Hz) showing that the substituent groups are α, α, or β, β orientated and since there is no deshielding of the OH group by interaction with the nitrogen nor axial-axial coupling between H(3) and H(4) it is concluded that the C(3) hydroxyl group is axial and the benzyl group equatorial (or α and α respectively).

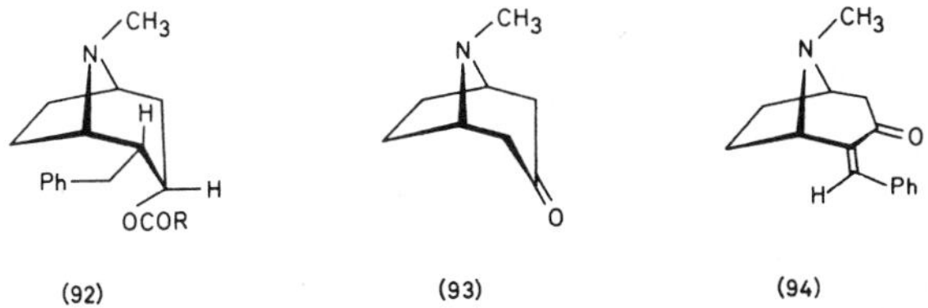

(92) (93) (94)

Proton noise decoupled ^{13}C spectra of this alkamine (and its esters) when compared with those of tropine, ψ-tropine and their benzoyl esters confirms the position and identity of the C(2) and the C(3) substituent groups. The 2-benzyl group causes deshielding of all the methine carbons [C(1), C(2), C(3)] except C(5), the other bridgehead carbon. Since it also causes shielding of C(7) but leaves C(4) unaffected it is concluded that the benzyl group is equatorially orientated and extends below the tropane ring (*cf.* ferrugine, I.R.C. Bick, J.W. Gillard and M. Woodroff, Chem. and Ind., 1975, 794). The chemical shifts of the *N*-Me and C(6), which are diagnostic for the tropine/ψ-tropine configuration, most resemble those of 1α*H*,5α*H*-tropan-3α-yl benzoate and the substituent groups are therefore both α-orientated. (M. Lounasmaa, P.M. Wovkulich and E. Wenkert, J. org. Chem., 1975, 40, 3694). The alkamine is synthesized from tropinone (93) and benzaldehyde, the *trans* mono benzylidine derivative (94) being isolated in 15% yield. Hydrogenation over Pd/C gives a mixture of 2α- and 2β-benzyl-1α*H*,5α*H*-tropan-3-ones and reduction with lithium aluminium hydride yields a mixture of the four possible 2-benzyl-3-hydroxy derivatives, one of which is identical to the natural alkamine, (M. Lounasmaa and C.J. Johanson, Tetrahedron Letters, 1974, 2509).

Work with *K. strobilina* leaves reveals the presence of two similar alkaloids, (-)-acetylknightinol (95; R = Ac) and (+)-knightinol (95; R = H), which have been shown by spectroscopic means to be 2-benzhydryl-3α-acetoxy-1α*H*,5α*H*-tropanes.

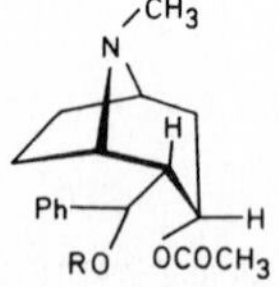

(95)

(c) 2,3,6-Trisubstituted tropanes

The leaves of *K. deplanchei* contain three crystalline bases of this type, C(96a), E (96b) and F (96c). Ester alkaloid C has been shown by mass spectroscopy to have a free C(3) hydroxyl group and on hydrolysis gives benzoic acid and an alkamine m/z = 247, a molecular weight 16 more than that of the alkamine of alkaloids A and B. The ^{1}H nmr spectra of both alkamines are sufficiently similar to conclude that the C(2) benzyl group δ = 2.27 and C(3) hydroxyl group (3β-H, δ = 3.84) have the same configuration. The H(6) proton appears as a double doublet ($J_{6ax-7ax}$ = 8Hz; $J_{6ax-7eq}$ = 3Hz), and since there is no observable coupling between H(5) and H(6) it implies that the C(6) hydroxyl group has the β-configuration. Analysis of the ^{13}C nmr spectra of C and its hydrolysis product confirms the 2α, 3α, 6α configuration of the substituted groups. The C-methine [C(2)], δ = 44.9, N-methines [C(1) and (5)], δ = 64.2 and 66.5, and the *O*-methines [C(3) , δ = 65.4, and C(6), δ = 80.5] of alkaloid C are distinguished by their increasing downfield shifts, the residual ^{13}C-^{1}H couplings in the single-frequency off-resonance decoupled spectra, and by comparison with spectra of tropine and its esters. Since one of the oxymethines has a shift almost identical to that of tropine C(3) (63.6 ppm) the signal at 80.5 ppm is assigned to the other oxymethine C(6). One of the aminomethine shifts resembles C(1) of tropine, and the downfield shift of the other (δ = 66.5) permits its differentiation and locates the benzoyloxy group at C(6). Benzoylation of the alkaloid allows the identification not only of the C(3) signal but the distinction of C(4) from C(7), the latter appearing unchanged at 32.7 ppm. Comparison with the spectra of alkaloids A and B shows that alkaloid C has the same C(2) C(3) stereochemistry. Examination of the alkamine of C shows strong shielding of C(2), C(4) and the *N*-methyl, a result which indicates *N*-methyl inversion, which can only be justified by hydrogen bonding between the nitrogen lone pair and the C(6) hydroxyl group, and thus the latter has the β *(exo)* configuration.

The substitution patterns of alkaloids E (96b) and F (96c) (*K. deplanchei*) have been established mainly by mass spectroscopy. Hydrolysis of E gives benzoic and cinnamic acids and an alkamine m/z 263 (16 more than C alkamine). Alkaloid F is tentatively assumed to have the same alkamine as E. By reasoning analogous to that outlined above, the identity and the configuration of the substituent groups of F has been established as 2α-benzhydryl, 3α-hydroxy, 6β-benzoyloxy with an equatorial *N*-methyl group.

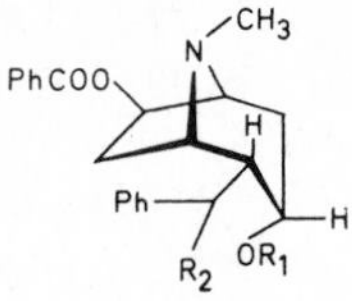

(96)

(97)

(a) R_1=H R_2=H

(b) R_1=PhCH=CHCO R_2=OH

(c) R_1=H R_2=OH

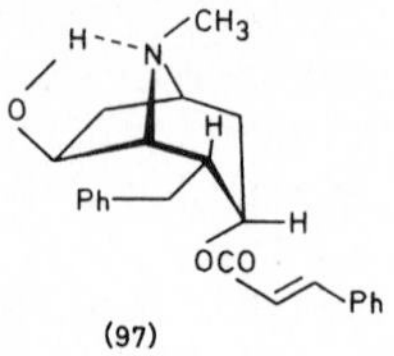

(d) 2,3,7-Trisubstituted tropanes

K. deplanchei leaf alkaloid D (97) is an optically inactive alkaloid and mass spectroscopy establishes that it is a C(2) benzyl, C(3) cinnamoyl, C(6) or C(7) hydroxytropane. Hydrolysis produces an alkamine with an identical mass spectrum (M^+ = 247) when compared with that of the alkamine from base C, but their melting points (192-5° and 170-2° respectively) and their ^1H nmr spectra differ. The ^{13}C nmr spectrum of the former alkamine shows by the position of the C(2), C(4) and *N*-methyl groups that the extra hydroxyl group occupies the 7β (*exo*) position. The proximal position of the C(7) hydroxyl exerts a β-effect on C(1) (δ = 71.6, *cf.* δ = 62.8 for C alkamine), but the C(2) and C(3) chemical shifts reflect the same configurational relationship as in the base C.

Leaves of *K. strobilina* contain two similar bases; F, (+)-knightoline (98), with a 2α-benzyl, 3α-acetoxy, 7β-hydroxytropane nucleus; and an amorphous base, K, (+)-knightalbinol (99), a 2α-benzhydryl, 3β-hydroxy, 7β-acetoxytropane, the first ψ-tropine-like base so far isolated.

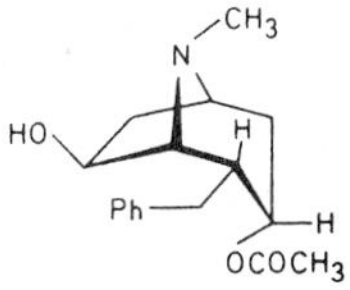

(98)

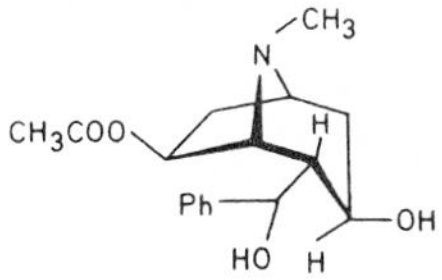

(99)

(e) 3,6-Disubstituted tropanes

Knightia strobilina shows a relationship, not only with *Bellendena*, but with solanaceous plants (particularly *Datura*) in that two of its bases, G (100a) and I (100b) are 1α*H*,5α*H*-tropan-3α,6β-diol derivatives. The former is the 6-benzoate the latter the 3-cinnamate but the absolute configurations have not been assigned.

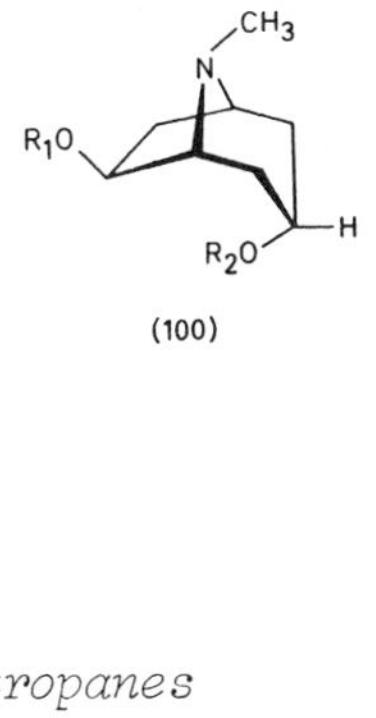

(a) R_1 = PhCO R_2 = H

(b) R_1 = H R_2 = PhCH=CHCO

(100)

(f) Pyranotropanes

The major alkaloid from *K. strobilina* is D, (+)-strobiline (101) and it is accompanied by its 5,6-dihydroderivative (alkaloid E) in the bark; the previously described alkaloids apparently being restricted to the leaves. Their structures were established by spectroscopic comparison with the known *Bellendena* and *Darlingia* bases.

(101)

(g) Erythroxylaceae

Leaves of *Erythroxylum coca* and *E. truxillense* are well known to contain cocaine and this sets the S. American species apart from other members of the genus which are widespread throughout Africa, Asia, Australia, Java, Sumatra and Indonesia. Cocaine forms 50-80% of the total alkaloid fraction in these plants, the remainder consisting of such bases as benzoylecgonine, cinnamyl cocaine, methylcocaine, methylecgonidine, the benzoyl esters of tropine and ψ-tropine and the dimeric truxillines.

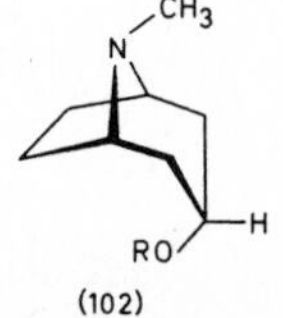

(102)

(a) $R=3,4,5(CH_3O)_3C_6H_2CO$

(b) $R=3,4,5(CH_3O)_3C_6H_2CH{=}CHCO$

Erythroxylum monogynum was first examined in 1938 (R.N. Chopra and N.N. Ghosh, Arch. Pharm., 1938, 276, 340) when the leaves were shown to contain (-)-ecgonine and cinnamyl cocaine. Subsequent investigation has shown the alkaloid pattern of the root bark to be more complex, several bases amounting to 0.85% dry weight being isolated by partition column technique (W.C. Evans and M.W. Partridge, Quart. J. Pharm. Pharmacol., 1948, 21, 126; J. Pharm. Pharmacol., 1952, 4, 769) and thin layer chromatography. Two tropine esters, the 3,4,5-trimethoxybenzoate (102a), and the 3,4,5-trimethoxycinnamate (102b) have been isolated and their structures established (J.T.H. Agar, W.C. Evans and P.G. Treagust, J. Pharm. Pharmacol., 1974, 26, 11P).

A third base (103) in quantity too small to measure the optical rotation contained two carbonyl groups (ν_{max} doublet at 1701 cm^{-1}) and on alkaline hydroylsis gave two acids, benzoic and 3,4,5-trimethoxycinnamic and 1α*H*,5α*H*-tropan-3α,6β-diol. Analysis of the mass spectrum of the base placed the cinnamoyl group at C(3) (prominent fragment at m/z 333) and the α-configuration was assigned because of the band width, $W_{\frac{1}{2}}$ 13Hz, of the H(3) nmr signal at δ 4.16. The racemic alkaloid (picrate m.p. 5° higher) is obtained by benzoylation of 6β-hydroxy-1α*H*,5α*H*-tropan-3-one, reduction over Raney nickel and esterification of the 3α-ol with trimethoxycinnamoyl chloride (J.T.H. Agar and W.C. Evans, J. chem. Soc., Perkin I, 1976, 1550). A fourth crystalline base (104), M^+ = 367, contained ester and hydroxyl functions (ν_{max} 1700 and 3400 cm^{-1}) and the ^{1}H nmr spectrum reveals an *N*-methyl δ 2.56; three methoxyl groups, δ 3.92; a characteristic downfield 2 proton singlet δ 6.73 for a 1,3,4,5-substituted aromatic system and other resonances characteristic of the teloidine nucleus.

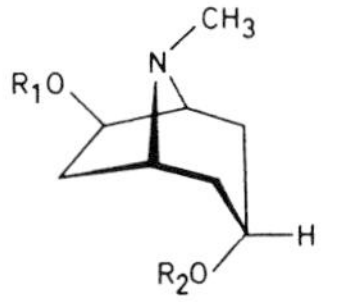

CH_3 N R_1O H R_2O

(103)

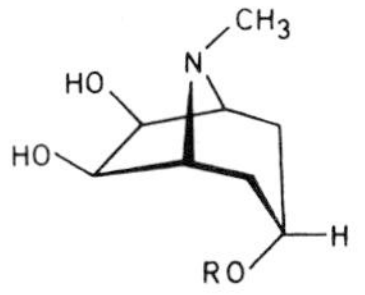

CH_3 N HO HO H RO

(104)

R_1 = PhCO

R_2 = 3,4,5$(CH_3O)_3C_6H_2CH{=}CHCO$

R = 3,4,5$(CH_3O)_3C_6H_2CO$

Teloidine and 3,4,5-trimethoxybenzoic acid are produced on hydrolysis and the position and configuration of the acid moiety follows from the presence of the ion at m/z 307 in the mass spectrum and the H(3) band width (δ 5.32, $W_{\frac{1}{2}}$ 14 Hz) in the nmr spectrum of the parent alkaloid. Tropine and ψ-tropine are also present. Apart from *Datura* and *Anthocercis* (Solanaceae) this is the first reported co-occurrence of tropine, tropan-3α,6β-diol and teloidine esters.

The shrub, *E. dekindtii* from Angola, the source of the West African drug, Olokuto, has been shown to contain at least five alkaloids (0.07%) in the root bark, (M.A.I. Al-Yahya, W.C. Evans and R.J. Grout, J. chem. Soc., Perkin I, 1979, 2130). The principal base of the root bark (0.02%), 1α*H*,5α*H*-tropan-3α-yl isovalerate (105a) is eluted from pH 6.8 partition columns in ether. From the picrate mother liquors of the isovalerate ester, a second new base 1α*H*,5α*H*-tropan-3α-yl phenylacetate (105b) is isolated, and has been identified by mass spectroscopy. Methylecgonidine is also obtained from the ether fraction. A minor base (0.013%) isolated from the chloroform fraction is unusual in that it gives a strong blue fluorescence in uv-light. Mass spectroscopy indicates that it is a tropine ester (m/z 42, 82, 83, 94, 95, 96) and the base peak m/z 124 (M^+-111) allows $C_4H_3O.COOH$ for the acid. The ^1H nmr spectrum reveals a widely spaced series of double doublets characteristic of the AMX system of 2-furoic acid, [H(3), δ 7.16, J = 3.6 and 0.8 Hz; H(4) δ 6.47, J = 3.6 and 0.8 Hz; H(5) δ 7.6, J = 1.8 and 0.5 Hz] and signals consistent with a tropine nucleus. The structure, 1α*H*,5α*H*-tropan-3α-yl 2-furoate (105c) has been confirmed by synthesis.

(105)

Also present are methylecgonidine (0.017%), valeroidine, 1α*H*,5α*H*-*nor*tropan-3α-yl isovalerate and 2-methylbutyrate, and tropine.

The Queensland tree *E. australe* contains meteloidine (106a) (0.014%) as its main leaf alkaloid, (S.R. Johns and J.A. Lamberton, Austral. J. Chem., 1967, 20, 1301). Extraction of the roots yields a single optically inactive base (0.001%) of molecular weight 277. The ^{1}H nmr spectrum shows an aromatic multiplet δ 8.0-7.5; H(3)tr δ 5.25, J = 6 Hz; H(6) and H(7) s δ 4.55; a broad two proton multiplet δ 3.1 H(1) and H(5) and an *N*-methyl at δ 2.52, similar to that of meteloidine [H(3) tr δ 5.4, J = 6 Hz; H(6) and (7) s δ 4.45; H(1) and (5) δ 3.1; NMe δ 2.55]. The mass spectrum has a prominent ion at m/z 217 (70%) corresponding to *N*-methylpyridinium benzoate and it is concluded that this base is 6β,7β-dihydroxy-1α*H*,5α*H*-tropan-3α-yl benzoate (106b).

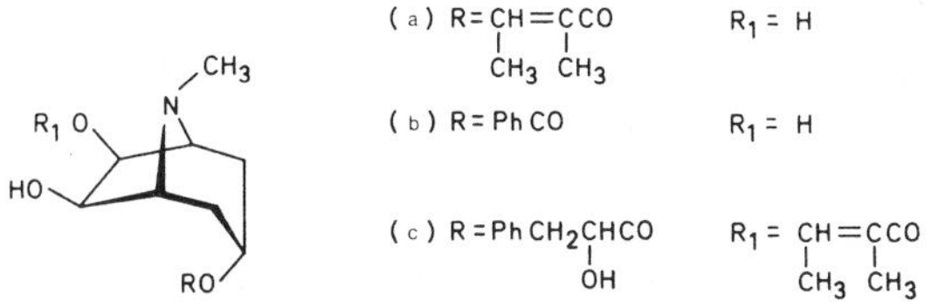

(a) R = CH=C(CH$_3$)CO (with CH$_3$ on CH) R$_1$ = H

(b) R = PhCO R$_1$ = H

(c) R = PhCH$_2$CH(OH)CO R$_1$ = CH=C(CH$_3$)CO (with CH$_3$ on CH)

(106)

The leaf alkaloids are more complex although meteloidine is the major constituent and in addition (±)-6β-hydroxy-1α*H*,5α*H*-tropan-3α-yl tiglate (107a) (0.001%) is present. Two other bases were dextrorotatory and showed certain similarities but no *N*-methyl signals were detected in their nmr spectra and pyridinium ions were prominent in their mass spectra. One alkaloid appears to be a tigloyl ester since the nmr spectrum shows a down-field one proton quartet at δ 6.85, J = 8 Hz and two C-methyls as a singlet at δ 1.8 (*cf.* C-Me signals of tigloyl residue of meteloidine at δ 1.85 and δ 1.8 ppm). Normally, however, the signals from the α-and β-methyls are well separated as in 6β-hydroxy-1α*H*,5α*H*-*nor*tropan-3α-yl tiglate (107b); α-methyl δ 1.85 and β-methyl δ 1.75 (W.C. Evans and V.A. Woolley, Phytochem., 1978, 17, 171).

Other signals are consistent with a *nor*tropan-3α,6β-diol nucleus and since the mass spectrum contains an ion at m/z 99 (4-hydroxypyridinium) but not at m/z 181 (pyridinium tiglate) the structure 1α*H*,5α*H*-*nor*tropan-3α,6β-diol 6-tiglate (107c) has been proposed. The second alkaloid (106c) (0.003%) M^+ 390 shows nmr signals resembling those of the meteloidine tigloyl moiety (6.75 and 1.8 ppm). A five proton aromatic multiplet 7.95-7.4 ppm in conjunction with other features is consistent with a phenyllactate residue. In particular, the triplet at δ 4.45, J = 8 Hz is mentioned [H(2) of phenyllactate] although this should give rise to a complex twelve line ABX system. The prominent ion at m/z 247 corresponding to pyridinium phenyllactate locates the aromatic acid at C(3) and the tigloyl group is therefore attached to the C(6) or the C(7) hydroxyl group.

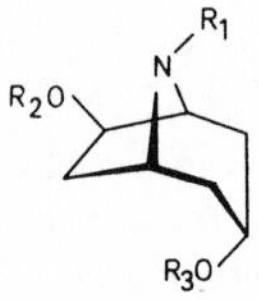

(a) $R_1 = CH_3$ $R_2 = H$ $R_3 = CH(CH_3){=}C(CH_3)CO$

(b) $R_1 = H$ $R_2 = H$ $R_3 = CH(CH_3){=}C(CH_3)CO$

(c) $R_1 = H$ $R_2 = CH(CH_3){=}C(CH_3)CO$ $R_3 = H$

(107)

The stem bark of the Queensland tree, *Erythroxylum ellipticum* contains 0.32% total alkaloid shown to be mainly 1α*H*,5α*H*-tropan-3α-yl 3,4,5-trimethoxycinnamate with a small proportion of the benzoyl ester. (S.R. Johns, J.A. Lamberton and A.A. Sioumis, Austral, J. Chem., 1970, 23, 421).

Three new 1α*H*,5α*H*-tropan-3α,6β-diol esters, Catuabins A (108a) B (108b) and C (108c) (0.01, 0.005 and 0.00064% respectively) have been isolated from *E. vacciniifolium* (E. Graf and W. Lude Arch. Pharm., 1978, 311, 139; 1977, 310, 1005). In their nmr spectra each has in addition to the usual tropane signals (NMe, 2.51 ppm ; H(1) and H(5), 3.37 ppm; H(2) and H(4), 1.72 ppm; H(7), 2.60 ppm), a triplet for the H(3) signal (A, 5.3 ppm; B, 5.36; C, 5.26) and a double doublet for H(6) (A, 5.82 ppm; B, 5.90 ppm, C, 5.97 ppm). The bases are therefore, substituted in the 3α position (R.J. Bishop, *et al.*, J. chem. Soc., (C), 1966, 74; J. Parello, *et al.*, Bull. Soc. Chim., France, 1963, 2, 2787; W.C. Evans and V.A. Major, J. chem. Soc., (C), 1968, 2775). Analysis of the H(6) signal shows couplings J_{1-7} = 8 Hz; $J_{6-7trans}$ = 4 Hz; J_{6-7cis} <1 Hz and that the C(6) substituent group is β. Catuabin A shows a second *N*-methyl group (δ 3.9) and other resonances characteristic of pyrrole-2-carboxylic acid [H(4') 6.11 ppm, dd, $J_{4'-3'}$ = 7 Hz, $J_{4'-5'}$ = 3 Hz; H(5), 6.77 ppm, a pseudotriplet, $J_{5'-4'}$ = 4 Hz, $J_{5'-3'}$ = 2 Hz] (*cf.*M.A.I. Al-Yahya, W.C. Evans and R.J. Grout, J. chem. Soc., Perkin I, 1979, 2130). The second acid moiety is 3,4,5-trimethoxybenzoic acid, easily identified by the six proton *O*-methyl singlet [C(3) and (5) 3.97 ppm] , the *O*-methyl singlet [C(4) at 3.92 ppm] and the down-field two proton singlet (7.28 ppm) for the remaining equivalent aromatic protons. Mass spectroscopy supports these findings and since there is a fragment (109) at m/z 307 the trimethoxybenzoyl group is located at C(3), the *N*-methyl-pyrrole-2-carboxy residue at C(6). All twenty one carbon singlets in the ^{13}C nmr spectrum can be identified by their shifts and by the multiplicity of the $^{13}C-^{1}H$ signals in the off-resonance spectra.

By analogous reasoning Catuabin B has a 6β-benzoyloxy and a 3α-(3,4,5-trimethoxy) benzoyl residue. Unlike Catuabin A, its mass spectrum does not have an ion at m/z 305.

Catuabin C resembles A having an *N*-methylpyrrole (m/z 124, 108, 80) and a pyrrole (m/z 94, 93, 66) 2-carboxylic acid residue, and other spectral data are consistent with the structure shown. It is not known whether the alkaloids have the 3*R*,6*R* or the 3*S*,6*S* configuration.

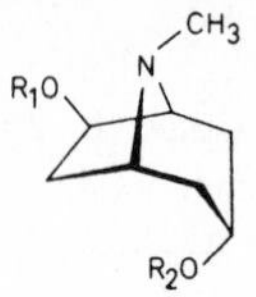

(108) 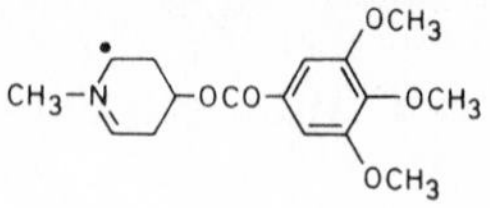 (109)

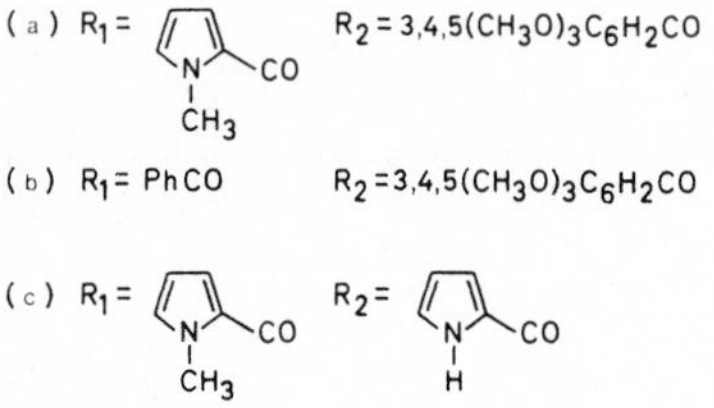

(a) R_1 = N-methylpyrrole-2-CO $R_2 = 3,4,5(CH_3O)_3C_6H_2CO$

(b) R_1 = PhCO $R_2 = 3,4,5(CH_3O)_3C_6H_2CO$

(c) R_1 = N-methylpyrrole-2-CO R_2 = pyrrole-2-CO

(h) Solanaceae

In the Solanaceae field several events appear to have foreshadowed recent work: the isolation of littorine, an alkaloid isomeric with hyosyamine; the discovery of heterodiesters of 1α*H*,5α*H*-tropan-3α,6β-diol and teloidine; the realisation that these alkaloids could exist as their *N*-oxides in nature.

(i) Littorine and related topics

Littorine, (-)-1α*H*,5α*H*-tropan-3α-yl 2-hydroxy-3-phenylpropionate, (110), was first obtained from the Australian *Anthocercis littorea* where it was accompanied by hyoscyamine, meteloidine, tropine and several other unknown bases (J.R. Cannon *et al.*, Austral. J. Chem., 1969, 22, 221), and from *Brugmansia* (previously *Datura* : T.E. Lockwood, Bot. Mus. Leaflet, Harvard University, 1973, 23, 273) *sanguinea*, a plant, indigenous to S. America (W.C. Evans and V.A. Major, J. chem. Soc., (C), 1968, 2775). The distribution of this base and others in *Datura* has been summarised (W.C. Evans, A. Ghani and V.A. Woolley, Phytochem., 1972, 11, 2527). Other species, *A. viscosa* and *A. fasciculata*, contain hyoscyamine

Alkaloids of *Anthocercis*

	A. Littorea		A. viscosa	
	Aerial parts	Roots	Aerial Parts	Roots
Total alkaloid %	0.16	0.10	0.11	0.12
Hyoscyamine	+	+	+	+
Apoatropine	+	+	+	+
Norhyoscyamine	+	+	+	+
Littorine	+	+	+	+
Hyoscine	+	+	+	+
Norhyoscine	+	+		
Meteloidine	+	+	+	+
1αH, 5αH-tropan-3α,6β,7β-triol 3,6-ditiglate		+		+
1αH, 5αH-tropan-3α,6β-diol 6-tiglate	+			
1αH, 5αH-tropan-3α-yl tiglate	+	+	+	+
Tigloidine		+		+
Tropine	+	+	+	+
ψ-Tropine	+	+	+	+
1αH, 5αH-tropan-3α,6β-diol			+	

+ implies presence of alkaloid

as the main alkaloid. A later study demonstrated that the alkaloid distribution pattern was more complex (W.C. Evans and P.G. Treagust, Phytochem., 1973, 12, 2505). *A. tasmanica* (total alkaloid 0.064%) appeared to be different from the others: hyoscine was the main base with nicotine, and littorine and hyoscyamine were also detected. (I.R.C. Bick *et al.*, Austral. J. Chem., 1974, 27, 2515). Recently, nicotine has been obtained from *A. frondosa* (W.C. Evans, and K.P.A. Ramsey, J. Pharm. Pharmacol., 1974, 31, 9P) and the situation perhaps resembles that of another Australian plant, *Duboisia*, some members containing tropane bases, others nicotine and *nor*nicotine.

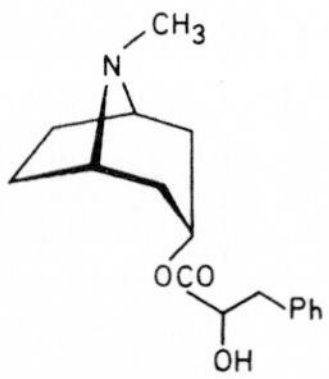

(110) 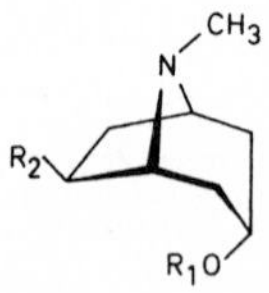 (111) 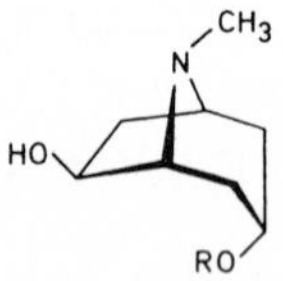 (112)

(a) $R_1 = CH_3CH_2CH(CH_3)CO$ $R_2 = H$

(b) $R_1 = (CH_3)_2CHCO$ $R_2 = H$

(c) $R_1 = (CH_3)_2CH_2CHCO$ $R_2 = OH$

(d) $R_1 = CH_3CH_2CH_2CO$ $R_2 = H$

$R = (CH_3)_2CHCO$

Alkaloids (0.05%) from the leaves of another Australian plant, *A. albicans*, proved to be quite different (W.C. Evans and K.P.A. Ramsey, Phytochem., 1981, 20, 497). In addition to known bases hyoscine, valtropine (1α*H*,5α*H*-tropan-3α-yl 2-methylbutyrate) (111a), 1α*H*,5α*H*-tropan-3α-yl isobutyrate (111b), valeroidine (111c), hyoscyamine and 1α*H*,5α*H*-tropan-3α-yl tiglate, it contained a new base (112), M^+ 227, with a fragmentation pattern consistent with a 1α*H*,5α*H*-tropan-3α,6β-diol nucleus (E. Blossey, *et al.*, Tetrahedron, 1964, 20 585;

W.C. Evans, V.A. Woolley, Phytochem., 1978, 17, 171). Lack of an ion m/z 113 indicated that it was a mono-ester esterified at C(3) with a $C_4H_8O_2$ acid. Apart from the normal signals of a dihydroxytropane (NMe, 2.5; H(3) tr, 4.95; H(6), dd. 4.5 ppm) the ^{1}H nmr spectrum showed a six proton doublet, δ 1.15, and alkaline hydrolysis of the base yielded isobutyric acid. The alkamine was shown to be (-)-1α*H*,5α*H*-tropan-3α,6β-diol, (i.e. 3*S*,6*S*, the same as valeroidine from *Duboisia*); a most interesting observation since both plants are in the *Salpiglossideae* tribe of the family. A second subspecies of *A. albicans* contains *apo*hyoscine, hyoscine, (main alkaloid), the acetyl and tigloyl esters of tropine, 1α*H*,5α*H*-tropan-3α,6β-diol 3-isobutyrate and a second new base 1α*H*,5α*H*-tropan-3α-yl n-butyrate (111d). *A. genistoides* from Western Australia contains hyoscine, *apo*hyoscine, *nor*hyoscine, meteloidine and the hitherto unknown *aponor*hyoscine (113). The related plants *Anthotroche myoporoides* and *A. walcottii* also contained *aponor*atropine (114).

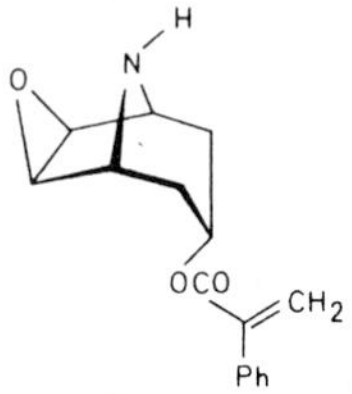

(113)

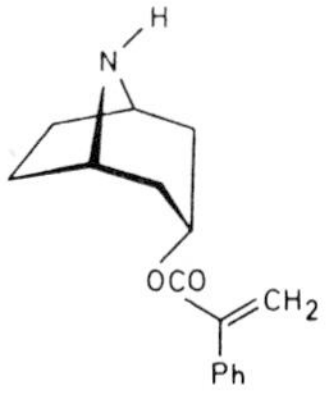

(114)

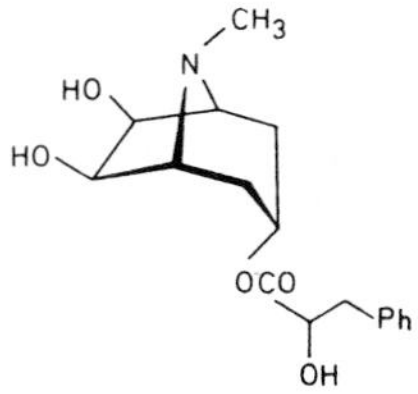

(115)

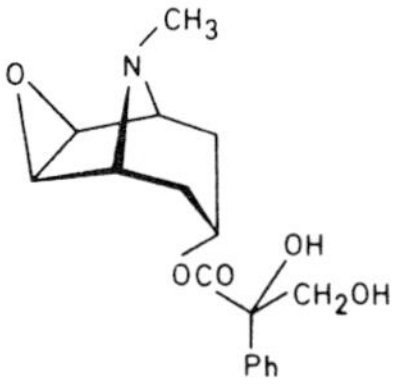

(116)

Alkaloids of *Solandra*

Alkaloidal content % x 10^2

	A		B			C		D		E	
	R	T	R	L	S	R	T	R	T	R	T
Total Alkaloid	64	16	13		14			36	26		
Hyoscine	0.4	0.2	0.7	+	1.5	+	+	0.3	1.0	+	+
Hyosyamine	40		2	+	0.4	+	+			+	+
Atropine		8						16	15		
Littorine	0.4		+					+			
Tigloidine	2.0	0.4	0.3		0.6	+		6.0	0.4	+	+
1αH, 5αH-tropan-3α-yl tiglate	3.0	0.2	1.0	+			+	1.6	0.4	+	+?
1αH, 5αH-tropan-3α-yl acetate	2.0	0.5	0.5					1.3	0.3	+	
Valtropine	3.0	0.4	0.2	+		+	+	2.0	0.4	+	+?
Noratropine		6.0	8.0	+	9.0		+	4.0	8.0	+	+
Norhyoscine					1.2						
Tropine	+	+	+	+	+	+	+	+	+	+	+
ψ-tropine	+	+	+	+	+	+	+	+	+	+	+

A, *Solandra grandifolia*; B, *S. guttata*; C, *S. hartwegii*; D, *S. hirsuta*; R, roots; T, aerial parts; L, leaves; S, stems.

+ implies presence of alkaloid

Littorine has also been detected in the genus *Solandra*, vine-like plants which, like *Datura*, are placed in the *Datureae* tribe (W.C. Evans, A. Ghani and V.A. Woolley, Phytochem., 1972, 11, 470).

Hyoscyamus sp also contain littorine in the roots (A. Ghani, W.C. Evans and V.A. Woolley, Bangladesh Pharm. J., 1972, 1, 12).

The alkaloid 6,7β-dihydroxylittorine (115) is a constituent (0.016%) of an Australian peach flowered *Brugmansia candida* as are hyoscine (major base), meteloidine, *nor*hyoscine, *nor*hyoscyamine and the previously unknown isometeloidine (1α*H*,5α*H*-tropan-3α,6β,7β-triol 6-tiglate).

Brugmansia sanguinea is the source of another interesting variation on tropic acid since α-hydroxyhyoscine (116) has been isolated from its leaves which are a commercial source of hyoscine. Hydrolysis of the base afforded α-phenylglyceric acid (C.F. Moorhoff, Planta Medica, 1975, 28, 106).

(j) Heterodiesters and related topics

Heterodiesters of hydroxytropanes were also first isolated some time ago from *Brugmansia sanguinea* when 1α*H*,5α*H*-tropan-3α,6β-diol 3-tiglate 6-acetate (117a) (0.05 - 0.1%) and the acetyl ester of tropine (0.02%) were extracted from the leaves. (W.C. Evans and V.A. Major, J. chem. Soc., (C), 1966, 1621). Later, 1α*H*,5α*H*-tropan-3α,6β,7β-triol 3-tiglate, 6-isovalerate (117b), was obtained from roots of the same source and for ^{1}H nmr comparison a series of mixed esters of teloidine were synthesized: 3,6-ditiglate, 7-isovalerate (117c); 3-tiglate, 6,7-di-isovalerate (117d); 3-tiglate, 6,7-di-(2-methylbutyrate) (117e); 3-tiglate 6-(2-methylbutyrate) (117f). (W.C. Evans and V.A. Major, J. chem. Soc., (C), 1968, 2775). The related plant *Brugmansia suaveolens* yielded 1α*H*,5α*H*-tropan-3α,6β-diol 3-tiglate 6-(2-methylbutyrate) from the leaves together with isometeloidine (117g) (W.C. Evans and J.F. Lampard, Phytochem., 1972, 11, 3293; *cf*. W.J. Griffin, Austral. J. Chem., 1976, 29, 2329). The only mixed ester to be obtained from other than a tree datura source is 1α*H*,5α*H*-tropan-3α,6β-diol 3-tiglate 6-propionate (117h) from *Datura innoxia* (P.J. Beresford and J.G. Woolley, Phytochem., 1974, 13, 1249).

Distribution of tropane alkaloids in *Datura* roots	*Datura* sp.								
	A	B	C	D	E	F	G	H	I
Littorine	+	+	+	+	+	+	+	+	+
1αH, 5αH-tropan-3α6β-diol ditiglate	+	+	+	+	+	+	+		+
1αH, 5αH-tropan-3α6β,7β-triol 3,6-ditiglate	+	+	+	+	+	+	+	+	+
Hyoscyamine and hyoscine	+	+	+	+	+	+	+	+	+
Meteloidine	+	+	+	+		+		+	+
Norhyoscine			+	+				+	
Norhyoscyamine				+		+	+		
Tigloidine			+		+	+	+	+	
1αH, 5αH-tropan-3α-yl tiglate		+		+		+		+	+
Oscine								+	+
1αH, 5αH-tropan-3α-yl acetate									+
1αH, 5αH-tropan-3α,6β,7β-triol 3-tiglate 6-isovalerate									+
Apoatropine	+		+		+				
Tropine and ψ-tropine	+	+	+	+	+			+	+
1αH, 5αH-tropan-3α,6β-diol				+			+		

A, *D. stramonium*; B, *D. ferox*; C, *D. innoxia*; D, *D. meteloides*; E, *D. metel* var. *fastuosa*; F, *D. leichhardtii*; G, *Brugmansia (Datura) cornigera*; H, *B. candida*; I, *B. sanguinea*.

+ implies presence of alkaloid

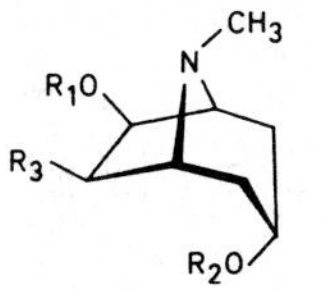

(117)

(a)	$R_1 = CH_3CO$	$R_2 = CH_3CH{=}C(CH_3)CO$	$R_3 = H$
(b)	$R_1 = (CH_3)_2CHCH_2CO$	$R_2 = CH_3CH{=}C(CH_3)CO$	$R_3 = OH$
(c)	$R_1 = CH_3CH{=}C(CH_3)CO$	$R_2 = CH_3CH{=}C(CH_3)CO$	$R_3 = (CH_3)_2CHCH_2COO$
(d)	$R_1 = (CH_3)_2CHCH_2CO$	$R_2 = CH_3CH{=}C(CH_3)CO$	$R_3 = (CH_3)_2CHCH_2COO$
(e)	$R_1 = CH_3CH_2CH(CH_3)CO$	$R_2 = CH_3CH{=}C(CH_3)CO$	$R_3 = CH_3CH_2CH(CH_3)COO$
(f)	$R_1 = CH_3CH_2CH(CH_3)CO$	$R_2 = CH_3CH{=}C(CH_3)CO$	$R_3 = OH$
(g)	$R_1 = CH_3CH{=}C(CH_3)CO$	$R_2 = H$	$R_3 = OH$
(h)	$R_1 = CH_3CH_2CO$	$R_2 = CH_3CH{=}C(CH_3)CO$	$R_3 = H$
(i)	$R_1 = CH_3CH{=}C(CH_3)CO$	$R_2 = CH_3CH{=}C(CH_3)CO$	$R_3 = H$
(j)	$R_1 = CH_3CH{=}C(CH_3)CO$	$R_2 = CH_3CH{=}C(CH_3)CO$	$R_3 = OH$

Two alkaloids, the 3,6-ditigloyl esters of 1α*H*,5α*H*-tropan-3α,6β-diol (117i) and 3α,6β,7β-triol (117j) are invariably found in *Datura* roots but not elswhere and it is of chemotaxonomic interest that the former base has been isolated from *Mandragora autumnalis* and *M. vernalis* (B.P. Jackson and M.I. Berry, Phytochem., 1973, 12, 1165).

(k) N-Oxides

Although *N*-oxides of alkaloids, particularly necine bases, are common, *N*-oxides of tropanes were not known in nature until recently. The axial (minor product) and equatorial (major product) N—O isomers of hyoscyamine and hyoscine have been prepared using *m*-chlorperbenzoic acid in chloroform at 0° (J. Cymerman Craig and K.K. Purushothan, J. org. Chem., 1970, 35, 1721) in preference to hydrogen peroxide (G. Fodor *et al.*, Canad. J. Chem., 1971, 49, 3258). The *N*-oxide hydrochlorides can be distinguished by nmr spectroscopy, hyoscyamine NMe(eq) appearing at δ 3.45, NMe(ax) at δ 3.56, in contrast to the signals from hyoscine NMe(eq) δ 3.62, NMe(ax) δ 3.45 caused by deshielding by the proximal C(6)(7) epoxide. By tlc it has been shown that both of the *N*-oxides

Alkaloids of *Hyoscyamus*

	Plant			
	A	B	C	D
Apohyoscine	+		+	
Apoatropine	+			
Tigloidine	+			
Hyoscine	+	+	+	
Norhyoscine			+	
1αH, 5αH-tropan-3α-yl tiglate	+			+
Littorine	+			+
Atropine	+			
Hyoscyamine		+	+	+
Tropine and ψ-tropine	+			

A, *Hyoscyamus albus*; B, *H. aureus*; C, *H. niger*; D, *H. pusillus*.

+ implies presence of alkaloid

of hyoscyamine occur in *Atropa belladonna*, *Datura stramonium*, *Hyoscyamus niger*, *Scopolia lurida*, *S. carniolica* and *Mandragora officinarum* (*cf*. B.P. Jackson and M.I. Berry, *loc. cit.*) whereas the equatorial N—O isomer of hyoscine was found in all but *Mandragora* (J.D. Phillipson and S.S. Handa, J. Pharm. Pharmac., 1973, 25, 117P; Phytochem., 1974, 14, 999). The *N*-oxide of 1α*H*,5α*H*-tropan-3α-yl tiglate is a constituent of *Physalis alkekengi* root (H. Yamaguchi, *et al*., Yakugaku Zasshi, 1974, 94, 1115) and 6β-hydroxyhyoscyamine *N*-oxide occurs in *Physochlaina alaica* (R.T. Mirzamatov, *et al*., Khim. prir. Soedin., 1974, 10, 540).

The genus *Schizanthus*, which like *Duboisia*, *Anthocercis* and *Anthotroche*, is in the *Salpiglossideae* tribe (R. Wettstein in A Engler and K. Prantl, Die Naturlichen Pflanzenfamilien, 1897, IV Teil, IIIb, p 4) is the source of several new alkaloids. *S. pinnatus* leaves gave schizanthin A (118) (0.0038%) a laevorotatory oily base and *laevo* schizanthin B (121) (0.014%) an amorphous powder. Both bases gave (+)-1α*H*,5α*H*-tropan-3α,6β-diol (3*R*,6*R*) on hydrolysis. Schizanthin A, M^+ 379, ($C_{20}H_{29}NO_6$), showed in its ^1H nmr spectrum an ethyl group δ 1.33, t. J = 7 Hz and δ 4.25, J = 7 Hz and several C-methyls, δ 1.81, 1.88, 1.93 and 2.18 (one too many for the structure allotted), which were shown by comparison with reference spectra to most resemble CMe chemical shifts of mesaconic and senecioic acids and the corresponding olefinic protons (multiplets) at 6.69 and 5.62 ppm respectively were also present. Other signals (one NMe) were characteristic of the dihydroxytropane nucleus but H(3) 5.0 ppm and H(6) 5.43 ppm are reported to be multiplets. Since the mass spectrum showed ions at m/z 253 corresponding to the 4-hydroxy-*N*-methylpyridinium ester of ethyl mesaconate (119), and one at

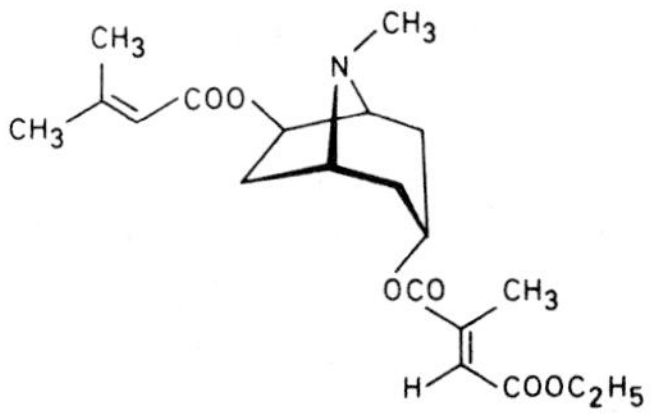

CH3 CH3 COO N CH3 OCO CH3 H COOC2H5

(118)

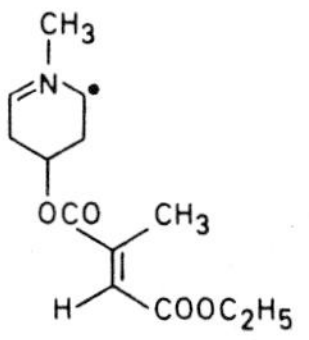

CH3 N OCO CH3 H COOC2H5

(119)

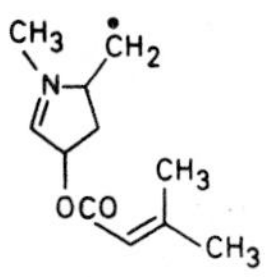

CH3 CH2 N OCO CH3 CH3

(120)

195 corresponding to the *N*-methylpyrrolidine derivative ester of senecioic acid (120) it was concluded that the alkaloid had the ethyl mesaconate residue at C(3), the senecioic moiety at C(6). However, because of the discrepancy of C-methyl signals, other isomers cannot be excluded. Schizanthin B ($C_{31}H_{44}N_2O_8$) was more unusual and its ^1H nmr spectrum displayed signals of senecioic acid (CMe at δ 1.90 and 2.18; olefinic proton, δ 5.67 m) and mesaconic acid (CMe at δ 1.83; olefinic proton, δ 6.82 m). As with the previous base H(3) and H(6) were multiplets. The mass spectrum had a prominent ion at m/z 446 (49%) corresponding to the di-ester of mesaconic acid with a two tropane fragment(122) and the structure shown (121) was, therefore, proposed. (H. Ripperger, Phytochem., 1979, 18, 171).

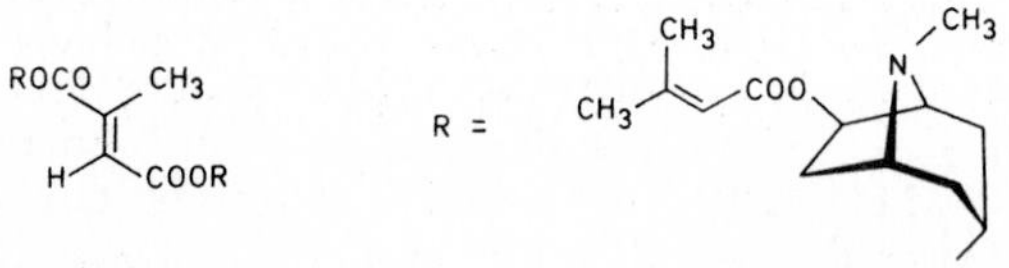

(121)

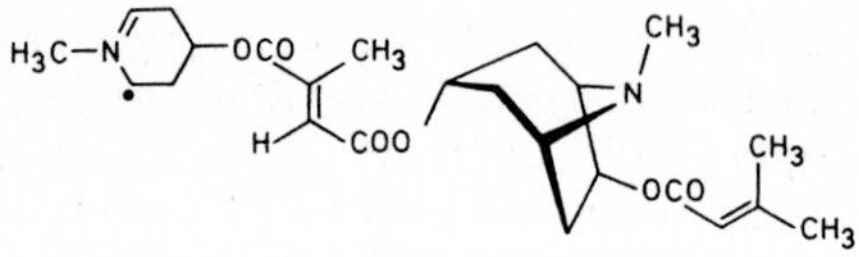

(122)

Schizanthus hookeri an indigenous Chilean plant also contains new bases (A. San Martin *et al.*, Phytochem., 1980, 19, 2007). The roots gave two new (+)-1α*H*,5α*H*-tropan-3α,6β-diols, the 3-senecioyl and the 6-angeloyl mono-esters. Tropine, (+)-ψ-hygroline (2*R*,2'*S*-hygroline) (123) and (-)-hygroline (2*R*,2'*S*-hygroline (124) were also isolated (see R. Lukes *et al.*, Coll. Czech. chem. Comm., 1960, 25, 483).

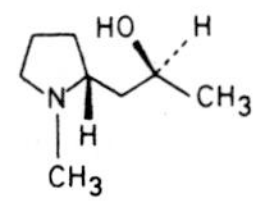

(123)

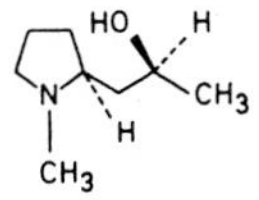

(124)

From *Duboisia leichhardtii* two *apo*hyoscine dimers (discopine esters of 1-phenyl-1,2,3,4-tetralin-1,4-dicarboxylic acid; *cf.* belladonnine) (125a and b), *apo*hyoscine, *apo*atropine, *nor*hyoscyamine and 1α*H*,5α*H*-tropan-3α-yl isobutyrate have been isolated (*cf.* W.C. Evans and K.P.A. Ramsey, Phytochem., 1981, 20, 497), and their structures established by spectroscopic means (K. Kagei, *et al.*, Yakugaku Zasshi, 1980, 100, 216).

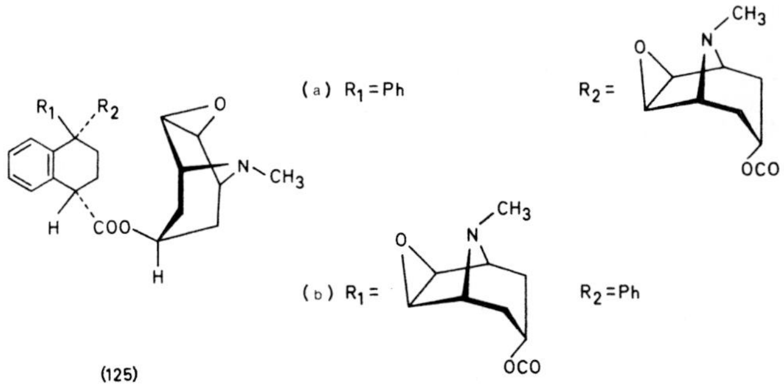

(a) R_1 = Ph

(b) R_2 = Ph

(125)

Tissue cultures from the same plant did not contain hyoscine or hyoscyamine, but nicotine, and the isobutyryl and isovaleryl esters of tropine were detected by glc-MS (K. Kagei, Yakagaku Zasshi, 1980, 100, 574).

(l) Secotropanes

Physalis peruviana roots have yielded the first secotropane alkaloids, (+) and (±)-physoperuvine (126a) and (+)-*N*,*N*-dimethyl-physoperuvine (126c), (A.B. Ray, *et al.*, Chem. Ind. (London), 1976, 454; M. Sahai and A.B. Ray, J. org. Chem., 1980, 45, 3265). The structures were established by spectroscopy (including ^{13}C nmr analysis) and *N*,*N*-dimethyl-physoperuvine has been synthesized by Michael addition of dimethylamine to cyclohept-2-en-3-one followed by quaternization with methyl iodide. Both (+)-physoperuvine and its benzoyl derivative (126b) show positive Cotton effects at 290 nm in the CD curves which resemble that of *R*-(+)-methylcycloheptan-3-one.

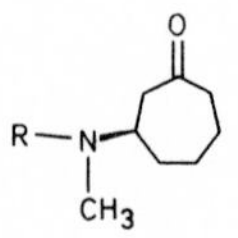

(a) R=H

(b) R=PhCO

(c) R=$(CH_3)_2$

(126)

3. Spectroscopy

The mass spectroscopy of the tropane alkaloids has been reported (E.C. Blossey, *et al.*, 1964, Tetrahedron, 20, 585).

Summaries of the ^{1}H nmr signals from the tropane ring are available (W.C. Evans and V.A. Major, J. chem. Soc., (C), 1966, 1621; 1968, 2775; G. Fodor, in Chemistry of the Alkaloids, ed. E. Pelletier, Van Nostrand Reinhold Co., 1970, p 431).

The following table gives ^{13}C nmr signals for a number of alkaloids.

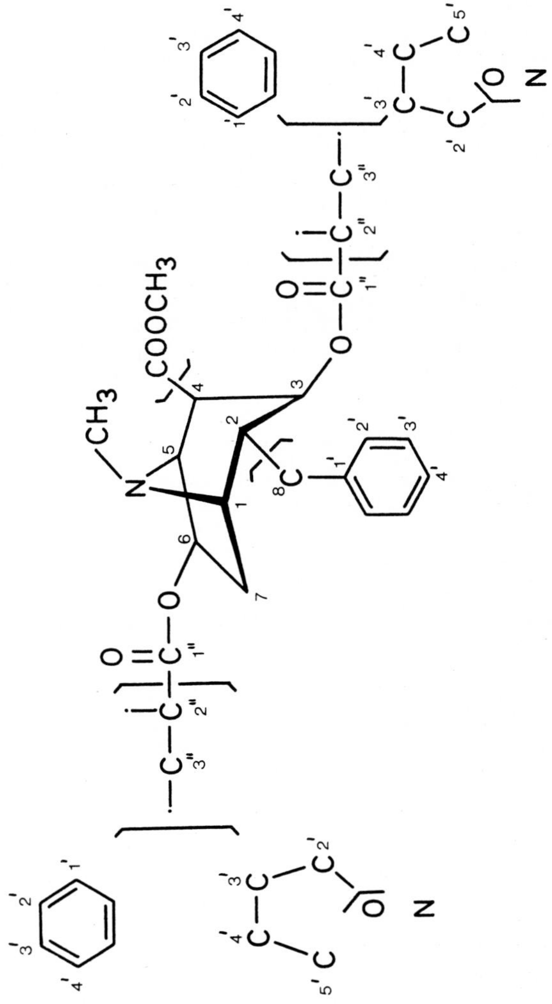

(127)

Carbon Nos. in (127)	1	2	3	4	5	6	7	8	NMe	C(2) CO OMe	1^1	2^1	3^1	4^1	5^1	1"	2"	3"
Nortropane[1]	54.7	32.9	17.2	32.9	54.7	29.0	29.0											
Tropane[1]	61.2	29.9	15.9	29.9	61.2	25.6	25.6		40.4									
Tropinone[1]	60.2	47.1	207.8	47.1	60.2	27.3	27.3		37.8									
Tropine[1]	59.8	39.1	63.6	39.1	59.8	25.7	25.7		40.0									
Tropine[4]	59.7	38.8	63.3	38.8	59.7	25.5	25.5		39.8									
ψ-Tropine[1]	60.1	38.3	62.7	38.3	60.1	26.7	26.7		39.2									
Benzoyl-tropine[1]	59.4	36.2	67.7	36.2	59.4	25.3	25.3		39.9									
Benzoyl-ψ-tropine[1]	60.0	35.3	67.5	35.3	60.0	26.3	26.3		38.3									
Hyoscy-amine[1]	59.3	35.7	67.6	35.7	59.3	24.8	24.8		39.7									
Hyoscy amine[2]	59.94	36.50	68.48	36.31	59.86	25.21	25.63		40.41		136.90	128.97	129.58	128.36		173.15	55.11	64.45
Hyoscy-amine[3]	59.4	36.1	67.8	35.9	59.3	25.2	24.7		40.0		135.7	128.6	128.0	127.4		172.0	54.5	63.8
Hyoscy-amine[4]	59.2	35.8	67.4	35.6	59.2	24.7	25.05		39.7		135.9	128.3	127.8	127.1		171.6	54.6	63.4
Hyoscy-amine[5]	59.4	35.8	67.6	36.0		24.8	25.2		40.0		135.9	128.5	127.9	127.3		171.8	54.6	63.1
Hyoscine[4]	57.3	30.2	66.3	30.4	57.3	55.4	55.8		41.6		135.4	128.3	127.6	127.3		171.2	54.0	63.3
Hyoscine[5]	57.3	29.9	66.1	30.1	57.3	55.4	55.8		41.3		135.5	128.4	127.6	127.3		171.7	54.1	63.3
Hyoscine[1]	58.2	34.7	66.6	34.7	58.2	55.9	55.9		43.4		135.9	128.5	127.9	127.4		171.7	54.5	63.7
Hyoscine[3]	57.4	30.4	66.4	30.3	57.3	56.0	55.6		41.6		135.5	128.5	127.7	127.5		171.3	54.1	63.5

Carbon Nos. in (127)	1	2	3	4	5	6	7	8	NMe	C(2) CO OMe	1^1	2^1	3^1	4^1	5^1	1"	2"	3"
Hyoscine[3]	57.4	30.4	66.4	30.3	57.3	56.0	55.6		41.6		135.5	128.5	127.7	127.5		171.3	54.1	63.5
Hyoscine N-oxide[1]	69.6	32.0	61.9	32.0	69.6	53.4	53.4		52.4									
Apohyoscine[5]	57.9	31.3	67.0	31.3	57.9	56.5	56.5		42.3		136.7	128.3	128.2	128.0		165.4	141.9	126.8
Norhyoscyamine[5]	53.0	36.8	68.4	37.0	53.0	28.8	28.5		-		136.0	128.6	128.1	127.5		172.0	54.8	63.6
Apoatropine[5]	59.9	36.4	68.1	36.4	59.9	25.5	25.5		40.3		136.9	128.4	128.1	128.1		165.9	142.0	126.6
Isobutyryloxytropane[5]	55.6	36.5	66.9	36.5	55.6	25.5	25.5		40.3							175.7	34.2	18.7
Cocaine[4]	64.5	49.9	66.6	35.2	61.3	25.0	25.0		40.7	170.2 50.8	130.2	127.9	129.3	132.5		165.5		
Cocaine[2]	62.13	50.78	67.50	35.94	65.37	25.80	25.81		41.38	171.63 51.51	130.50	131.00	129.10	133.60		167.11		
1-Phenyl-1,2,3,4-tetralin-1,4-dicarboxylic acid discopine[1][5]	57.7	30.9	67.0 67.5	31.3	57.9	56.2	56.5		42.2 42.5									
" [2]	57.6	30.7	66.8 67.4	31.0 31.2	57.8	56.0	56.2 56.4		42.0 42.2									
Alkaloid A (K.deplanchei)[6]	63.2	45.8	69.7	37.0	59.6	25.3	21.9	35.2	40.3		130.3 139.0	129.2 127.9	128.1 128.7	132.7 125.7		165.8		
Alkaloid B (K. deplanchei)[6]	63.3	45.4	69.3	37.0	59.8	25.3	21.6	35.1	40.4		139.2	128.1	128.7	125.8		169.9	21.3	
Alkamine A and B	63.8	46.3	65.6	39.9	60.0	25.3	21.8	35.4	40.3		140.1	127.9	128.7	125.5				
Alkaloid C (K. deplanchei)[6]	64.2	44.9	65.4	36.9	66.5	80.5	32.2	35.2	40.6		130.3 140.0	128.1	128.1 128.8	132.7 125.7		165.8		

	1	2	3	4	5	6	7	8	NMe	C(2) CO OMe	1^1	2^1	3^1	4^1	5^1	1"	2"	3"
Alkamine C	62.8	40.0	66.0	32.7	67.4	76.2	37.1	35.1	36.8		139.8	128.3	128.7	125.8				
Alkaloid D (K.deplanchei)[6]	70.7	38.6	69.3	31.0	58.3	40.9	72.8	35.0	36.7		134.0 138.9	128.7 127.9	128.3 128.7	226.0				
Alkamine D	71.6	40.1	66.1	34.4	58.8	40.8	73.3	35.1	37.1		139.8	128.3	128.7	125.8				
Alkaloid E (K.deplanchei)6	62.5	49.4	64.6	36.0	66.3	80.3	33.1	74.2	40.3		130.3 142.7	129.2 126.4	128.1 128.4	132.7 127.8		165.8		
Catuabin A[7]	60.40	33.95	67.75	35.25	60.05	78.75	36.55		40.60		125.60	106.85	153.30 OMe 56.40	142.50 OMe 60.95		165.45		
											- NMe 36.66	122.70	107.95	117.90	129.65	160.95		
N Benzoylphysoperuvine[8]	60.9	43.4	213.3	40.2	28.2	21.6	33.3		55.9		136.6	129.4	128.5	126.6		171.0		

Explanatory Notes: Nos 1,2,3 etc refer to tropane ring; Nos $1^1,2^1,3^1$ etc refer to aromatic rings numbered from the bridgehead carbon or heterocyclic rings numbered from the heteroatom; Nos 1",2",3" etc refer to the ester carbonyl (1") and carbons before the aromatic or heterocyclic rings, with di- or tri-esters, the 3 substituent group preceeds the 6; where aromatic or heterocyclic acid moieties are themselves substituted, then the substituent group, eg. OMe appears in the same column with the substituted atom, (see Catuabin A).

[1]E. Wenkert *et al.*, Acc. Chem. Res., 1974, 7, 46; [2]V.I. Stenberg and N.K. Narain, J. heterocyclic.Chem., 1977, 14, 225; [3]E. Leete *et al.*, J. Amer. chem. Soc., 1975, 97, 6826; [4]A.M. Taha and G. Rucker, Egypt. J. pharm. Sci., 1977, 18, 59; [5]K. Kagei *et al.*, Yakugaku Zasshi, 1980, 100, 216; [6]M. Lounasmaa *et al.*, J. org. Chem., 1975, 40, 3694; [7]E. Graf and W. Lude, Arch. Pharm., 1978, 311, 139; [8]M. Sahai and A.B. Ray, J. org. Chem., 1980, 45, 3265.

Chapter 12

FIVE-MEMBERED MONOHETEROCYCLIC COMPOUNDS (continued)

THE PYRROLE PIGMENTS

The manuscript for this Chapter was not available for publication in this Volume. So some key references are listed below.

The great expansion of activity in research into porphyrins, chlorins, and the cobalamines since the early 1970's has been accompanied by the appearance of several substantial publications which offer a wealth of readily accessible information; refs. 1-4 present information in increasing detail. Much of this recent research has been connected with investigation of the biosynthesis of the tetrapyrrole pigments, and this has been clearly reviewed in refs. 5 and 6.

1. K.M. Smith in "Comprehensive Organic Chemistry", eds. Sir Derek Barton and W.D. Ollis, Pergamon Press, Oxford, 1979, Vol.4, Chap. 17.2, pp.321-355.

2. "Porphyrins and Metalloporphyrins", ed. K.M. Smith, Elsevier, Amsterdam, 1975.

3. "The Porphyrins", ed. D. Dolphin, Academic Press, New York, 1979, Vols.I-VII.

4. "B_{12}", ed. D. Dolphin, John Wiley and Sons, New York, 1982, Vols. I and II.

5. M. Akhtar and P.M. Jordan in "Comprehensive Organic Chemistry", eds. Sir Derek Barton and W.D. Ollis, Pergamon Press, Oxford, 1979, Vol.5, Chap. 30.2, pp.1121-1166.

6. 'The Biosynthesis of Porphyrins, Chlorophylls, and Vitamin B_{12}', by F.J. Leeper in Natural Product Reports, 1985, *2*, 1.

Chapter 13

FIVE-MEMBERED MONOHETEROCYCLIC COMPOUNDS (continued)

AZAPORPHYRINS; BENZOPORPHYRINS; BENZOAZOPORPHYRINS; PHTHALOCYANINES AND RELATED STRUCTURES

The manuscript for this Chapter was not available for publication in this Volume. So some key reviews of relevant material are listed below.

1. 'Azaporphyrins', by A.H. Jackson in "The Porphyrins", ed. D. Dolphin, Academic Press, New York, 1979, Vol.1, Chap. 9.

2. "The Phthalocyanines", by F.H. Moser and A.L. Thomas, CRC Press, Boca Raton, Florida, 1983, Vols. I and II.

3. "Coordination Compounds of Porphyrins and Phthalocyanines", by B.D. Berezin, Transl. by V.G. Vopian, John Wiley and Sons, Chichester, U.K., 1981.

Chapter 14

FIVE-MEMBERED MONOHETEROCYCLIC COMPOUNDS: THE INDIGO GROUP

M. SAINSBURY

Introduction

The commercial importance of indigo (indigotin) has received a stimulus in the last decade through the increased popularity of certain types of clothing wherein part fading of the dyestuff is deemed fashionable. As a result interest in the chemistry of indigoid dyes has revived to a degree, and some of the structures long thought satisfactory have been re-examined and reassigned.

1. Indigo and its derivatives

(a) Methods of synthesis

(i) Three new syntheses of indigo have been reported by J. Gosteli (Helv., 1977, 60:1980), the first of these utilizes isatoic anhydride (1, R=H) as starting material. This with dimsyl sodium in dimethylsulphoxide affords 2-amino-ω-methylsulphinylacetophenone (2, R=H) which, when warmed with hydrochloric acid, undergoes a Pummerer rearrangement and loss of methylthiol to give indigo.

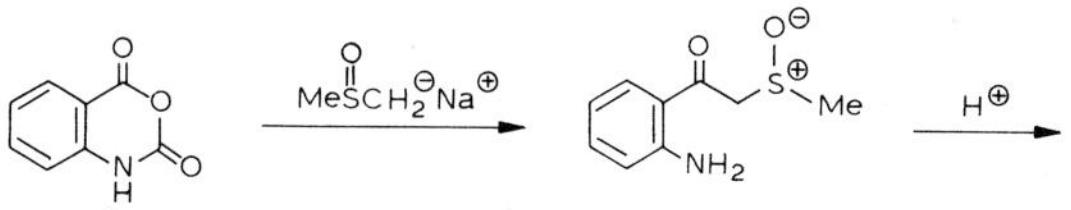

$MeS(O)CH_2^{\ominus}Na^{\oplus}$; $H^{\oplus}$

(1) (2)

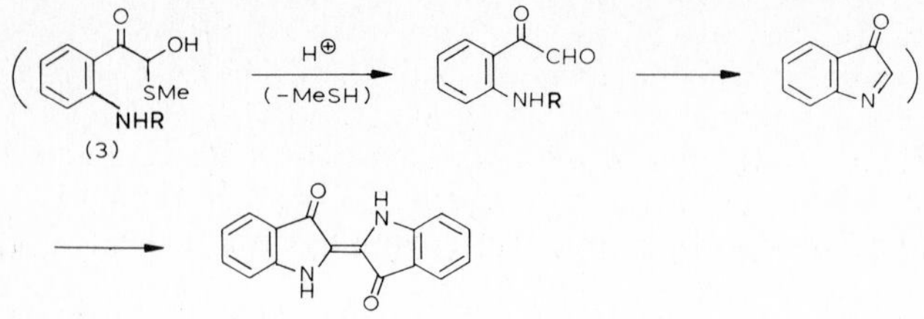

(3)

(4)

A more satisfactory yield (77% *vs* 39%) is obtained by acetylating the amine (2), and in this case the presence of the rearrangement product 2-acetamido-ω-methylmercapto-ω-acetoxyacetophenone (3, R=Ac) is detected in the mother-liquors. When the acetylation is conducted under vigorous conditions (not specified) the rearrangement product predominates. On treatment with hydrochloric acid (3) is transformed into indigo (4) (*cf.* M. von Strandtmann *et al.*, U.S. patent, 1976, 3,937-704).

In the second approach isatoic anhydride is reacted with nitromethane to give 2-amino-ω-nitroacetophenone (5). This compound on warming with hydrochloric acid yields indigo. It is suggested that a Nef rearrangement is involved in this transformation, in the course of which the nitromethyl group of the nitroacetophenone is converted into a carboxaldehyde function. Again the synthesis is improved by the *N*-acetylation of the starting material (5), the final step then occurs in 88% yield.

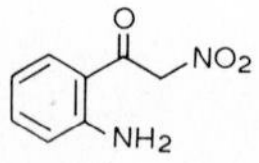

(5)

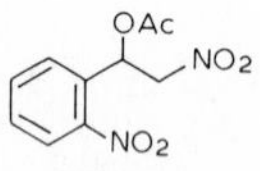

(6)

Styrene reacts with acetic anhydride and nitric acid, together with a catalytic amount of concentrated sulphuric acid, to give the dinitroacetate (6). Treatment of this with methanol and toluene-p-sulphonic acid gives the corresponding alcohol, which is a known compound having been made many years ago by J. Thiele (Ber., 1899, 32:1293). Incidentally, Thiele noted that this compound gives a blue colour when it is exposed to ferrous ion. The alcohol is converted into indigo (59% yield) by sodium dithionite in aqueous sodium hydroxide.

(ii) *N*-Methyl- and *N*,*N*-dimethyl-indigo have been prepared by the oxidative dimerisation of indoxyl derivatives using the one electron oxidant 2-chloro-2-nitropropane. *N*-Methylindigo, for example, is obtained together with indigo, (molar ratio ∿ 2:1) by the reaction of *O*-acetylindoxyl and *N*-methyl-*O*-acetylindoxyl with this reagent in methanol solution containing sodium methoxide (G. Kaupp, Ber., 1970, 103:990). *N*,*N*-Dimethylindigo results from a similar experiment, this time using as starting material *N*-methyl-*O*-acetylindoxyl.

(iii) 1,3-Dimethyl-2,1-benzisoxazolium perchlorate (7) adds cyanide, azide or methoxide ion to give 3-substituted 2,1-benzisoxazolines of the general type (8). Pyrolysis of the cyano derivative (8, X=CN) affords *N*-methylindoxyl (10) and *N*,*N*-dimethylindigo (11) (N.F. Haley, J. org. Chem., 1978, 43: 1233). It is considered that the exocyclic methylene compound (9), formed by elimination of hydrogen cyanide from 3-cyano-1,3-dimethyl-2,1-benzisoxazoline, rearranges to *N*-methylindoxyl, as a key step in this sequence of reactions. Air oxidation of the last product then yields *N*,*N*-dimethylindigo.

Me, $X^{\ominus}$, O, $N^{\oplus}$, Me, $ClO_4^{\ominus}$ → Me, X, O, N, Me → Δ

(7, X = CN, N_3, OMe) (8)

CH_2, O, N, Me → O, N, Me → [O] → O, Me, N, N, Me, O

(9) (10) (11)

(iv) 4,4',5,5',6,6',7,7'-Octahydroindigo (13), m.p. >300°, λ_{max} (log ε) nm 317 (4.60), 524 (4.05) (methanol) has been synthesised by G. Pfeiffer and H. Bauer (Ann., 1980, 564) by the oxidative coupling of the tetrahydroindoxyl (12) using potassium ferricyanide.

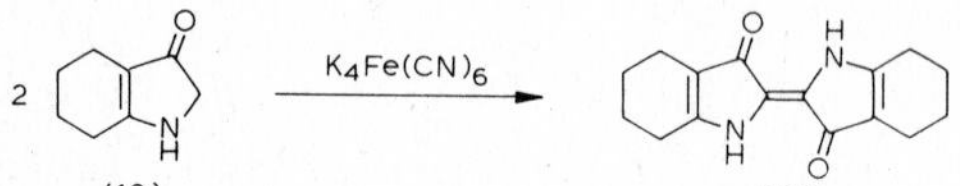

$K_4Fe(CN)_6$

(12) (13)

(v) 4,4'-Dibutyl-5,5'-dimethylpyrrolindigo (14), m.p. >300°, λ_{max} (log ε) nm 317 (4.20), 529 (3.98) (methanol), and di(cyclopentan[b]pyrrolindigo) (15), λ_{max} 313.5, 505 nm, are prepared by similar reactions.

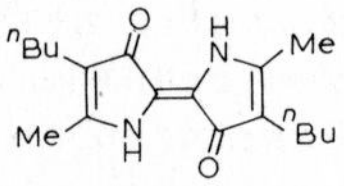

(14) 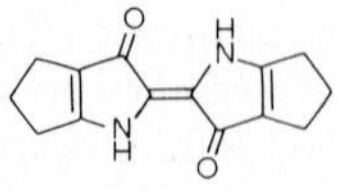 (15)

(vi) Tetrahydroindigo (17), m.p. >300°, λ_{max} (log ε) nm 315 (4.47), 565 (4.32) is made by combining 3-methoxyindolin-3(2H)-one (16) with tetrahydroindoxyl (12), and *N*-methyl and *N,N*-dimethyl derivatives of octahydroindigo and tetrahydroindigo have also been synthesised. 1-Methyl-4,4',5,5',6,6',7,7'-octahydroindigo, m.p. 195° (begins to decompose at ∿140°), λ_{max} (log ε) nm 315 (4.28), 552 (4.00); 1,1'-dimethyl-4,4',5-5', 6,6', 7,7'-octahydroindigo, m.p. 90-92° (decomp.), λ_{max} (log ε) nm 328 (4.27), 621 (4.17); 1-methyl-4,5,6,7-tetrahydroindigo, m.p. 193°, λ_{max} (log ε) m, 318.5 (4.28), 584 (4.13); 1'-methyl-4,5,6,7-tetrahydroindigo, m.p. 153°, λ_{max} (log ε) nm 319 (4.23), 595 (4.15); 1,1'-dimethyl-4,5,6,7-tetrahydroindigo, m.p. 90-92° (decomp.), λ_{max} (log ε) nm 330 (4.34), 582 (3.97).

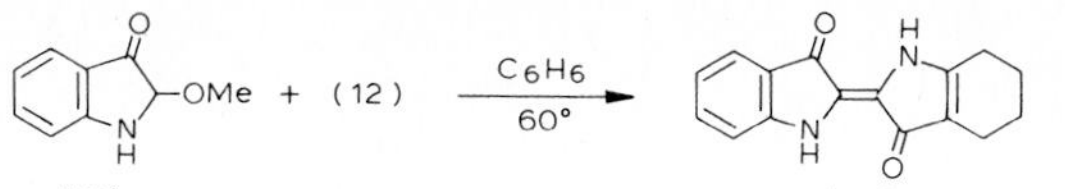

(16) + (12) $\xrightarrow[60°]{C_6H_6}$ (17)

(vii) 4,5,6,7-Tetrahydroindirubin (20) m.p. >300°, is the product of a reaction between isatin (18) and tetrahydroindoxyl (12). If acid is excluded the intermediate adduct (19) can be isolated as yellow solid, m.p. 145° (decomp.). When the adduct is heated in benzene solution containing a catalytic amount of trifluoroacetic acid the colour of the reaction mixture turns red and tetrahydroindirubin is then formed.

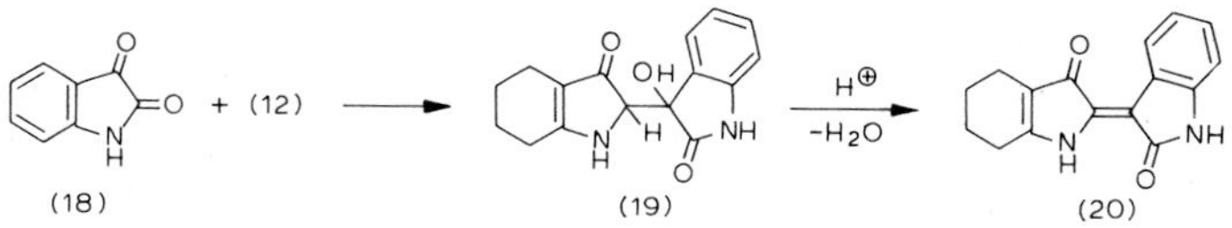

(18) + (12) → (19) $\xrightarrow[-H_2O]{H^{\oplus}}$ (20)

(viii) The series of the six possible combination isomers of indigo has been completed by the synthesis of 3-oxo-2-(3-oxo-1-isoindolinylidene)indoline (phthalorubin) (23) and 2-oxo-3-(3-oxo-1-isoindolinylidene)indoline (phthalaurin) (25) (E. Wille and W. Lüttke, Ber., 1973, 106:3257). These two compounds are prepared by the reaction of monothiophthalimide (21) with *N,O*-diacetylindoxyl (22) and oxindole (24) respectively.

Phthalorubin sublimes at 275° and melts between 310-320° (decomp.), λ_{max} (log ε) nm 516 (4.16). Phthalaurin sublimes at 265° and has m.p. 299-303°, λ_{max} (log ε) nm 455 (3.85).

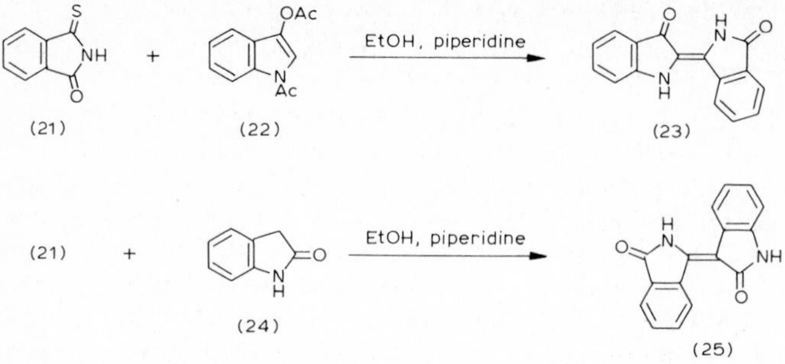

(21) (22) (23)

(21) (24) (25)

(ix) The synthesis of polymeric indigos has been extended by E. Ziegler *et al.* (Monatsh., 1970, 101:923) using a method previously described (Ziegler and Th. Kappe, Angew Chem., 1964, 76:921; Ziegler, H.G. Foraita and Kappe, Monatsh., 1967, 98: 324). The new compounds have structures (26), (27) and (28) (n unspecified).

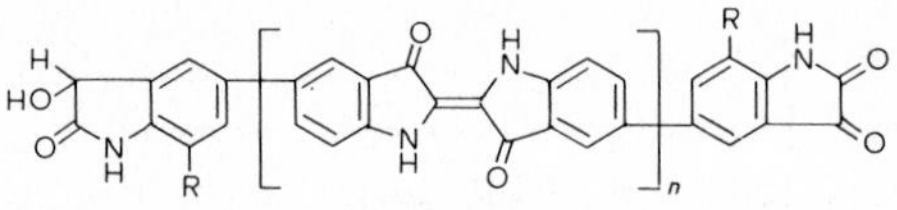

(26) R = H
(27) R = Me

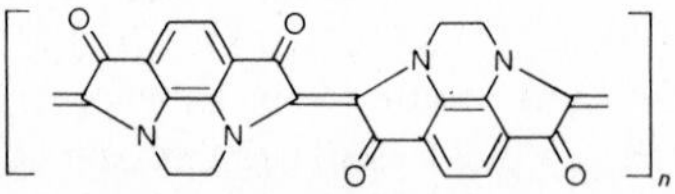

(28)

(x) Minor amounts of thioindigo (36) are often observed in the reaction products of thiochromone and thiochromanone derivatives (R.M. Christie *et al.*, J. chem. Res., 1980, 8), but an 80% yield may be obtained when 2,3-dibromo (or chloro) thiochromone-*S*-oxide (29) is heated with sodium acetate (2 mol. equiv.) in boiling acetic acid. No thioindigo is formed if a 2,3-dihalogenothiochromone is used as starting material (N.E. MacKenzie and R.H. Thomson, Chem. Comm., 1980, 559). The following scheme is proposed to account for the generation of the dyestuff. Nucleophilic displacement of the halogen at C-2 affords the acetate (30) which rearranges via the intermediate (31) to the *S*-acetate (32). Intramolecular displacement of the acetate group then occurs to yield the thiiranium species (33) which reacts with water to give thioindoxyl-2-carboxylic acid (34). Decarboxylation generates 2-bromothioindoxyl (35) which is a known precursor of thioindigo.

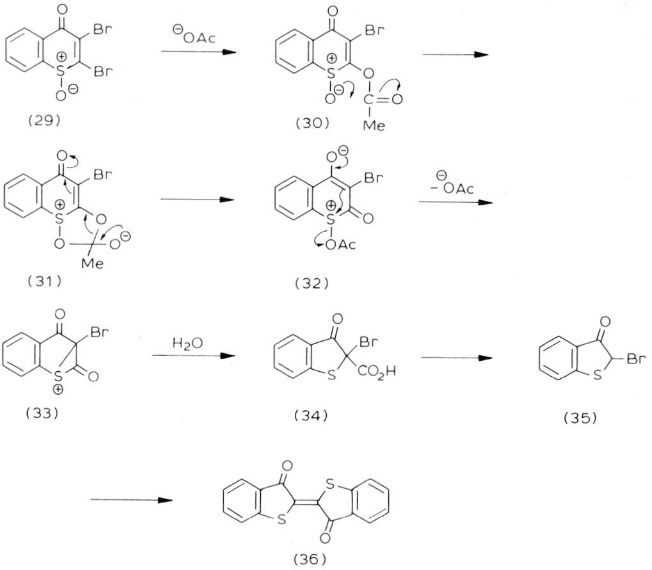

(29) (30) (31) (32) (33) (34) (35) (36)

(b) *Spectroscopy and* E-Z *isomerism of indigo and thioindigo derivatives*

(i) The properties of the indigo chromophore have been the subject of much theorectical and experimental commentary in the last decade principally through the researches of groups lead by Lüttke and by H. Bauer. Some of these studies are to be found in papers already cited, but a useful review is also available (Bauer and G. Pfeiffer, Chem. Zeit., 1976, 100:373).

(ii) The electronic absorption maxima of indigo in various media have been compared (A.R. Monahan and J.E. Kuder, J. org. Chem., 1972, 37:4182). The lowest π-π^* transition occurs at 590 nm (CCl_4), 604 nm ($CHCl_3$), 610 nm (EtOH), 640 nm (amorphous solid), 668 nm (crystalline solid). These differences are shown to be due to variations in hydrogen bonding interactions, and quantum chemical calculations support the conclusion that in the solid state associated species occur. It is pertinent to note that in related structures where the NH groups are replaced by O, S or Se similar changes in spectral properties are not observed and indigoid derivatives bearing bulky substituents at positions 4, 5 and/or 7 do not show hydrogen-bonding shifts between solution and solid phases. In the latter case the size of the substituents inhibits interactions between NH and CO groups in neighbouring molecules.

(iii) The photochemical *cis-trans*-isomerization of *N,N'*-diacetylindigo (1, R=Me) has been studied extensively (W.R. Brode, E.G. Pearson and G.M. Wyman, J. Amer. chem. Soc., 1954, 76:1034; Wyman and A.F. Zenhäusern, J. org. Chem., 1965, 30:2348) and thioindigo (D.A. Rogers, J.D. Margerum and Wyman, J. Amer. chem. Soc., 1957, 79:3464; A.J. Henry, J. chem. Soc., 1946, 1156).

Although *cis*-thioindigo can be isolated by chromatography of the photostationary *cis-trans*-mixture, *cis-N,N'*-diacylindigos are normally very unstable and readily change into the corresponding *trans*-isomers. However, Y. Omote, S. Imada and H. Aoyama (Chem. Ind., 1979, 415; Bull. chem. Soc. Japan, 1979, 52:3397) have observed that the *cis*-isomers of *N,N'*-diacetyl, distearoyl, dibenzoyl and 3,5-dinitrobenzoyl-indigo crystallise from solutions of the corresponding *trans*-isomers that have been irradiated with ultraviolet light. The relative rates of the *cis*→*trans* isomerism of these four *cis-N,N'*-diacylindigos are 7.9, 11.0, 1.0 and 5.1 respectively. Physical data for the isomers are summarised in table 1.

TABLE 1

Properties of *cis*- and *trans*-isomers of *N*,*N*'-diacylindigo (37)

R		mp (decomp)°	$\nu^{KBr}_{C=O}$ cm^{-1}	$\lambda^{C_6H_6}$	ε_{max}
CH_3	*cis*	244-246	1720, 1690	438	4500
	trans	256-257	1690, 1680	562	7000
$CH_3(CH_2)_{16}$	*cis*	101-103	1720, 1700	437	3900
	trans	101-102	1690, 1670	567	7100
C_6H_5	*cis*	246	1725, 1675	460	3900
	trans	256-257	1695, 1660	574	7700
$3,5-(NO_2)_2C_6H_3$	*cis*	205-210	1730, 1660	436	2800
	trans	249-252	1700, 1680	552	6200

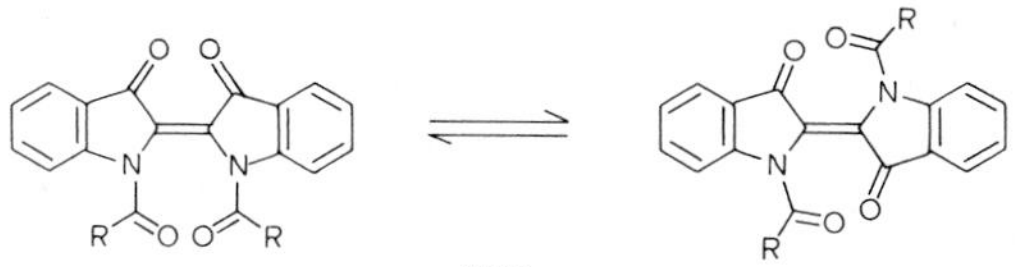

cis-form (37) *trans*-form

(iv) R. Hasenkamp, V. Luhmann and Lüttke (Ber., 1980, 113: 1708) have prepared a number of new thioindigo dyes with interrupted peripheral conjugation. As expected the electronic absorption maxima of these compounds occur at shorter wavelengths than those of the parent compound although the extinction coefficients are similar. Examples of these compounds are collected in table 2 where some of their spectral properties are compared.

TABLE 2

Some new derivatives of thioindigo

Structure	m.p.°	λ_{max}(log ε)nm	ν C=O	ν C=C cm^{-1}
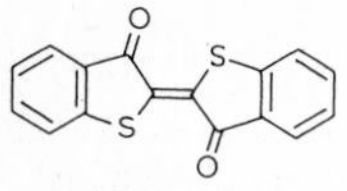 (thioindigo)	>280	541 (4.19)	1658	-
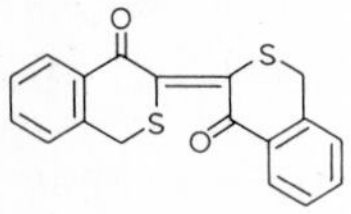	235	477 (4.03)	1627	-
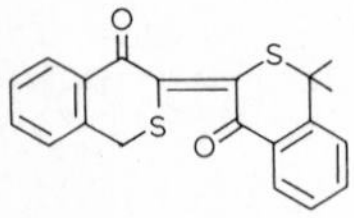	186	454 (3.96)	1658	-
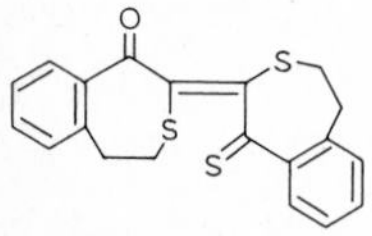	213	388 (3.97)	1676, 1637	1500

(c) *Tyrianpurple and its biological precursors*

The dyestuff Tyrian purple has a very long history. Members of the genera *Murex* from the families *Muricidae* and *Thaisidae* concentrate precursors of the dye in their hypobrachial glands. The occurrence of the dye or its precursors is not restricted to gastropod molluscs, however, since Tyrian purple has been identified in the hemichorate *Ptychodera flava layanica* Spengel(T. Higa and P.J. Scheuer, Heterocycles, 1976, 4:227). J.T. Baker and M.D. Sutherland (Tetrahedron Letters, 1968, 43) and Baker and C.C. Duke (Austral. J. Chem., 1973, 26:2153) have identified the precursor of Tyrian purple in *Dicathais orbita* Gmelin as tyrindoxyl sulphate (38) and the counter cation has been shown to be choline, or its ester, in *D. orbita* and *Mancinella keineri* Deshayes (Baker and Duke, Tetrahedron Letters, 1976, 1233).

The *in vivo* transformation of tyrindoxyl sulphate involves an enzymic hydrolysis to give tyrindoxyl (39) which in turn is oxidised to the corresponding dehydro derivative (40). This product and tyrindoxyl are assumed to form a quinhydrone-type complex (tyriverdin), i.e. a charge transfer complex with the elemental composition $C_{18}H_{14}Br_2N_2O_2S_2$, $\frac{1}{2}H_2O$ or $C_{36}H_{28}BrN_4O_4S_4$, H_2O, which is the immediate precursor of Tyrian purple (Baker, Endeavour, 1974, 32:11).

$OSO_3^{\ominus}$ SMe Br N H (38) —sulphatase→ OH SMe Br N H (39) —[O]→

O SMe Br N (40) - - - - - - → O H N Br Br N H O Tyrian purple

Tyriverdin can be isolated (Baker, Pure appl. Chem., 1976, 48:38), but it fails to exhibit a molecular ion peak in conventional mass spectrometric analysis. However, field desorption techniques can be applied and lead to the molecular formula $C_{18}H_{14}Br_2N_2O_2S_2$ (C. Christopherson *et al.*, Tetrahedron Letters, 1977, 1747; Tetrahedron, 1978, 34:2779). On this basis tyriverdin is allocated structure (42), a conclusion supported by the fact that it is formed when 2,2'-diacetoxy-6,6'-dibromoindigotin (41) reacts with methanethiol in pyridine solution containing triethylamine.

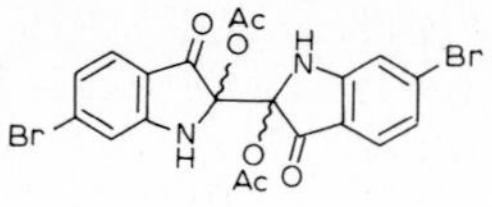

(41) 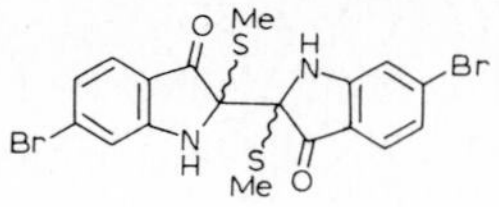 (42)

5,5',7,7'-Tetrabromo-6,6'-dimethoxyindigotin (43), m.p. >300°, and 5,6',7-tribromo-6-methoxyindigotin (44), m.p. >300° also occur in P. *flava* (Higa and Scheuer, *loc. cit.*) accompanied by 5,7-dibromo-6-methoxyindole (45).

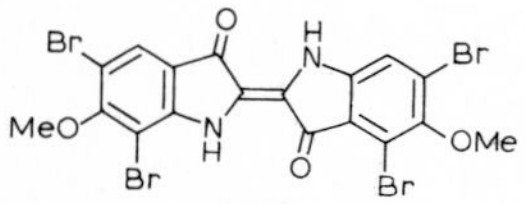

(43) 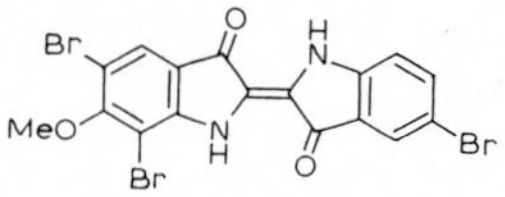 (44) 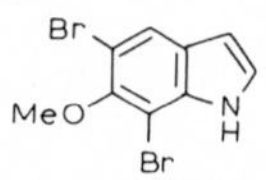 (45)

(d) Deoxyindigo (structural reassignment)

Deoxyindigo, obtained by treating indigo with hydrazine and sodium hydroxide in ethanol followed by aeration, was assigned structure (46) by W. Borsche and R. Meyer (Ber., 1921, 54:2854). A secondary product given the name dehydrodeoxyindigo and thought to have the orthoquinonoid structure (47) was also isolated in a similar reaction conducted by P. Seidel (*ibid.*, 1944, 77:788; 1950, 83:20).

These compounds have been re-examined by J. Bergman, B. Egestad and N. Eklund (Tetrahedron Letters, 1978, 3147). Since deoxyindigo was isolated directly from an alkaline medium, it was argued that an anionic form would be favoured. Such a salt should be very susceptive to oxidative coupling and a high resolution mass spectrometric study of deoxyindigo showed that in reality deoxyindigo has the molecular formula $C_{32}H_{20}N_4O_2$.

The structure has been reformulated as (50, R=H) and confirmed by an independent synthesis: the 3,3'-biindoyl lithium (49) reacts with dehydroindigo (48) to form the di-*N*-sulphonyl derivative of deoxyindigo (50, R=SO_2Ph). This, when hydrolysed, affords deoxyindigo.

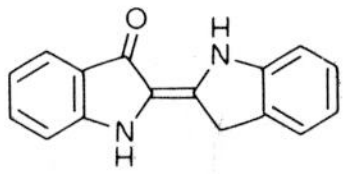

(46)

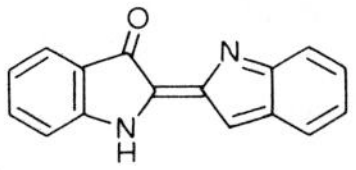

(47)

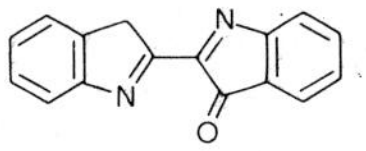

(48)

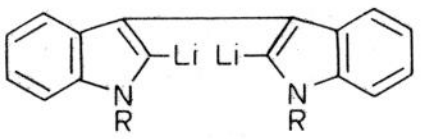

(49)

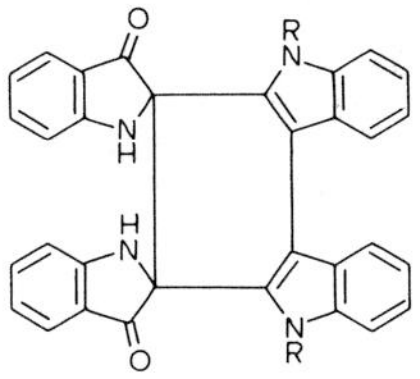

(50)

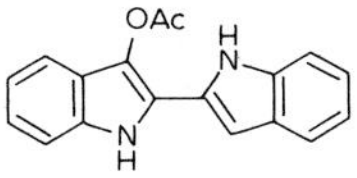

(51)

The *O*-monoacetate (51) of the structure originally proposed for deoxindigo is obtained by the addition of acetic anhydride (air being excluded) to Borsche's reaction mixture. Whereas addition of allyl chloride gives the allyl derivative (52), m.p. 171-173°. Interestingly the isomer (54) has been obtained by the addition of allyl bromide to an alkaline solution 2-(3-indolyl)indoxyl assumed to contain the anion (53) (E. Houghton and J.E. Saxton, J. chem. Soc. (C), 1969, 595).

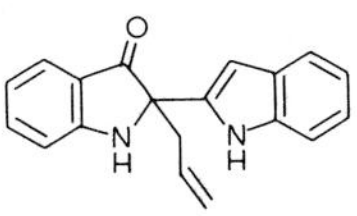

(52)

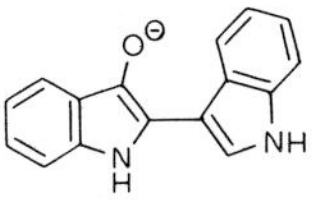

(53)

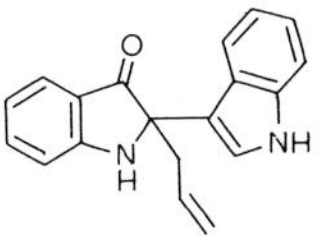

(54)

(e) *Reaction between indigo and hydrazine*

When indigo reacts with anhydrous hydrazine at 35° a product $C_{16}H_{13}N_5O$, m.p. 227-229°, is formed, whereas at 100° a compound $C_{16}H_{12}N_4O$, m.p. 218-220°, is obtained. An X-ray diffraction analysis shows that the latter compound has structure (55, R=NH_2). Treatment of this compound or the lower temperature product with Raney nickel gives the *N*-deamino derivative (55, R=H), m.p. 318-320°. The product of the low temperature reaction is assumed to be the zwitterion (56).

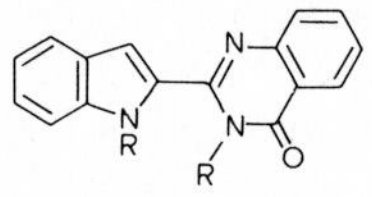

(55)

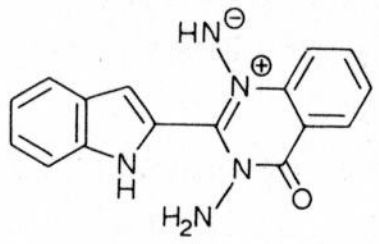

(56)

(f) *Indigo diimine*

"Indigo diimine" (W. Madelung, Ber., 1913, 46:2259; Ann., 1914, 405:58) does not exist as the symmetrical form (57), but rather as the tautomeric structure (58), both in solution and in the solid state (H. Sieghold and Lüttke, Angew. Chem. internat. Edn., 1975, 14:52).

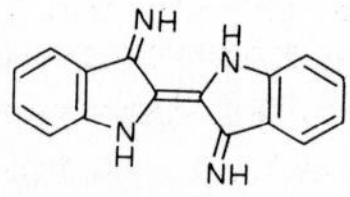

(57)

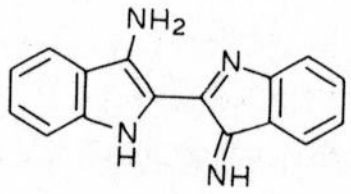

(58)

Chapter 15

CYANINE DYES AND RELATED COMPOUNDS

D.J. FRY

Since the contribution in the Second Edition was written most of the work on cyanine dyes has been concerned with investigations of their physical properties rather than on the synthesis of new dyes or on new synthetic methods. The main reason for this emphasis arises from attempts to explain how the dyes act as spectral sensitizers for silver halide and other semi-conductors. An excellent review of the chemistry and physical properties of cyanine dyes has been written by D.M. Sturmer ("The Chemistry of Heterocyclic Compounds", ed. A. Weissberger and E.C. Taylor, John Wiley and Sons, New York, 1977, Chapter VIII, Syntheses and Properties of Cyanine and Related Dyes). Much the same ground is covered by Sturmer and D.W. Heseltine in "The Theory of the Photographic Process", 4th Edition (ed. T.H. James, Macmillan Publishing Company, New York, 1977, Chapter 8, Sensitizing and Desensitizing Dyes). A good review on the physical properties of cyanine dyes has been produced by S. Dähne (Phot. Sci. Eng., 1979, 23, 219) and A.H. Herz has reviewed the aggregation of dyes with particular reference to cyanines (Adv. Colloid. Interface Science, 1977, 8, 237). Cyanine dyes from thiazolium compounds are reviewed by H. Larivé and R. Denilauler ("The Chemistry of Heterocyclic Compounds", ed. A. Weissberger and E.C. Taylor, Vol. 34, Thiazole and its Derivatives, Part 3, John Wiley and Sons, New York, 1979, page 23).

Most of the work published in patents deals with special combinations of substituents in known dye structures, or with combinations of dyes, to give improved technological properties, and is not of great scientific importance.

S. Dähne (Science [Washington], 1978, 199, 1163) has defined the energetically stabilized polymethine state where a chain of n atoms is occupied by $(n \pm 1)$ π-electrons, with the electrons being delocalised as symmetrically as possible across the chain (1).

$$\cdots\cdots\ (n \pm 1)\ \pi \cdots\cdots$$
$$X - (CR)_{n-2} - X^1$$

(1)

In the cyanine dyes $X = X^1 = N^{\frac{1}{2}+}$, in merocyanines $X = N^{\frac{1}{2}+}$, $X^1 = O^{\frac{1}{2}-}$ and in oxonols $X = X^1 = O^{\frac{1}{2}-}$.

There is a theoretical basis for this definition of the polymethine state (J. Fabian and H. Hartmann, J. molec. Structure, 1975, 27, 57). The polymethine state is associated with intense light absorption at long wavelengths and with an alternation of π-charge density distribution along the chain. On excitation, the charge density distribution is reversed (Dähne, S. Kulpe, K.-D. Nolte and R. Radeglia, Phot. Sci. Eng., 1974, 18, 410).

The alternation of π-charge will lead to a rehybridization of the σ-electron system of individual atoms to different degrees and cause an alternation in bond angles. On excitation of the molecule by light the change in the alternation of charge density will also lead to a change in the bond angles and the molecule will bend. The alternation of the bond angles in the ground state was first shown by X-ray analysis (S. Kulpe *et al.*, J. prakt. Chem., 1973, 315, 865) and confirmed by ^{13}C-H coupling constants in the nmr spectrum (R. Radeglia, *ibid.*, 1973, 315, 1121). Although there is an alternation of bond angles, the bond lengths are equal. These data are shown for the simple bis-dialkylamino cyanine cation (2).

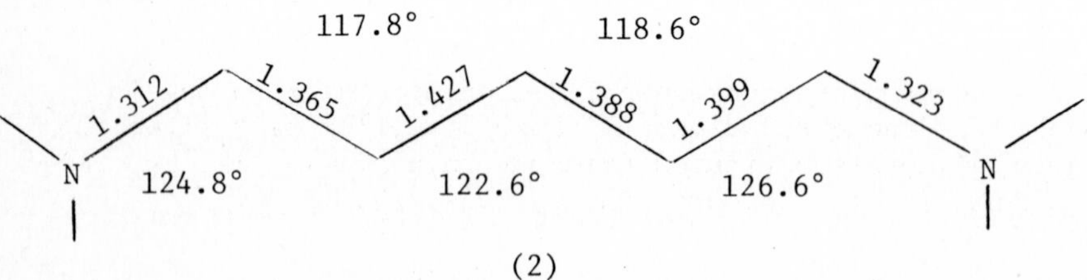

(2)

(bond lengths are in Ångstrom units)

1. *Cyanines*

1.1 *Fluoro-substituted dyes*

Cyanines containing two or three fluorine atoms in the chain are prepared by the following scheme (L.M. Yagupol'skii, Yu.L Yagupol'skii, Zh. org. Khim., 1977, 13, 1996).

S CH + ClF = CF_2 → S C - CF = CF_2 N Me N Me (3)

(3) + S C - CH_2R N Me BF_4^-

$4\text{-}CH_3C_6H_4NEt_2/CH_3CN$

S C - CHR = CF - CF = C S N N Me BF_4^- Me

(4)

R = H, λ_{max} 545 nm

R = F, λ_{max} 578 nm

Condensation between 1,2-dichloro-hexafluorocyclopentene and a 2,3-dimethyl benzothiazolium salt gives the tetrafluoro-substituted bridged-chain dye (5) (A. Ya. Il'chenko *et al.*, Zh. org. Khim., 1979, 15, 1532).

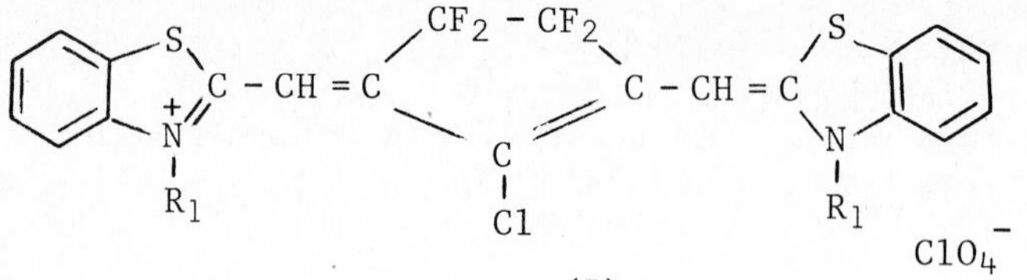

(5)

Substitution of F for H gives a 10 nm bathochromic shift. A meso-F-substituted pentamethine dye is obtained by the reaction between a quaternary salt with an active methyl group and the bis-aldehyde $O=CH-CF-CH=O]^{-}Na^{+}$. An alternative synthesis is to use $Me_2\overset{+}{N}=CH.CFCHO$ as the intermediate (M.M. Kul'chitskii, Ukr. Khim. Zh. [Russ. Ed.] 1979, 45, 872).

1.2 *Long-chain dyes*

The very long chain trideca- and pentadeca-methine dyes (6) are obtained by reaction between the appropriate heterocyclic quaternary salt and the bis-di-*N*-alkylanil salts from undecatetraenedial and tridecapentaenedial respectively. The condensations are effected in acetic anhydride in the presence of a tert. aliphatic amine at a low temperature (N.V. Monich *et al.*, USSR Pat. 503,408/1973; Chem. Abs., 1977, 87, 103371).

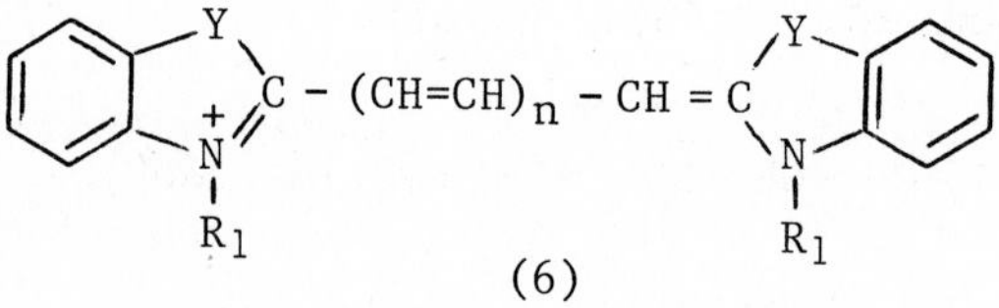

(6)

Y = S, Se R_1 = alkyl, aryl n = 6 or 7.

The bridged-chain dyes (8) are obtained from the intermediate (7) synthesized via Vilsmeier formylation of cyclopentanone, cyclohexanone or cycloheptanone (Yu. L. Slominskii *et al.*, Zh. org. Khim., 1978, 14, 2214).

$$PhN^{+}H=CH-C(=C(Cl)-)(CH_2)_n-C=CHNHPh \quad Cl^-$$

n = 2 - 4

(7)

$$\text{benzazolium}(Y,\ N^+R)-CH=CH-CH... \quad ClO_4^-$$

(8)

The presence of the meso-chlorine atom gives a bathochromic shift of 17-31 nm compared to the meso-H dye. The chlorine atom is reactive because of the charge alternation along the chain and readily undergoes nucleophilic displacement, e.g. by $CH_2(CN)_2$.

Another intermediate for bridged-chain dyes is obtained by the reaction scheme (9) (Y. L. Slominskii *et al.*, Ukr. Khim. Zh. [Russ. Ed.], 1980, 46, 61).

$$PhNMe-CH=CH-CH=\overset{+}{N}MePh \;\; I^- \; + \text{ cyclic aldehyde}$$

$$\downarrow$$

$$Ph\overset{+}{N}Me=CH-CH-C(CH_2-X-CH_2)C=CH-NMePh$$

X = direct link, CH_2, O

(9)

The resultant dyes (10) show a bathochromic shift of 45, 128 and 15 nm for X = direct link, CH_2 and O respectively.

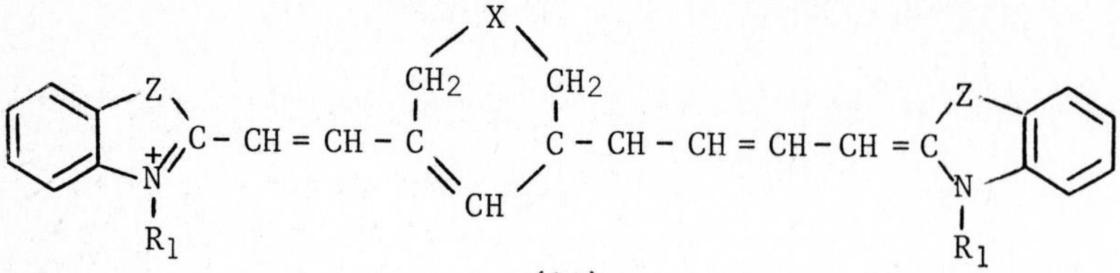

(10)

X = direct link, CH_2, O   Z = S, CMe_2

Long chain dye bases (11) with a benzene ring in part of the chain are obtained by the Wittig reaction (I.N. Zhmurova *et al.*, Zh. org. Khim., 1975, 11, 2160).

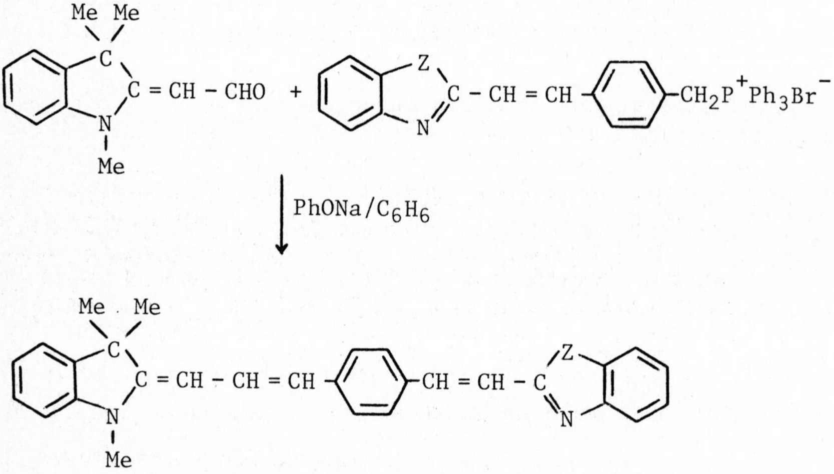

$PhONa/C_6H_6$

(11) Z = S or -CH = CH-

The base (11) is unstable to acid but can be quaternized to form the corresponding dye salt.

1.3 *Physical properties*

Much attention has been given to the physical properties of cyanine dyes in order to elucidate the mechanism of spectral sensitization. As a consequence, it is now the general opinion that the sensitization process involves the direct transfer of an electron from the dye to the conduction band of the semi-conductor (S. Dähne, Phot. Sci.

Eng., 1979, 23, 219) although the alternative mechanism of a radiationless transfer of energy may operate in special circumstances (R. Steiger *et al.*, Phot. Sci. Eng., 1980, 24, 185).

The most important data to be deduced for deciding the mechanism of spectral densitization are estimates of the relative location of the energy levels of the dye molecules which can be obtained from half-wave potentials, spectral data and quantum mechanical calculations. The highest occupied molecular orbital (HOMO) can be correlated with the anodic oxidation potential or the ionization energy and the lowest unoccupied molecular orbital (LUMO) can be correlated with the cathodic reduction potential or the electron affinity. The energy differences between these two levels is the excitation energy which can be obtained from spectral data. Although the spectral data can be determined accurately, experimental difficulties attend the determinations of half-wave potentials, ionization energies and electron affinities.

Table 1 shows the redox potentials (against the standard hydrogen electrode) of a number of dyes and the correlation with the excitation energy of the light absorption of longest wavelength (all dyes (12) carried N-ethyl groups on the heterocyclic ring). (R.F. Large, "Photographic Sensitivity", ed. R.J. Cox, Academic Press, London 1973, p. 241).

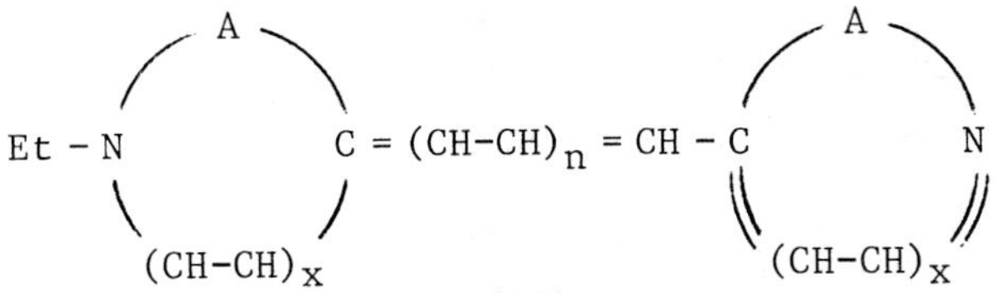

Et - N(A)(CH-CH)$_x$ C = $(CH\text{-}CH)_n$ = CH - C(A)(CH-CH)$_x$ N

(12)

In all the 2,2'dyes x = 0

2,2'-thia(oxa, selena) cyanines A = (benzo ring with X) X = S, O, Se

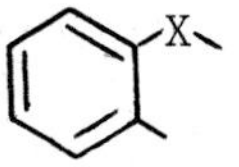

2,2'-cyanines A = (benzo ring with CH=CH)

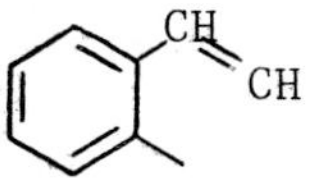

4,4'-cyanines A = 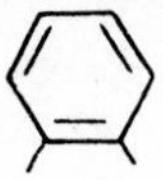 x = 1

Et N Cl

5,6-dichlorobenzimida-2,2'-cyanines A =

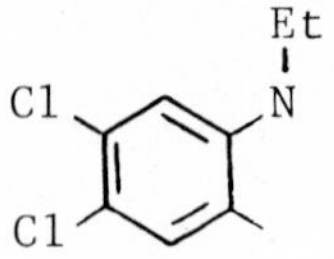

Cl

TABLE I

Dye (12)	$E_{ox}V$	$E_{red}V$	$E_{ox}-E_{red}$	$E_{h\nu}eV$	$E_{h\nu} \overset{\Delta}{-} E_{redox}$
2,2'-thiacyanines					
n = 0	>1.2	-1.38	>2.58	2.94	<0.36
n = 1	0.75	-1.00	1.75	2.23	0.48
n = 2	0.48	-0.83	1.31	1.91	0.50
n = 3	0.26	-0.72	0.98	1.63	0.65
2,2'-cyanines					
n = 0	0.99	-1.03	2.02	2.34	0.32
n = 1	0.58	-1.10	1.68	2.05	0.37
n = 2	0.28	-0.98	1.26	1.75	0.49
4,4'-cyanines					
n = 0	0.64	-1.14	1.78	2.09	0.31
n = 1	0.19	-0.99	1.18	1.76	0.58
2,2'oxacyanines					
n = 0	>1.0	-1.69	>2.69	3.33	<0.64
n = 1	0.94	-1.26	2.20	2.56	0.36
n = 2	0.49	1.16	1.65	2.14	0.51
5,6-dichloro-2,2'-benzimida cyanines					
n = 0	0.43	-1.84	2.27	2.51	0.24
n = 1	0.58	-1.50	2.08	2.41	0.33
n = 2	0.27	-1.28	1.55	2.06	0.51
2,2'-selenocyanines					
n = 0	>1.0	-1.58	>2.58	2.89	<0.31
n = 1	0.62	-1.02	1.64	2.18	0.54
n = 2	0.46	-0.84	1.30	1.88	0.58
n = 3	0.30	-0.64	0.94	1.61	0.67

Table II gives some experimental ionisation energies (R.C. Nelson and R.G. Selsby, Phot. Sci. Eng., 1970, 14, 342) and table III some electron affinities (J.W. Trusty and R.C. Nelson, Phot. Sci. Eng., 1972, 16, 421).

Table II

Dye	Ionisation energy eV
2,2'-cyanine	
n = 0	5.20
n = 1	4.66
n = 2	4.49
2,2'-thiacyanine	
n = 2	4.61
n = 3	4.49

Table III

Dye	Electron affinity eV
2,2'-thiacyanine	
n = 1	2.70
n = 2	2.73
n = 3	2.74

All the dyes carry N-ethyl groups. In the thiacyanine series the differences between ionization energy and electron affinity are 1.88 eV ($n = 2$) and 1.75 eV ($n = 3$) and these correlate well with the energy difference from spectral data of 1.91 eV ($n = 2$) and 1.63 eV ($n = 3$).

The protonation of cyanines has been studied by A.H. Herz (Phot. Sci. Eng., 1974, 18, 207). Some of his data are given in Table IV for a series of 2,2'-thiacyanines (13).

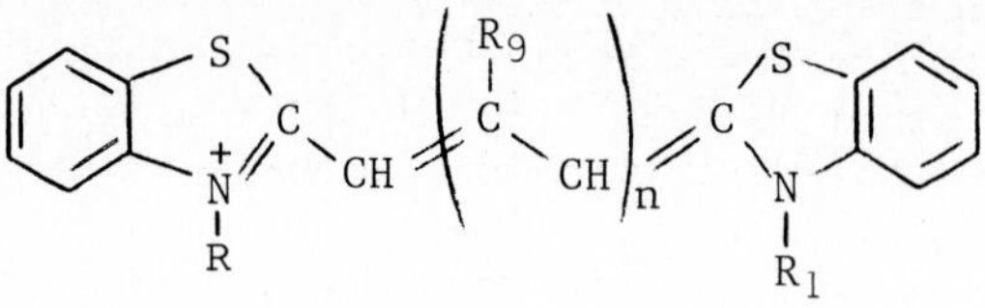

(13)

Table IV

n	R_9	R	R_1	$pK_a \pm 0.15$	λ_{max} nm
0	H	Et	Et	- 2.1	420
1	H	Et	Et	0.3	552
2	H	Et	Et	2.1	646
3	H	Et	Et	3.9	755
1	Me	Et	Et	3.0	539
1	Me	Et	$(CH_2)_4SO_3^-$	2.9	542
1	Me	Et	$(CH_2)_2COO^-$	2.9	539
1	Ph	Me	Me	1.8	554

Benzimidazole dyes (14) are more highly basic than the benzothiazole analogues (Table V).

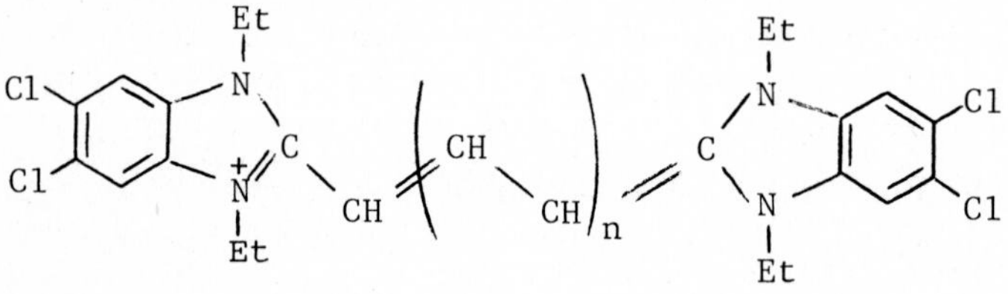

(14)

TABLE V

Dye (14)	pK_a	λ_{max}
n = 1	7.9	507
n = 2	8.5	610
n = 3	∿10.2	710

Although these data show that the shift in λ_{max} is proportional to the change in the pK_a for each vinylogous series, this is not always the case. An exception is found in the 2,2'-cyanine series and this may be due to the twisted nature of the shortest chain dye with a concomitant increase in the basicity. When dye molecules are aggregated, as is frequently found when they are adsorbed to a silver halide surface, the basicity is decreased.

Quantum-mechanical calculations have given values for redox potentials, ionization energies and electron affinities which agree well with the experimental values (T. Tani, Phot. Sci. Eng., 1972, 16, 258; Y. Ferre, H. Larivé and E.J. Vincent, *ibid.*, 1974, 18, 457; K.D. Nolte and S. Dähne, J. prakt. Chem., 1976, 318, 993).

X-ray studies of dye crystals have been reviewed by D.L. Smith (Phot. Sci. Eng., 1974, 18, 309). In general the crystals are formed of planes of dye molecules packed plane to plane and end to end like a tilted pack of cards: the inter-planar distances are 3.3 to 3.6Å.

Variable temperature Fourier transform proton nmr has been used to study further the *cis-trans* equilibria of meso-substituted trimethine cyanines (P.M. Henricks and S. Gross, J. Amer. chem. Soc., 1976, 98, 7169). In general the *mono-cis* form is preferred. The normal state of bis(3-ethylbenzoxazole-2)pentamethine cyanine is the *di-cis* form (15), which is converted to the all-trans form (16) on irradiation (G.R. Fleming *et al.*, Chem. Phys. Letters, 1977, 49, 1).

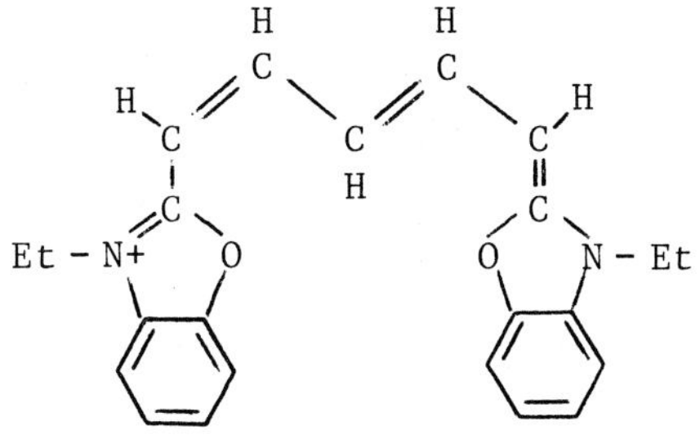

(15)

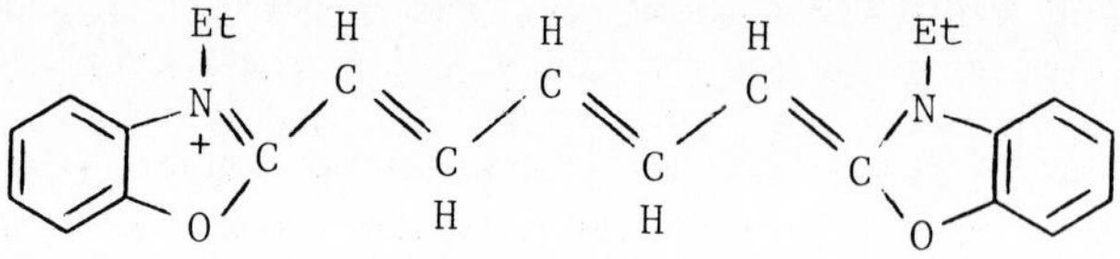

(16)

T. Tani and Y. Sano have studied the electron spin resonance properties of irradiated cyanine dyes (J. phot. Sci., 1979, 27, 231). No signals are detected when the dye is in a gelatin layer but they are found for dyes adsorbed to silver halide. The intensity and the decay time of the signals increase with increase in the height of the HOMO of the dye. The signals are ascribed to the formation of dye positive holes, presumably after donation of the excited electron to the silver halide lattice.

Electrophoretic studies show that anionic and zwitterionic dyes having acid-substituted N-alkyl groups are adsorbed to silver halide with the acid groups towards the solution phase (W.L. Gardner, D.P. Wrathall and A.H. Herz, Phot. Sci. Eng., 1977, 21, 325).

1.4 *Reaction of cyanines*

The alternation of π-electron density along the chain of cyanine dyes (17) leads to electrophilic reactions at atoms with high electron density and nucleophilic reactions at atoms with low electron density.

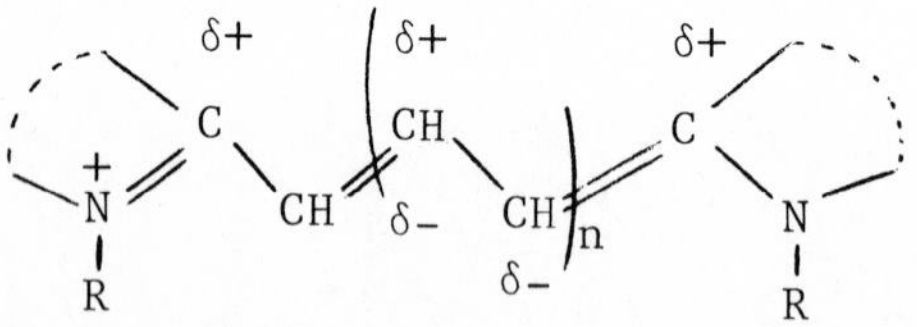

(17)

Examples of electrophilic reactivity are the H - D exchange of the α-methine groups (R. Radeglia *et al.*, J. prakt. Chem., 1978, 320, 539), bromination and nitration at the same positions (D. Lloyd and H. McNab, Angew. chem. internat. Edn. 1976, 15, 459) and the coupling of strongly basic monomethine cyanines (derived from benzimidazole, quinoline or benzothiazole) with diazonium salts of high electrophilicity (A. Klaproth and K. Dimroth, Ann., 1975, 1839).

Examples of nucleophilic reactions are the replacement of a meso-alkylthio group by the residue of an aromatic amine or a ketomethylene compound (C.C.C., 2nd Edition, Vol. IVB, p. 380).

Nucleophilic reactions also result in chain shortening (18) or ring closure to benzene derivatives (19) (H.E. Nikolajewski *et al.*, Angew. Chem. internat. Edn., English, 1965, 5, 1044).

$$\mathrm{Me_2\overset{+}{N}=CH-CH=CH-CH=CH-CH=CH-CH=CH-NMe_2}\quad ClO_4^-$$

$$\Big\downarrow \text{Boil + excess } NHMe_2$$

$$\mathrm{Me_2\overset{+}{N}=CH-CH=CH-CH=CH-CH=CH-NMe_2}\quad ClO_4^-$$

$$+\ \mathrm{Me_2N-CH=CH_2}$$

(18)

$$Me_2N{-}CH{=}CH{-}CH{=}(CH{-}CH{=})_m\overset{+}{N}Me_2$$

m = 2 — heat in aqueous solution

+H⁺

(19)

$-H_2O$

When m = 3, cyclisation leads to cinnamaldehyde (H. Althoff, B. Bornowski and S. Dähne, J. prakt. Chem., 1977, 319, 890).

Photo-oxidation of dyes proceeds with low quantum yield but the reaction can be sensitized with methylene blue. The ease of the reaction generally correlates directly with the electrochemical oxidation potentials but the rate is increased five-fold by meso-alkyl substitution. The products are the ketones (20) and (21) (G.W. Byers, S. Gross and P.M. Henricks, Photochem. Photobiol., 1976, 23, 37).

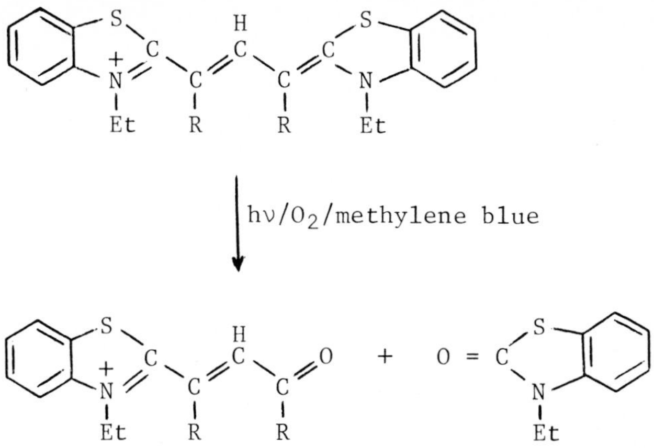

$h\nu/O_2$/methylene blue

(20) (21)

When cyanine dyes are subjected to mass spectrometry the first reaction is the loss of one of the *N*-substituents together with the anion to leave the cyanine dye base: no molecular ion is observed. The chain is then cleaved followed by cyclisation to structures like (22) and (23) (K.K. Zhigulev *et al.*, Chem. Abs., 1975, 82, 126598).

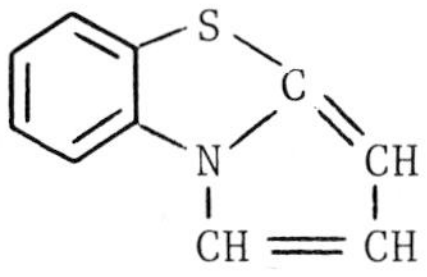

(22) 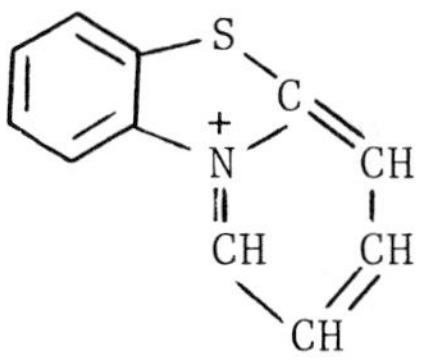 (23)

Cyanine dyes form 1:1 charge transfer complexes with strong electron acceptors like chloranil. These complexes absorb at long wavelengths and have been used to estimate ionization potentials (J. Nys and W. Van den Heuval, Phot. Korr., 1966, 102, 37). Strongly basic benzimidazole trimethine cyanines also form complexes with tetrahalogeno-p-benzoquinone in which five dye molecules are combined with one quinone. ESR measurements reveal the presence of acceptor and donor radicals, suggesting a certain amount of charge separation. Surprisingly, in view of the poor light stability of cyanines, the dye in the 5:1 complex is very stable to light (R. Steiger, J. phot. Sci., 1980, 28, 156).

1.5 *Miscellaneous syntheses and uses*

The synthesis of monomethine cyanines by the action of pentyl nitrite on a quaternized heterocycle having a reactive methyl group was described previously (C.C.C., 2nd Edition, Vol. IVB, 374). J.D. Mee has now described an alternative route to the intermediate 2-hydroxyimino-methyl heterocycle (25). It is of particular value for the synthesis of monomethine cyanines (24) where R and/or R^1 are sulphoalkyl groups (Research Disclosure, 1979, 18234).

Z_1 ring (N^+–R) C–CH_3 + ON–C_6H_4–NMe_2 $\xrightarrow{\text{base}}$ Z_1 ring (N^+–R) C–CH=N–C_6H_4–NMe_2

$\downarrow H_3O^+$

Z_1 ring (N^+–R) C–CH=N–OH $\xleftarrow{Na_2OH.HCl}$ Z_1 ring (N^+–R) C–CHO

(25)

(25) + Z ring (N^+–R^1) C–CH_3

[Z is a ring residue]

$\downarrow$ $(CH_3CO)_2O$, NEt_3

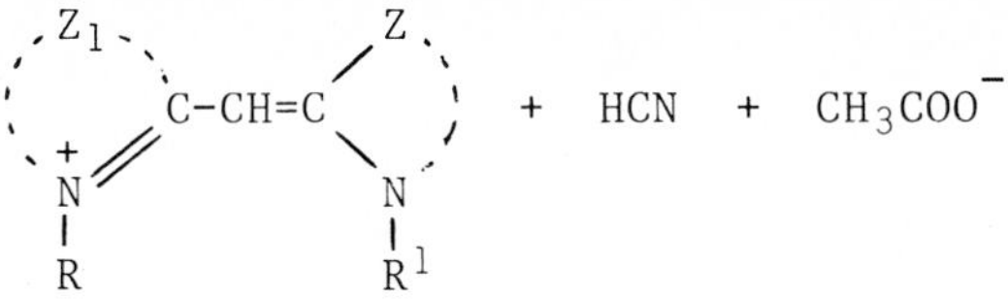

$$\text{(24)}\quad \overset{+}{N}(R)=C(Z_1)-CH=C(Z)-N(R^1) + HCN + CH_3COO^-$$

(24)

Cyanines have been linked to gelatin by making use of one of the groups like 2,4-dichloro-1,3,5-triazine used to link dyes to textiles (R. Steiger *et al.*, B.P. 1,529,201; U.S.P. 4,138,551/1975).

In addition to the widespread use of cyanines as photosensitizers and Q-switched lasers, they have also been patented as antioxidants for methyl linoleate (K. Fukuzumi and N. Ikeda, Japan Pat. 72,42922; Chem. Abs. 1976, 85, 126228).

2. *Acetylenic Dyes*

A double bond in the conjugated chain of a cyanine dye can be replaced by a triple bond (J.D. Mee, J. Amer. chem. Soc., 1974, 96, 4712; J. org. Chem., 1977, 42, 1035). The simplest compound (27) may be prepared by the elimination of hydrogen chloride from the compound (26).

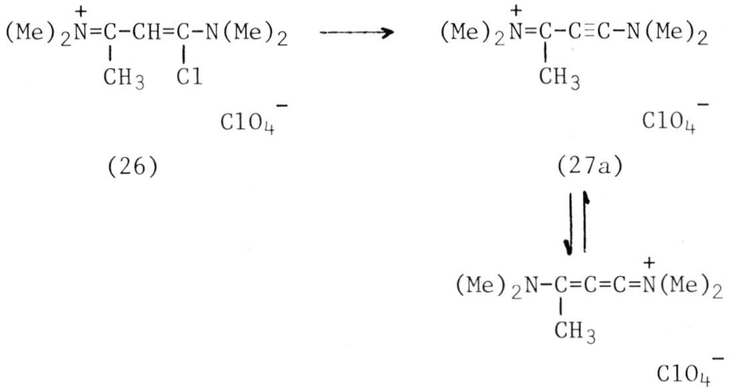

$$(Me)_2\overset{+}{N}=C(CH_3)-CH=C(Cl)-N(Me)_2\ ClO_4^- \longrightarrow (Me)_2\overset{+}{N}=C(CH_3)-C\equiv C-N(Me)_2\ ClO_4^-$$

(26) (27a)

$$\rightleftharpoons (Me)_2N-C(CH_3)=C=C=\overset{+}{N}(Me)_2\ ClO_4^-$$

(27b)

(27) has λ_{max} 309 nm (in CH_3CN).

Analogously, the *meso*-chlorotrimethine dye (28) gives the acetylenic dye (29).

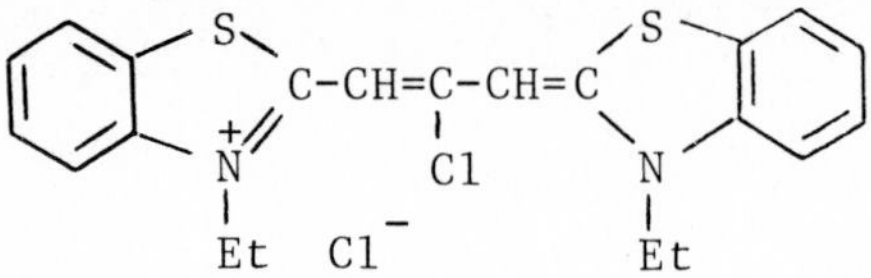

(28)

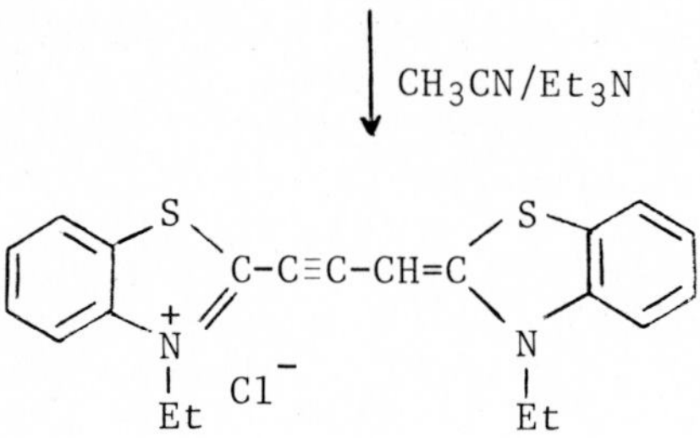

CH_3CN/Et_3N

(29a)

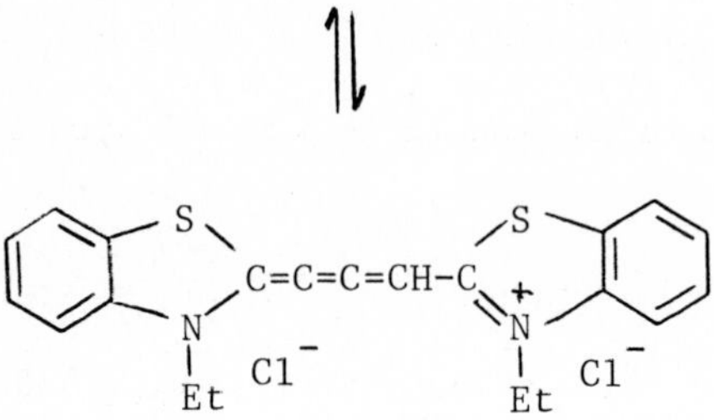

(29b)

The acetylenic structure (29a) is in equilibrium with the cumulene structure (29b). The non-equivalence of these structures results in a hypsochromic shift in the absorption maximum compared with that of the corresponding trimethine dye. The dyes decompose on heating without a definite melting point.

If the *meso*-chloro-substituted dye has two different heterocyclic end groups the elimination of hydrogen chloride takes place in two senses to give different, isomeric, acetylenic dyes. These can be synthesized unambiguously by the following synthetic scheme.

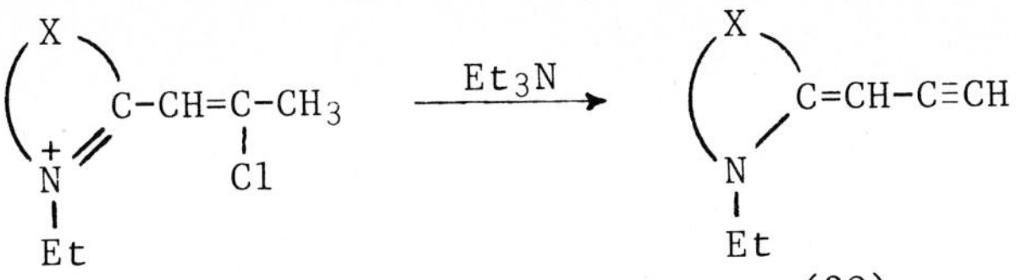

$$\text{X/N}^+(\text{Et})\text{=C-CH=C(Cl)-CH}_3 \xrightarrow{Et_3N} \text{X/N(Et)-C=CH-C}\equiv\text{CH}$$

(30)

$$(30) + \text{Z-C(Y)=N}^+\text{Et} \longrightarrow \text{X/N(Et)-C=CH-C}\equiv\text{C-C(Y)=N}^+\text{Et}$$

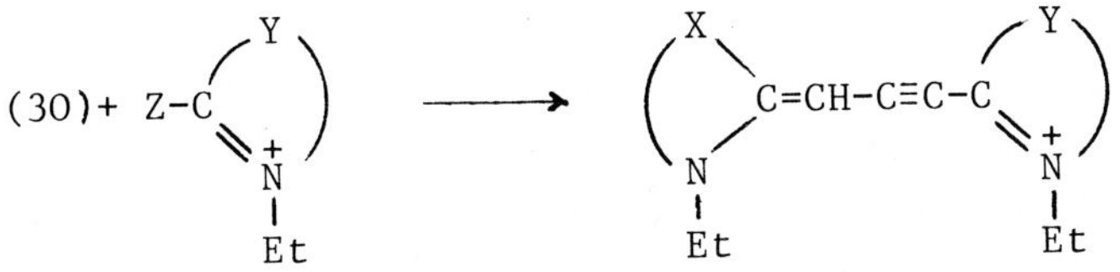

($Z = SO_3^-$, S alkyl)

X and Y complete a heterocyclic ring of the type customarily used in cyanine syntheses. The ability to isolate the intermediate (30) depends on the nature of the ring X. Examples of an isomeric pair of dyes are (31) and (32).

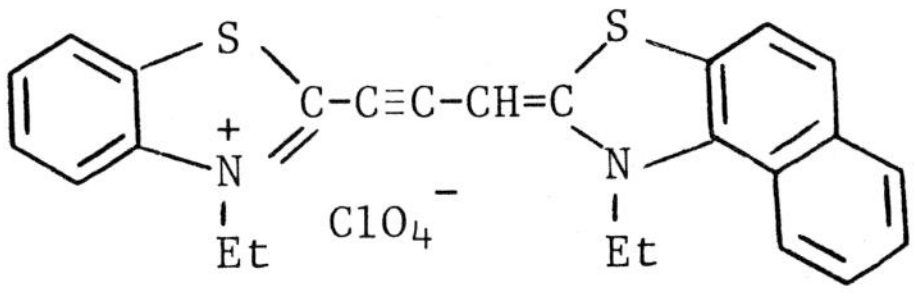

λ_{max} 535 nm (CH_3CN)

ε_{max} 8.4 x 10^4

(31)

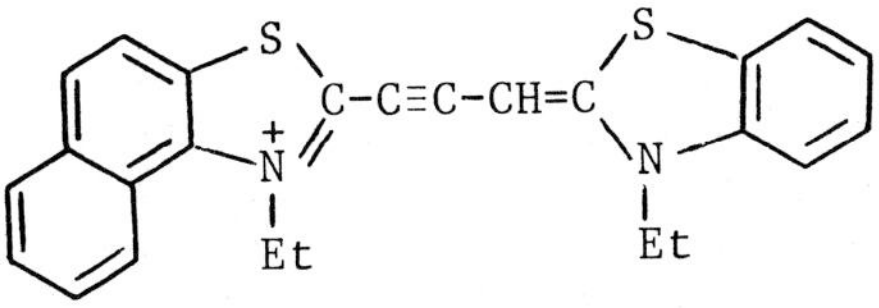

λ_{max} 526 nm (CH_3CN)

ε 8.4 x 10^4

(32)

Addition of hydrogen chloride to the triple bond occurs readily. Under suitable conditions an acid-catalysed interconversion of the two isomeric forms (31) and (32) can take place, e.g. by the action of acetic acid on an acetonitrile solution of the dyes to give an equilibrium mixture of the two forms. This interconversion is presumably achieved via a common dication formed by protonation of either of the isomers.

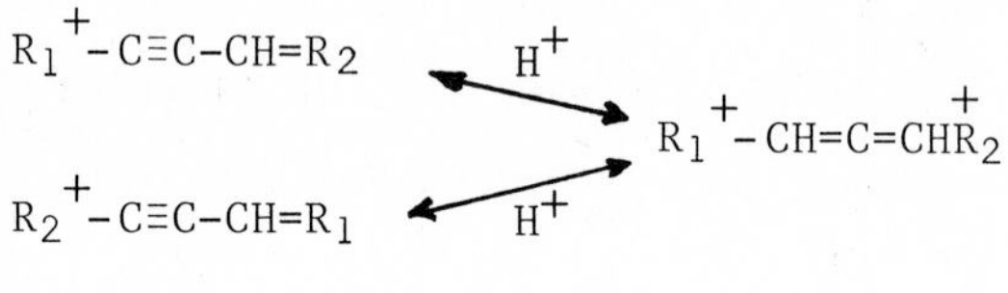

$$R_1^+{-}C{\equiv}C{-}CH{=}R_2 \xrightleftharpoons{H^+} R_1^+{-}CH{=}C{=}CH\overset{+}{R}_2 \xrightleftharpoons{H^+} R_2^+{-}C{\equiv}C{-}CH{=}R_1$$

(R_1 = R_2 = heterocycle)

In the equilibrium mixture, the dye having the shorter λ_{max} is the predominant one (Mee and D.M. Sturmer, J. org. Chem., 1977, 42, 1041).

Other molecules can add to the triple bond such as phenylthiol to give a *meso*-phenylthio-substituted trimethine cyanine or 1,3-diethyl-2-thiobarbituric acid to give the allopolar dye (33).

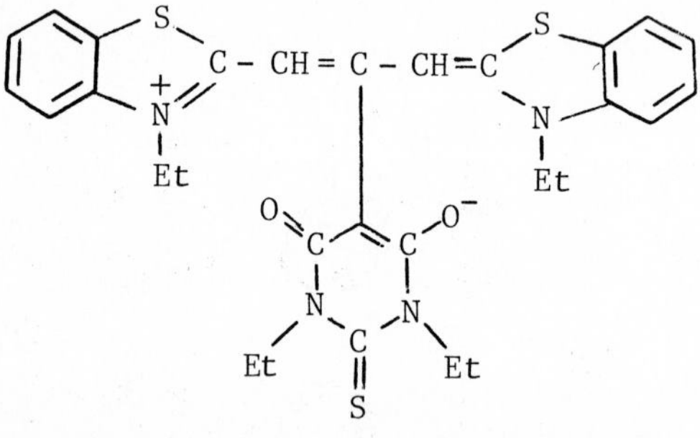

(33)

The acetylenic dyes are spectral sensitizers for silver halide (Kodak, B.P. 1,460,094; U.S.P. 4,025,349/ 1973).

Benzimidazole dyes containing a triple bond in the conjugated chain (36) have been prepared by the reaction between 1,3-dimethyl-2-methylene benzimidazoline (35) and tetrachloromethane in boiling benzene (J. Bourson, Bull.

soc. chim. France, 1975, 644. The reaction is thought to proceed via the intermediate (34).

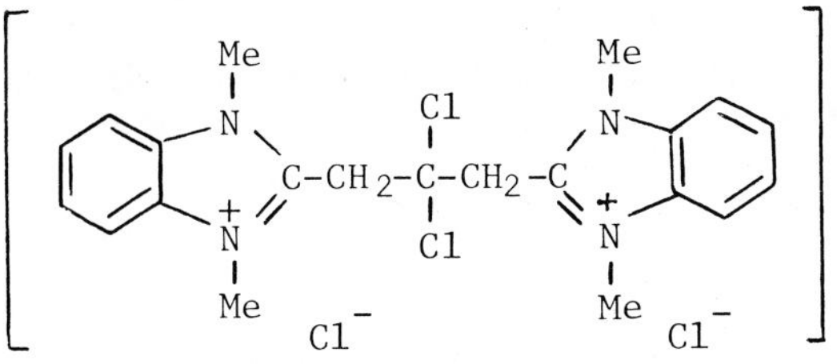

(34)

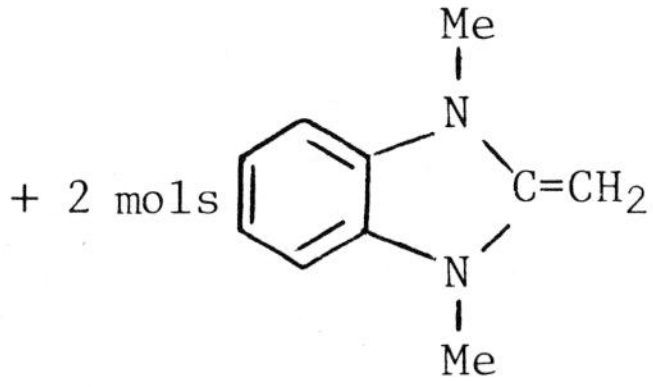

+ 2 mols

(35)

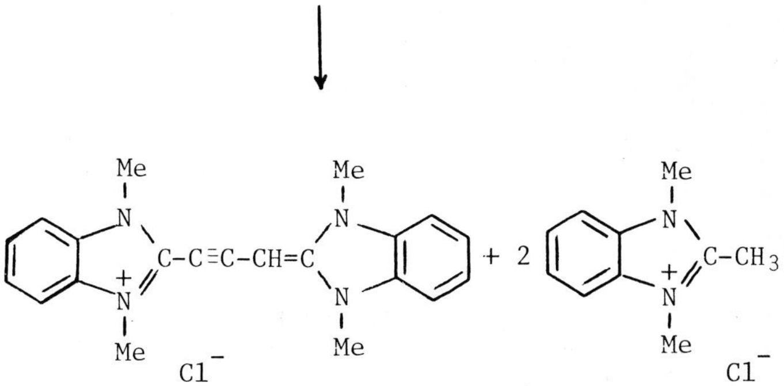

(36)

(36) had λ_{max} 459 nm, ε_{max} 6.21 x 10^4.

3. *Aza-, Phospha- and Arsa-Cyanines*

3.1 *Azacyanines*

Benzimidazole triazatrimethine cyanines (38) have been prepared by the reaction of the 2-azido quaternary salt (37) with sodium azide in dimethyl formamide (H. Balli and R. Maul, Helv., 1976, 59, 148). Some absorption data are given in Table I.

Dyes may be prepared similarly from a 2-azidothiazolium salt (Balli and Löw, Helv., 1976, 59, 155).

Et N R C-N$_3$ N Et —NaN_3/DMF→ R N C=N-N=N-C N Et Et R

(37) (38)

Dye (38)

Table I

R	λ_{max}	$\varepsilon_{max} \times 10^{-4}$
H	430	3.04
Me	436	3.26
CF_3	427	2.96
NO_2	443	3.90

meso-Phenyl tetra-azapentamethine cyanines (39) are green dyes for acrylic fibres (A. Förster *et al.*, E. Ger. Pat. 117081/1974; Chem. Abs. 1976, 85, 22770).

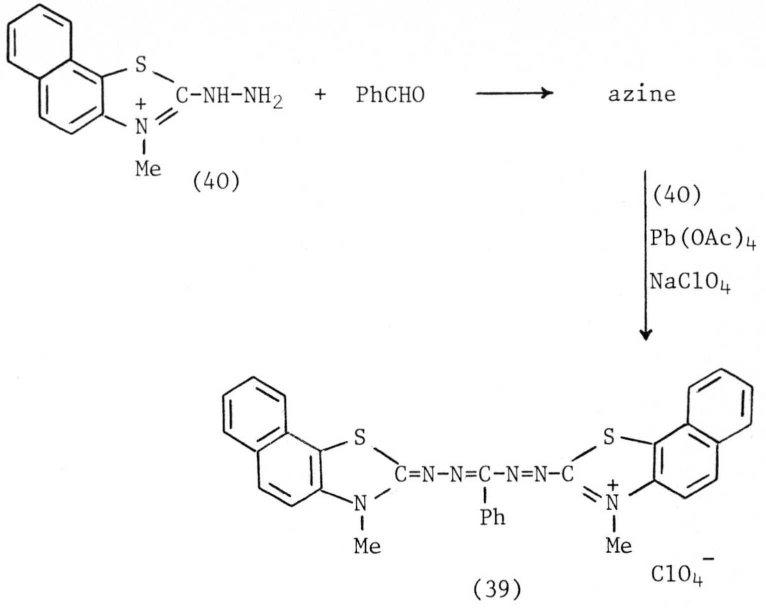

(40) + PhCHO → azine

(40), $Pb(OAc)_4$, $NaClO_4$ ↓

(39) ClO_4^-

meso-Mercaptotetra-azapentamethine cyanine dye bases (41) are prepared by the reaction scheme as shown below (R.G. Dubenko, I.M. Bazarova, P.S. Pel'kis, Chem. Abs., 1972, 76, 101193).

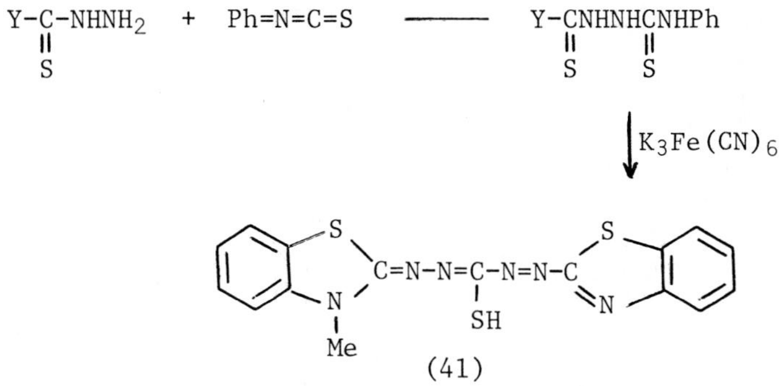

Y-C(=S)-$NHNH_2$ + Ph=N=C=S —— Y-C(=S)NHNHC(=S)NHPh

↓ $K_3Fe(CN)_6$

(41)

Y =

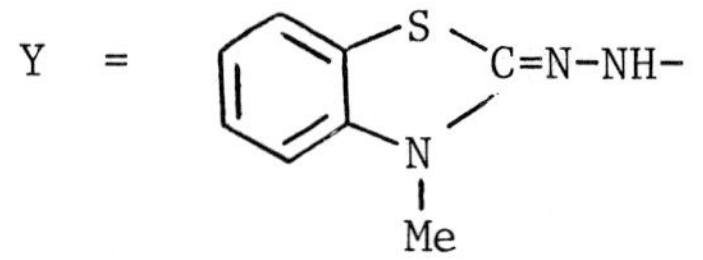

3.2 *Phospha- and arsa-cyanines*

The β-phosphatrimethine cyanine (42) is made from 2-methylene-1,3,3-trimethylindoline and PBr_3 which form bromobis(1,3,3-trimethylindolin-2-ylidenemethyl)phosphane from which the bromine is removed by treatment with trimethyloxonium tetrafluoroborate (N. Gamon and C. Reichardt, Angew. Chem. internat. Edn., 1977, 16, 404).

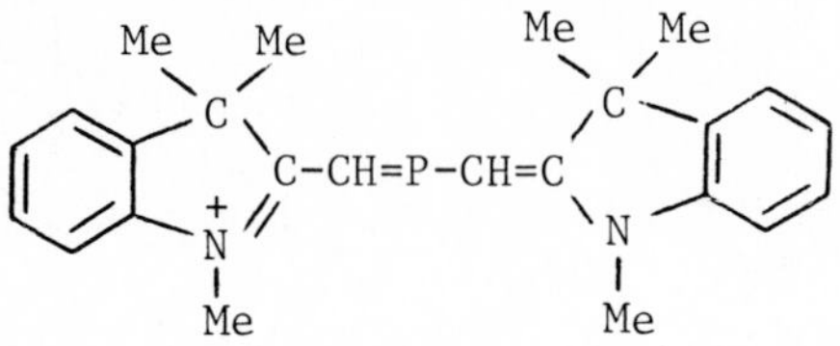

(42)

Compound (42) has λ_{max} 586 nm in chloroform and is very sensitive to hydrolysis. The absorption in chloroform of the analogous trimethine cyanine is 553 nm and of the β-aza dye 602 nm. The β-arsatrimethine indolinene dye is made analogously by reaction between arsenic tribromide and 2-methylene-1,3,3-triethylindoline. It has λ_{max} 664 nm in chloroform. It is likewise extremely sensitive to hydrolysis (Gamon and Reichardt, Tetrahedron Letters, 1979, 225).

4. *Merocyanines*

The solvatochromism of merocyanines (C.C.C., 2nd Edn., Vol. IVB, p. 407) has been studied further in relation to the electronegativity of the end groups X and X^1 (43).

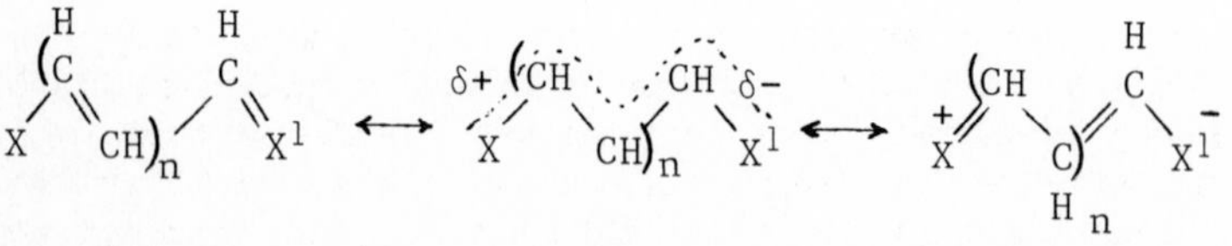

(a) (b) (c)

Electronegativity	$X^1 < X$	$X^1 = X$	$X > X^-$
	(a)	(b)	(c)

(43)

In dyes which show a bathochromic shift in absorption maximum in a solvent series of increasing polarity the ground state moves from the non-polar polyenic structure (43a) towards the polymethine structure (43b): dyes which show a hypsochromic shift in λ_{max} in a solvent series of increasing polarity move in the ground state from the state (43b) towards the dipolar polyenic state (43c) (P. Scheibe *et al.*, Ber. Bunsenges. physik. Chem., 1976, 80, 630; M. Wähnert and S. Dahne, J. prakt. Chem., 1976, 318, 321).

Some dyes (44) which can be considered as merocyanines are useful for the production of greenish-yellow colours on polyester fibres (Sandoz, B.P. 843,644; U.S.P. 2,850,520/1956).

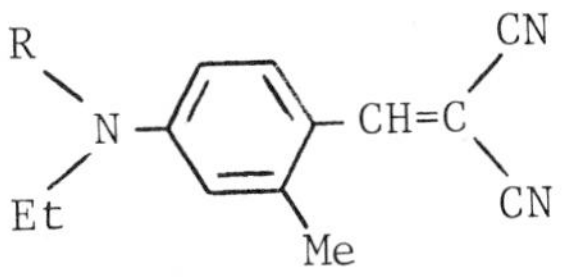

R(Et)N–C6H3(Me)–CH=C(CN)2

(44)

Increase in the size of the group R increases the resistance of the dye to sublimation, e.g.

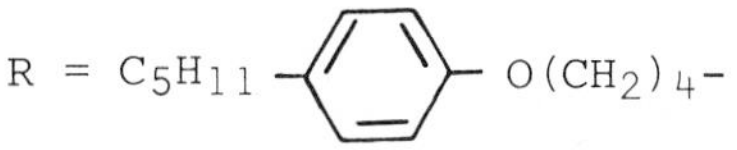

$$R = C_5H_{11}-C_6H_4-O(CH_2)_4-$$

(Bayer, B.P. 1,436,056; U.S.P. 3,920,720/1973).

5. *Oxonols*

Oxonol dyes continue to attract interest as filter and anti-halo dyes in photographic materials because they are readily bleached during processing. Water-insoluble mono- and tri-methine oxonols from 1,3-dialkyl(aryl)pyrazol-5-ones containing at least two free carboxylic acid groups are used in a highly dispersed form in a photographic emulsion or an anti-halo layer (R.G. Lemahieu, H. Depoorter and W.J. Vanassche, B.P. 1,563,809/1976).

The oxonol dyes (45) have been produced (S.W. Bland B.P. 1,278,621; U.S.P. 3,681,345/1969).

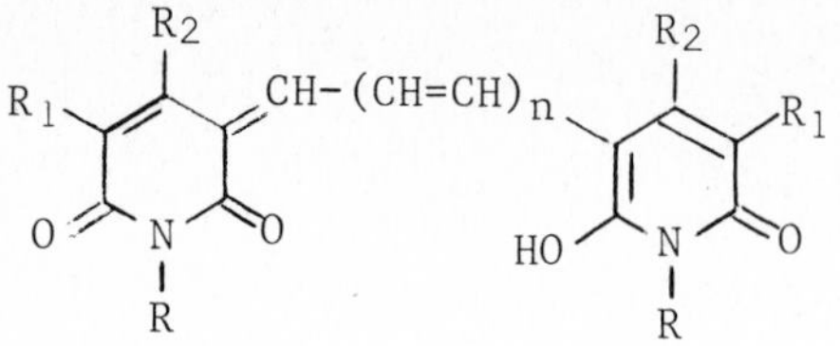

(45)

n = 0 - 2. R = H, Et, Ph, $PhCH_2$

R_1 = H, CN, $CONH_2$

R_2 = Me, COOH

A synthesis of trimethine oxonols involves a chain rupture of the type discussed in section 1.4 (R. Stolle, G. Bach, J. fur Signalaufzeichnungsmaterialen, 1979, 7, 59). A mixture of one mol of a monomethine oxonol and one mol of a pentamethine oxonol in alcoholic solution is boiled with a primary or a secondary amine to give two mols of the trimethine oxonol.

6. *Styryl Dyes*

Substitution of either of the chain H-atoms in a styryl dye (46); R^1 = H, n = 0) leads to a hypsochromic shift in the absorption maximum because the substituents destroy the planarity of the chromophoric system. In longer chain dyes (46) n = 1, A = -CH=CH-CH=CH-, X = R^1 = H) substitution of the β-hydrogen atom by phenyl gives a bathochromic shift of 110 nm (Z.F. Khalil, J. appl. Chem. Biotechnol., 1978, 28, 341).

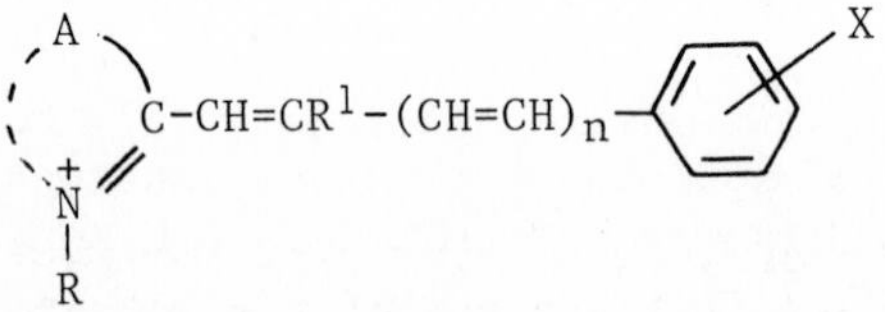

(46)

n = 1, R^1 = X = H, A = -CH=CH-CH=CH. λ_{max} = 390 nm.

n = 1, R^1 = Ph, X = H, A = -CH=CH-CH=CH-. λ_{max} = 500 nm.

A styryl dye has been prepared from the aldehyde intermediate N(4-formylphenyl)aza-[15]crown-5 and 1,4-dimethylpyridinium iodide. The absorption maximum of 470 nm is shifted towards the blue by 6.5, 1.5, 91 and 75 nm when the dye is complexed with Li, Na, Ca and Ba ions respectively (J.P. Dix and F. Vögtle, Ber., 1980, 113, 457.

If, in the synthesis of trimethine dyes substituted by F in the chain, the base condensing agent is N,N-diethylaniline instead of 4-methyl-N,N-diethylaniline (see p.   ), the reaction product is not the trimethine dye but the styryl dye (47).

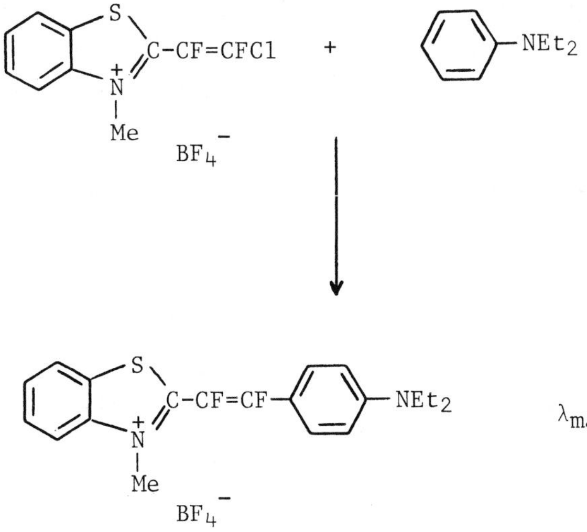

S C-CF=CFCl + N Me BF_4^- + C_6H_5-NEt_2

→

S C-CF=CF-C_6H_4-NEt_2 N Me BF_4^- λ_{max} 532 nm

(47)

7. *Pyrilium Dyes*

The stability of heptamethine pyrilium dyes is enhanced by including all but two of the methine groups in rings.

2,4-Diphenyl-5,6,7,8-tetrahydro-1-benzopyrilium perchlorate when treated with an intermediate (48) gives dyes (49) (G.A. Reynolds and K.H. Drexhage, J. org. Chem., 1977, 42, 885). The dyes are more stable than the analogue with an unbridged pentamethine chain. Replacement of R = H by R = Cl gives still higher stability and a slight red shift in λ_{max} (49), R = Cl, n = 2, λ_{max} 1145 nm, $\varepsilon = 1.43 \times 10^5$).

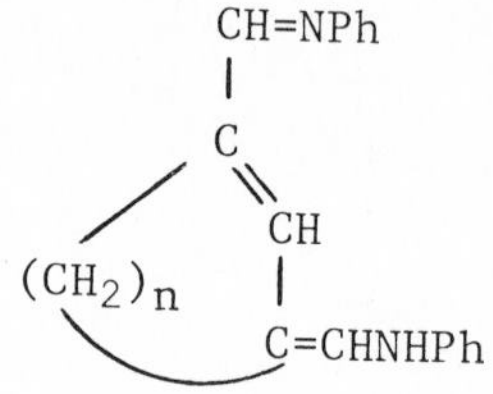

(48) n = 2 or 3

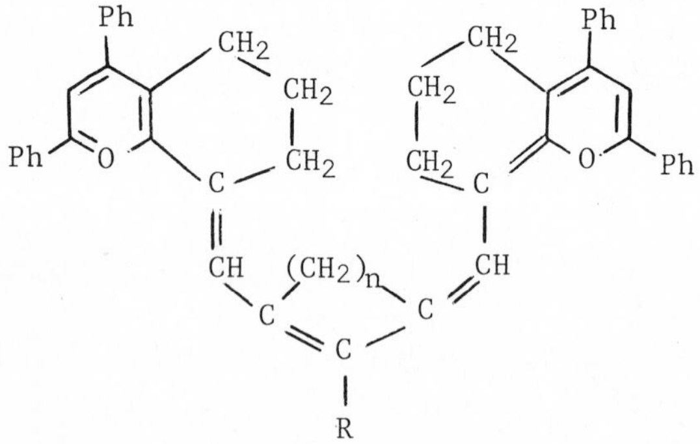

(49)

R = H, n = 2. λ_{max} 1138 nm $\varepsilon = 0.7 \times 10^5$

R = H, n = 3. λ_{max} 1090 nm $\varepsilon = 1.4 \times 10^5$

8. *Reaction Mechanisms*

A. Guillaume and A. Bruylants have studied the reaction between 2-methylene-1,3,3-trimethylindoline (50) and a series of 2-(2-ethylthiovinyl)heterocyclic quaternary salts (51). The reaction is first order with respect to (50) (Bull. Cl. Sci. Acad. R. Belg., 1976, 62, 177). Tertiary bases are inefficient catalysts but halide anions are weak catalysts. When the base (50) is prepared *in situ* from a 1,2,3,3-tetramethyl-1-H-indolium salt the halide ions are good catalysts.

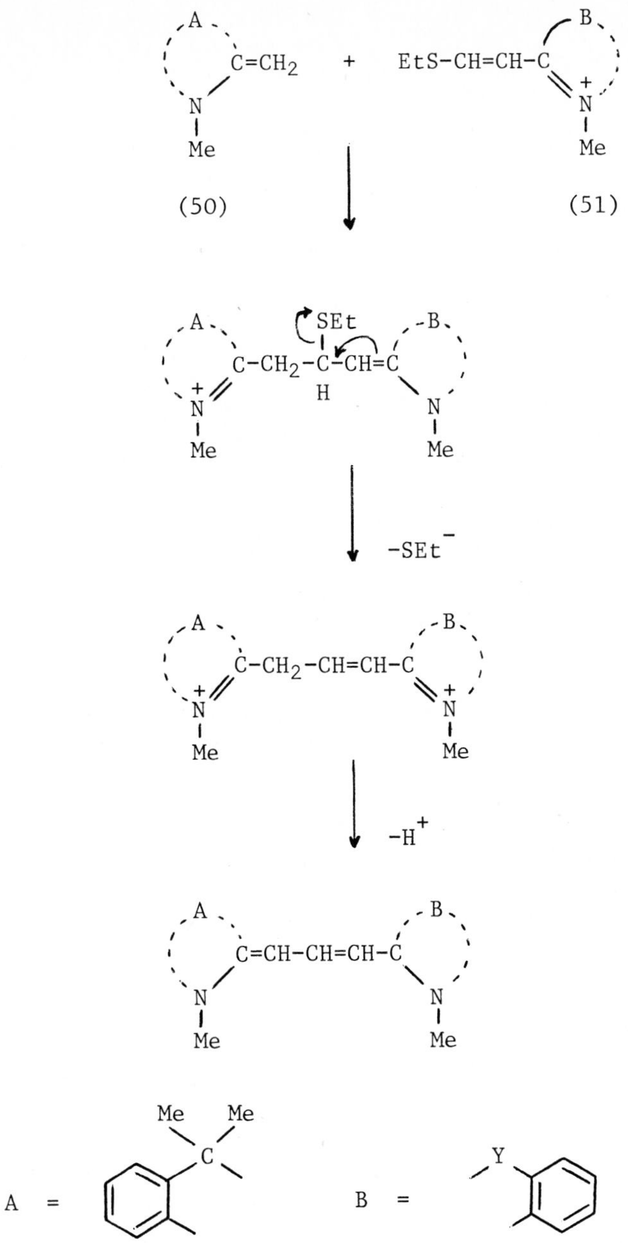

(50) + (51)

$-SEt^-$

$-H^+$

A =

B =

The reaction velocity constants for different groups Y at a concentration of 4×10^{-1} moles L^{-1} are given below.

Y =	O	S	Se	CMe_2	-CH=CH-
10^2K =	17.6	9.15	5.34	1.61	0.15

Provided the reactants (50) and (51) are sufficiently soluble at room temperature dye formation takes place rapidly and the dye is isolable in good yield. Higher temperatures can be harmful because of undesirable side reactions.

Guide to the Index

This index is constructed in a similar manner to the volume indexes of the first edition of the Chemistry of Carbon Compounds. However, to make the index easier to use, more descriptive entries have been made for the commonly occurring individual, and groups of chemicals.

The indexes cover primarily the chemical compounds mentioned in the text, and also include reactions and techniques, where named, and some sources of chemical compounds such as plant and animal species, oils, etc.

Chemical compounds have been indexed alphabetically under the names used by authors, editing being restricted to ensuring uniformity of entries under the same heading. In view of the alternative nomenclature that can often be used, a limited amount of cross-referencing has been done where it is considered to be helpful, but attention is particularly drawn to Convention 2 below.

For this and the succeeding volumes, the indexing conventions listed below have been adopted.

1. *Alphabetisation*

(a) The following prefixes have not been counted for alphabetising:

n-	*o-*	*as-*	*meso-*	D	*C*
sec-	*m-*	*sym-*	*cis-*	DL	*O-*
tert-	*p-*	*gem-*	*trans-*	L	*N-*
	vic-				*S-*
		lin-			*Bz-*
					Py-

Some prefixes and numbering have been omitted in the index, where they do not usefully contribute to the reference.

(b) The following prefixes have been alphabetised:

Allo	Epi	Neo
Anti	Hetero	Nor
Cyclo	Homo	Pseudo
	Iso	

(c) A letter by letter alphabetical sequence is followed for entries, firstly for the main entry, followed by the descriptive entry. The only exception to this sequence is the placing of plural entries in front of the corresponding individual entries to prevent these being overlooked by a strict alphabetical sequence which could lead to a considerable separation of plural from individual entries. Thus "butanes" will come before *n*-butane, "butenes" before 1-butene, and 2-butene, etc.

2. *Cross references*

In view of the many alternative trivial and systematic names for chemical compounds, the indexes should be searched under any alternative names which may be indicated in the main body of the text. Only a limited amount of cross-referencing has been carried out, where it is considered that it would be helpful to the user.

3. *Esters*

In the case of lower alcohols esters are indexed only under the acid, e.g. propionic methyl ester, not methyl propionate. Ethyl is normally omitted e.g. acetic ester.

4. *Derivatives*

Simple derivatives are not normally indexed if they follow in the same short section of the text.

5. *Collective and plural entries*

In place of "– derivatives" or "– compounds" the plural entry has normally been used. Plural entries have occasionally been used where compiunds of the same name but differing numbering appear in the same section of the text.

6. *Main entries*

The main entry of the more common individual compounds is indicated by heavy type. Multiple entries, such as headings and sub-headings over several pages are shown by "–", e.g., 67–74, 137–139, etc.

INDEX

Abrus precotorius, 81
Acacia simplicifolia, 80
Acanthaceae, 6
Accedinine, 145, 146
Accedinisine, 145, 146
2-Acetamido-ω-methylmercapto-ω-acetoxyacetophenone, 254
O-Acetoacetylclivonine, 165
2-Acetoacyltrop-2-ene, 221
Acetone dicarboxylic acid, 201, 207, 208
4-Acetoxy-1-(3,4-methylenedioxyphenyl)cyclohexyl propionyl isocyanate, 191
3α-Acetoxytropane, 202
O-Acetylclivatine, 165
O-Acetylclivonine, 165
O-Acetylcrotaverrine, 53, 62
7-Acetylechinatine, 39, 45
Acetylenes, cyclisation, 21
Acetylene dialdehyde tetraethylacetals, 208
Acetylenic dyes, 283—287
—, as spectral sensitizers, 286
3-*O*-Acetyl-3-epicorymine, 109
O-Acetylindoxyl, 255
7-Acetylintermedine 40, 45
Acetylketene, 165
Acetylknightinol, 224
Acetyllasiocarpine, 39, 45
7-Acetyllycopsamine, 41, 45
N-Acetylnorloline, 67, 68
N'-(Acetylnortropan-3'α-yloxycarbonyl)-nortropan-3α, 6α-oxide, 217
O-Acetylpetasinecic acid, 63
O-Acetylscopine, 210
O-Acetylsenecic acid, 63
O-Acetylsenkirkine, 52, 62
Acetylstriatic acid, 34
N-1-Acetylstrychnosplendine, 117
Acetylsyneilesine, 53, 62
O-Acetyltropoyl chloride, 211
N-Acetyltryptamine, 79
N-Acetyl-1-tryptophan, 81
Acrolein, addition to 1-nitrobut-3-ene, 216
Acrostalagmus cinnabarimus, 136
Acrylic ester, 185, 217
Acrylic fibres, 288
Acylic diesters, 39—45
2-Acylindoles, 112
3-Acylindoles, 132
N-Acylpyrrolidines, 9—13
N-Acylpyrrolidones, 13, 14
N-Acyltryptamines, 79
Adina rubescens, 98
Adirubin, 98
Aflatrem, 73
Aflavinine, 72, 73
Aglaia roxburghiana, 12
Ajmalicine, 100, 101
Ajmaline-sarpagine alkaloids, 111—114
Akagerine, 115
Akuammigine, 100, 101
Alalakine, 123
Alatovenine, 104, 105
Alkaloid A and B, 225, 249, 250
Alkaloid C, 249, 250
Alkaloid D, 226, 249, 250
Alkaloid E, 225, 227, 249, 250
Alkaloid F, 225
Alkaloids I, J, L, M, N, 114
Alkaloid P, 8
Alkaloids, ajmaline-sarpagine, 111—114
—, *Amaryllidaceae*, 151—197
—, *Anthocercis*, 235
—, *Aspidosperma*, 71
—, bis-indole, 71, 135—149
—, — from *Vinca rosea*, 141
—, Calabar bean, 84
—, carbazole derivatives, 76
—, carboline, 115
—, clavine, 86
—, containing the Iboga unit, 127—137
—, *Convolvulaceae*, 200
—, *Darlingia*, 4, 5
—, diketopiperazine, 135
—, eburnamine, 124
—, ergot, 71, 85, 86
—, *Erythroxylaceae*, 200, 228—233
—, eserine types, 84
—, *Euphorbiaceae*, 200
—, from *Crinum* species, 159
—, from *Knightia deplanchei*, 223
—, halogenated ergot, 85
—, heteroyohimbine, 99—106
—, *Hyoscyamus*, 242
—, in *Datura* roots, 240
—, indole, 71—149
—, indoline, 94

—, indoxyl, 130
—, *Knightia*, 222, 223
—, littorine, 234
—, lycorine, 168—176
—, monoester, 35—38
—, monoterpene indole, 67
—, otonecine, 15
—, *N*-oxides, 241
—, oxindole, 104, 105, 114, 139
—, pharmacology, 151
—, picraline-type, 107—110
—, *Proteaceae*, 200, 219—222
—, pseudo-yohimbine, 105
—, pyranotropanes, 227
—, pyrrolidine, 1—14
—, pyrrolizidine, 15—69
—, *Rhizophoraceae*, 200
—, sarpagine, 146
—, seco-strychnine, 116
—, secotropanes, 246
—, secoyohimbine, 99—106
—, sesquimeric bisindole, 138
—, simple indoles, 72—75
—, simple tryptamine derivatives, 79
—, simple tryptophan derivatives, 79
—, *Solanaceae*, 200, 202, 234
—, *Solandra*, 238
—, *Strychnos* 116—119
—, synthesis, 151
—, tropane, 199—250
—, vincamine-eburnamine, 71
—, with *Aspidosperma* units, 121—127
—, with Corynanthe-Strychnos units, 97—119
—, with isoprene derived moiety, 85—96
—, without isoprene derived moiety, 79—83
—, without tryptamine units, 72—78
—, with seco-type units, 120
—, with terpene derived moiety, 97—134
—, with tryptamine units, 79—134
—, yohimbine, 99—106
Alkamines A-D, 249, 250
Alkenes, dispolar cycloaddition to nitrones, 23
Allo-catharanthines, 125, 126
5-Allyl-1-pyrroline 1-oxide, 216
Alstonia lanceolifera, 107, 111
Alstonia quaternata, 108
Alstonia scholaris, 110
Amanita sp., 200
Amaryllidaceae alkaloids, 151—197
—, biosynthesis, 151—154
—, crystal structure determination, 166
—, photochemical reactions, 168
—, spectroscopy, 167
Ambelline, 159
2-Amino-ω-methylsulphinylacetophenone, 253
2-Amino-nitriles, 147
2-Amino-ω-nitroacetophenone, 254
2-Amino-1αH, 5αH-tropan-3β-yl benzoate, 204
Amsonia elliptica, 123
Anacampta disticha, 130
Andranginine, 125, 126
Angelic acid, 38, 45
7-Angelylheliotrine, 40, 45
14, 15-Anhydrocapuronidine, 128, 129
14, 15-Anhydro-1,2-dihydrocapuronidine, 128
Anhydroecgonine chloride, 221, 222
Anhydrovinblastine, 143, 144
Anodic oxidation potential, cyanines, 273
Anodic oxidative coupling reactions, 151
Anthocercis, 230, 243
Anthocercis albicans, 236, 237
Anthocercis alkaloids, 235
Anthocercis fasciculata, 234
Anthocercis frondosa, 236
Anthocercis genistoides, 237
Anthocercis littorea, 234, 235
Anthocercis tasmanica, 236
Anthocercis viscosa, 234, 235
Anthotroche, 243
Anthotroche myoporoides, 237
Anthotroche walcottii, 237
Antibiotics, from *Acrostalagmus cinnabarimus*, 136
Anit-halo dyes, 291
Antioxidants, cyanines, 283
Apoatropine, 218, 235, 240, 242, 245, 249
Apocynaceae, 15
Apogalanthamine, synthesis, 184
Apohaemanthamine, 160
Apohyoscine, 237, 242, 245, 249
Apohyoscine dimers, 245
Aponoratropine, 237
Aponorhyoscine, 237
Aristone, 133
Aristotelia sp., 132
Aristotelia chilensis, 132, 133

Aristotelia peduncularis, 67, 134
Aristotelia serrata, 132
Aristoteline, 132, 133
Aristotelinine, 133
Aristotelinone, 132
Aristotelone, 132
Arndt Eistert reaction, 177
Arnica montana, 13
Arolycorcidine, 156
Arolycoricidinol, 156
Aromatic esters, pyrrolizidine alkaloids, 65, 66
Arsacyanines, 290
β-Arsatrimethine indolinene dyes, 290
Arundo donax, 81
Aspergillus amstelodami, 89, 90
Aspergillus caesopitosus, 92
Aspergillus clavatus, 92
Aspergillus flavus, 135
Aspergillus fumigatus, 92
Aspergillus ruber, 89, 91
Aspergillus rubus, 87
Aspergillus ustus, 88
Asphodelus microcarpus, 1
Aspidofractines, 127
Aspidosperma album, 123
Aspidosperma alkaloids, 71, 121
Atanisatin, 77
Atropa sp., 199
Atropa belladonna, 243
Atropine, 199, 200, 218, 238, 242
Aurechinulin, 90
Auroglaucin, 91
Austamides, 87, 88
Axillarine, 33, 49
Azacyanines, 288
Azaporphyrins, 252
2-Azido quaternary salts, 288
2-Azidothiazolium salts, 288
Azoisobutyronitrile, 215

Babylonia japonica, 83
Balansia epichloe, 75
Belladonnine, 245
Bellendena sp., 227
Bellendena montana, 219
Bellendine, 202, 220—222
Benzaldehyde, 224
2-Benzamido-1α*H*,5α*H*-tropan-3β-ol, 204
Benzenes, from cyanines, 279
2-Benzhydryl-3α-acetoxy-1α*H*,5α*H*-tropanes, 224
2α-Benzhydryl-3β-hydroxy-7β-acetoxytropane, 226
Benzimidazole, 278
Benzimidazole dyes, 276, 286
Benzimidazole triazatrimethine cyanines, 288
Benzimidazole trimethine cyanines, 281
2,1-Benzisoxazolines, 255
Benzoazaporphyrins, 252
Benzoic acid, 204, 225, 229
Benzoporhyrins, 252
Benzothiazole, 278
Benzothiazole dyes, 276
Benzoylecgonine, 204, 228
N-Benzoylnorecgonine, 204
O-Benzoylnorecgonine, 204
Benzoylphysoperuvine, 246, 250
2-Benzoyltropane, 222
Benzoyltropine, 248
Benzoyl-ψ-tropine, 248
N-Benzoyltryptamine, 79
2-Benzyl-3-hydroxy-1α*H*,5α*H*-tropan-3-ones, 224
2-Benzyltropanes, 222
2-Benzyl-1α*H*,5α*H*-tropan-3-ones, 224
N-Benzyltropinone, 213
Betonicine, 2
3,3′-Biindoyl lithium, 264
Biosynthesis, *Amaryllidaceae* alkaloids, 151—154
Bis-aldehydes, reaction with quaternary salts, 270
Bis-dialkylamine cyanine cations, 268
Bis-di-*N*-alkylanil salts, 270
3,6-Bis(γ,γ-dimethylallyl)indole, 72
Bis(3-ethylbenzoxazole-2)pentamethine cyanine, 277
12,12′-Bis-11-hydroxycoronaridine, 137
Bisindoles, 126
Bis-indole alkaloids, 71, 135—149
—, from *Vinca rosea*, 141
O,*N*-Bis(trimethylsilyl)acetamide, 176
Bonafousia tetrastachya, 137, 148
Bonafousine, 148
Borreline, 80
Borreria sp., 80
Borreria verticillata, 140
Borrerine, 140
Borreverine, 140
Botrytis cinerea, 8
Brevianamides, 87, 88
Brevianamide-austamides, 94

Bromobis(1,3,3-trimethylindolin-2-ylidenemethyl)phosphane, 290
5-Bromo-*N*,*N*-dimethyltryptamine, 79
2-Bromothioindoxyl, 259
6β-Bromo-1α*H*,5α*H*-tropan-3α-yl acetate, 209
Brugmansia candida, 239
Brugmansia sanguinea, 234, 239
Brugmansia suaveolens, 239
Bufotenine, 79
Bufotenine-*N*-oxide, 79
Bulgarsenine, 29, 57, 64
But-3-enoic ester, 217
t-Butyldimethylsilyl chloride, 193
2-*t*-Butyl-4-hydroxy-5-methylphenyl sulphide, 176
n-Butyllithium, 212, 214
2-Butylmalic acid, 64
Buxomeline, 123

Cacalia yatabei, 51
Caccinia glauca, 65
Cadaba fruticosa, 2
Cadabine, 2
Calabar bean alkaloids, 84
Calcium acetone dicarboxylate, 206
Capparidaceae, 2
Capparis spinosa, 1
Capuronetta elegans, 128, 146
Capuronine acetate, 146
Capuvosidine, 146, 147
Capuvosine, 146, 147
Carbazoles, biosynthesis from *Rutaceae*, 76
—, cyclisation with terpenoid units, 77
—, from *Hyella caespitosa*, 78
—, indole alkaloids, 76
Carboline alkaloids, 115
2-Carbomethoxy-7-(δ-carbomethoxy-β-hydroxypropyl)hexahydropyrrolo-[1,2-b]isoxazole, 217
Carbomethoxy dimethyl acetals, 217
δ-Carbomethoxy-β-hydroxypropyl nitrones, 217
Carbomethoxy isoxazoles, 217
Carbomethoxy nitrones, 217
Carbomethoxypyrrole, 210, 211
Carbomethoxypyrrolo-isoxazoles, 217
N-Carbomethoxy-1α*H*,5α*H*-tropan-6-en-3-one, 211
N-Carbomethoxy-1α*H*,5α*H*-tropan-3-one, 211
5-Carboxystrictosidine, 98
Caribine, 155, 160, 162
Carinatine, 155, 158
Carvone, 215
Cassipourea gerrardii, 3
Catharanthine-*N*-oxide, 144
Catharanthus lanceus, 121
Catharanthus ovalis, 121, 122, 143
Catharanthus roseus, 97, 99, 141
Catharanthus trichophyllus, 121
Catharine, 141, 142
Cathenamine, 99
Cathodic reduction potential, cyanines, 273
Cathofoline, 108
Cathophylline, 121
Cathovalinine, 121
Catuabin A, 233, 250
Catuabins B and C, 233
Celastraceae, 15
Chaetocin, 136
Chaetomium cochliodes, 137
Chetomin, 137
Chloranil, 281
Chlorins, 251
2-Chlorocyclohepta-2,6-dienone, 214
Chloroformic ester, 172
Chlorohyellazole, 78
2-Chloro-6-methylcyclohepta-2,6-dienone, 215
2β-Chloro-5-methyl-1α*H*,5α*H*-tropan-3-one, 215
2-Chloro-2-nitropropane, 255
3-Chloroperbenzoic acid, 124, 169, 171, 173, 175, 188, 217, 241
Chlorophylls, 251
8-Chlororugulovasines, 85
Chlorotrimethine dyes, 284
Chlorotrimethylsilane, 214
2α-Chlorotropanes, 215
2β-Chloro-1α*H*,5α*H*-troapn-3-one, 214
Cimiciphytine, 148
Cinnamaldehyde, 4, 280
Cinnamic acid, 225
N-Cinnamoylnorcuscohygrine, 9
Cinnamyl cocaine, 228, 229
Clausanitin, 77
Clausena anisata, 77
Clausena heptaphylla, 76, 77
Clausena indica, 76
Claviceps paspali, 72, 86
Claviceps purpurea, 81, 85
Clavine alkaloids, 86
Clazamycins A, 67, 68
Clivacetine, 164, 165, 178

Clivatine, 164
Clivia miniata, 164, 165
Clividine, synthesis, 177
Clivimine, 164, 165
—, synthesis, 178
Clivonine, 164
—, synthesis, 177
Cobalamines, 251
Coca, 204
Cocaine, 200, 204, 205, 217, 222, 228, 249
Cocculidine, crystal structure, 166
Cocculine, crystal structure, 166
Cocculus laurifolius, 166
Codonopsine, 3, 4
Codonopsinine, 3
Codonopsis clematides, 3
Combretine, 2
Combretum micranthum, 2
Convolvulaceae alkaloids, 200
Convolvulus erinacius, 1
Coordination compounds, porphyrins, 252
—, phtalocyanines, 252
Coromandalin, 26, 35, 38
Coronaridine, 127, 128
Coronaridinol, 127
Corydallis pallida, 8
Corydolactam, 8
Corymine, 109
Corynantheine, 103
Corynanthe-Strychnos alkaloids, 97—119
Cotton effect, 195, 246
p-Coumaryltryptamine, 80
Courbonia glauca, 1
Criglaucidine, 159
Criglaucine, 159
Crinamine, 159, 160, 163
—, irradiation, 168
—, synthesis, 186
Crinane, synthesis, 151
Crinane bases, 151
Crinatine, 160
Crinidine, 163
Crinine, 163
—, synthesis, 184
Crinine ring-systems, 152
Crinum alkaloids, 159
Crinum asiaticum, 160, 164
Crinum glaucum, 159
Crinum jagus, 159
Crinum natans, 159, 160
Crinum ornatum, 159
Criocerine, 124
Crispatic acid, 27
Criwelline, 159, 195
—, synthesis, 186
Croalbidine, 24, 47, 61
Croalbinecine, 24, 61
—, physical properties, 25
Croburhine, 47
Crocandine, 46, 61
Cromaduric acid, 27, 34, 61
Cromadurine, 27, 46, 61
Cronaburmic acid, 28
Cronaburmic lactone, 34, 61
Cronaburmine, 28, 48, 61
Crotafoline, 28
Crotalaria albida, 47
Crotalaria axillaris, 49
Crotalaria burhia, 47
Crotalaria candicans, 46
Crotalaria grahamiana, 48
Crotalaria laburnifolia, 52
Crotalaria madurensis, 46
Crotalaria nana, 48, 50
Crotalaria striata, 50
Crotalaria verrucosa, 53
Crotalaria walkeri, 52, 53
Crotalaric acid, 32, 34, 61
Crotalarine, 32, 47, 61
Crotananic acid, 28, 34, 61
Crotananine, 28, 50, 61
Crotastriatine, 50, 61
Crotaverric acid, 32, 34
Crotaverrine, 32, 53, 62
Cryptoechinulins A, B, C and D, 90, 91
Cryptoechinulins E, F and G, 91
Crystal structure determination, *Amaryllidaceae* alkaloids, 166
Curassavine, 26, 35, 38
Curassavine *N*-oxide, 38
Curculigo orchiodes, 157
Curtius rearrangement, 177, 185, 204
Cuscohygrine, 1
Cyanines *see also* Cyanine dyes
—, anodic oxidation potential, 273
—, as antioxidants, 283
—, as photosensitizers, 283
—, bridged-chain dyes, 270—272
—, cathodic reduction potential, 273
—, π-electron density along chain, 278
—, electron spin resonance, 278
—, electrophilic reactivity, 278
—, electrophoretic studies, 278
—, energy levels, 273

—, fluoro-substituted, 269
—, half-wave potential, 273
—, ionization energy, 273, 275
—, linked to gelatin, 283
—, long-chain dyes, 270—272
—, photo-oxidation, 280
—, protonation, 275
—, physical properties, 272—278
—, reactions, 278
—, reaction mechanisms, 294—296
—, redox potential, 273, 274
—, ring closure to benzenes, 279
—, X-ray studies, 277
—, zwitterionic dyes, 278
2,2′-Cyanines, 273, 274
—, ionisation energies, 275
4,4′-Cyanines, 274
Cyanine dyes *see also* Cyanines, 267—283
—, as spectral sensitizers, 267, 272, 273
—, 1:1 charge transfer complexes, 281
—, from thiazolium compounds, 267
—, mass spectrometry, 281
3-Cyano-1,3-dimethyl-2,1-benzisoxazoline, 255
Cyanomethylphosphonic ester, 185
Cyclamidomycin, 6
Cyclic diesters, 46—64
Cyclohepta-2,6-dienone, 213
Cyclohepta-2,6-dione, 212
Cycloheptanone, Vilsmeier formylation, 270
Cycloheptanone ethylene ketal, 212
Cyclohept-2-en-3-one, addition to dimethylamine, 246
Cycloheptenone aminoesters, 21
Cyclohexanone, Vilsmeier formylation, 270
Cyclohex-2-enone, 214
Cyclomahanimbine, 77
Cyclopentanone, Vilsmeier formylation, 270
Cyclopropanes, intramolecular opening, 19, 22
Cyclopropylimines, acid-catalysed rearrangement, 19
—, rearrangement to pyrrolidines, 2
Cyclositsirikine, 103
Cyclostachine A and B, 11
Cyphomandra betacea, 1

Darlingia sp. 221, 227
Darlingia alkaloids, 4, 5
Darlingia darlingiana, 4, 222
Darlingia ferruginea, 222
Darlingianine, 4, 5
Darlingine, 221, 222
Darlinine, 5
Datura sp., 1, 201, 204, 221, 227, 230, 239, 241
Datura alkaloids, 240
Datura innoxia, 240
Datura meteloides, 203
Datura sanguinea, 202, 234
Datura stramonium, 243
Deacetylbowdensine, 159
Decarbomethoxydihydrovindoline, 145
19-Decarbomethoxy-14′,15′-dihydrovindolinine, 145
Deformo-akuammiline, 108
20,21-Dehydroajmalicine, 99
12,13-Dehydrobrevianamide, 87
Dehydrodarlingianine, 5
Dehydrodarlinine, 5
Dehydrodeoxyindigo, 264
21-Dehydrogeissoschizine, 99, 100
Dehydroheliotridine, 24
6-Dehydrohyoscyamine, 211, 212
Dehydroindigo, 264
Dehydroisosenaetnine, 55, 63
Dehydrosenaetnine, 55, 63
6-Dehydrotropine, 211
6-Dehydro-ψ-tropine, 211
Dehydrotryptophans, 89
Dehydrotyrindoxyl, 263
N-Demethyldihydrogalanthamine, 164
N-Demethylepimacronine, 155, 162
N-Demethylgalanthamine, 164
—, crystal structure, 166
O-Demethylgalanthamine, 164
O-Demethylgalanthine, 158
Demethylhomolycorine, 156
O-Demethyllycoramine, 156
Dendrobium chrysanthum, 9
Dendrobium lohohense, 2
Dendrobium pierardii, 2
Dendrodoa grossularia, 81
Dendrodoine, 81
Deoxybrevianamide E, 87, 88
Deoxycapuvosine, 146, 147
4-Deoxy-*ribo*-hexapyranuronate, 75
Deoxyindigo, 264
Deoxyisostrychnine, 149
Deoxypretazettine, 195
Deoxytazettine, 195
Derris elliptica, 3
Desacetyldesformo picraline, 107
Desdanine, 6

Deserpidine, 104
Desethylibophyllidine, 130
O-Desmethylcrinamine, 160
N-Desmethylgalanthamine, 160
Desmethyl marcfortine B, 96
Desmodium caudatum, 79
Desmodium triflorum, 1
Desoxycordifoline, 98, 99
2,2′-Diacetoxy-6,6′-dibromoindigotin, 263
3,17-Di-*O*-acetyl-3-epicorymine, 109
O,O-Diacetylhamayne, 160
N,N′-Diacetylindigos, *cis-trans* isomerisation, 260
N,O-Diacetylindoxyl, 257
N,O-Diacetylruspolinone, 6
N,N′-Diacylindigos, properties of *cis*- and *trans*-isomers, 261
1,3-Dialkylpyrazol-5-ones, 291
1,3-Diarylpyrazol-5-ones, 291
1,8-Diazabicyclo[5,4,0]undecene-7, 188
Diazonium salts, 278
Dibenzo[*c,e*]azocines, 184
Dibenzoylindigo, 260
2α,4α-Dibromo-*N*-carbomethoxy-1α*H*,5α*H*-tropan-6-en-3-one, 210
5,6-Dibromo-*N,N*-dimethyltryptamine, 79
αα′-Dibromoketones, 212
5,6-Dibromo-6-methoxyindole, 264
2,3-Dibromothiochromone *S*-oxide, 259
Dibromotropanes, 211
4,4′-Dibutyl-5,5′-dimethylpyrrolindigo, 256
Dicathais orbita, 263
5,6-Dichlorobenzimida-2,2′-cyanines, 274
α,α-Dichlorocyclopropanol, 214
2,3-Dichloro-5,6-dicyanobenzoquinone, 11
1,2-Dichlorohexafluorocyclopentene, 269
2,3-Dichlorothiochromone *S*-oxide, 259
2,4-Dichloro-1,3,5-triazine, 283
N,N-Dicyclohexylcarbodiimide, 193
Di(cyclopentan[b]pyrrolindigo), 256
Diels-Alder reaction, 12, 172, 175
2,5-Diethoxy-2,5-dihydrofuran, 207
N,N-Diethylaniline, 293
Diethyl phosphorochloridate, 180
10-Diethylphosphoroxygalanthamine, 181
1,3-Diethyl-2-thiobarbituric acid, 286
2,3-Dihalogenothiochromones, 259
2,3-Dihydrobellendine, 221
5,6-Dihydrobellendine, 221
Dihydrocaranine, 171
α-Dihyrocaranone, 174
Dihydrocarbazoles, 139
Dihydro-β-carbolines, 82
Dihydrocatharine, 142
Dihydrocleavamines, 146
Dihydroclivamine, 178
Dihydrocyclomahanimbine, 77
Dihydrodarlingianine, 5
5,6-Dihydrodarlingine, 221
Dihydroepimacronine, 155, 163
Dihydrofuran, cycloaddition to pyrroline oxide, 21
Dihydrogalanthamine, synthesis, 182, 183
12,13-Dihydro-12-hydroxyaustamide, 88
2,3-Dihydroisobellendine, 221
Dihydrolycorine hydrobromide, 167
Dihydromancunine, 105
2,7-Dihydropleiocarpamine, 145
Dihydropycnanthine, 145
Dihydropyrrolizinone, 24
3,12-Dihydroroquefortine, 95
Dihydrositsirikine diol diacetate, 105, 106
Dihydrovobasine, 112, 147
4,5-Dihydrowisanidine, 11
3,4-Dihydroxy-2,5-dihydroxymethylpyrrolidine, 3
Dihydroxyheliotridane, 22
6,7β-Dihydroxylittorine, 239
2,3-Dihydroxy-3-methylpentanoic acid, 26
6,7-Dihydroxyteloidine, 209
Dihydroxytropanes, 237, 243
6α,7β-Dihydroxy-1α*H*,5α*H*-tropan-3-one, 209
6β,7β-Dihydroxy-1α*H*,5α*H*-tropan-3α-yl benzoate, 231
1α,4β-Diiodo-2α-acetoxytropan-3-one, 218
Diisobutylaluminium hydride, 185, 192, 201, 211
Diisopropylamine, 214
Diketopiperazines, 96
Diketopiperazine alkaloids, 135
Diketopiperazine units, 87, 89
2,5-Dimethoxy-2,5-dihydrofuran, 207, 208

Dimethoxyethane, 185
2,2-Dimethoxypropane, 188
6,7β-Dimethoxyteloidine, 209
2-(1,1-Dimethylallyl)indole, 89
Dimethylamine, addition to cyclohept-2-en-3-one, 246
Dimethylanthracene, Diels-Alder adduct, 23
1,3-Dimethyl-2,1-benzisoxazolium perchlorate, 255
2,3-Dimethylbenzothiazolium salts, 269
Dimethylformamide, 188
3,3-Dimethylglutaric anhydride, 33
N,N-Dimethylindigo, 255
1,3-Dimethyl-2-methylene benzimidazoline, 286
1,1′-Dimethyloctahydroindigo, 256
N,N-Dimethyloctahydroindigo, 256
N,N-Dimethylphysoperuvine, 246
1,4-Dimethylpyridinum iodide, 293
3,10-Dimethyltetrahydrocyclohepta[*b*]pyran-5,8-imine-4-one, 202
1,1′-Dimethyltetrahydroindigo, 256
N,N-Dimethyltetrahydroindigo, 256
N,N-Dimethyltryptamines, 79
N,N-Dimethyltryptophan methyl ester, 81
3,5-Dinitrobenzoylindigo, 260
Diphenyl diselenide, 188
Diphenyl selenide, 169
2,4-Diphenyltetrahydro-1-benzopyrilium perchlorate, 293
Dipterine, 79
Distearoylindigo, 260
2,3-Disubstituted tropanes, 223
3,6-Disubstituted tropanes, 227
Ditryptophenaline, 135, 136
Dolichantoside, 98
Donaxaridine, 81
Donaxarine, 81
Doronenine, 29, 57, 64
Doronicum macrophyllum, 54
Doronine, 54, 63
Dregamine, 112, 147
Duboisia sp., 201, 221, 236, 237, 243
Duboisia leichhardtii, 245
Dyes, acetylenic, 283—287
—, chlorotrimethine, 284
—, cyanine *see also* Cyanines, 267—283
—, merocyanines, 291
—, oxonol, 291, 292
—, pyrilium, 293, 294
—, styryl, 292, 293
—, zwitterionic cyanines, 278
Dysidea herbacea, 14

Eburnamine alkaloids, 124
Eburnamine-vincamine alkaloids, 71
Ecgonine, 204, 229
—, absolute configuration, 205
Ecgonine methyl ester, 217
Ecgoninic acid, 205
Echimidinic acid, 45
Echinulins, 87
Eglandine, 127, 128
Eglandulosine, 127, 128
Ehretia aspera, 65
Ehretinine, 65, 66
Eicosanoic acid, 9
Electron affinity, thiacyanines, 275
Electron spin resonance, cyanines, 278
Electrophilic reactivity, cyanines, 278
Elswesine, synthesis, 191
Emmons-Horner reaction, 176
Energy levels, cyanines, 273
17-Epi-akagerine, 115
3-Epicorymine, 109
Epidarlinine, 5
5,6-Epidihydrobellendine, 221
1-Epi-δ-dihydrocaranine, 174
6a-Epi-3-epimacronine, 195
17-Epihomolycorine, 157
Epimacronine, 163, 195
Epimaritidine, 180
16-Epi-19-oxokopsinine, 127
3-Epipretazettadiol, 195
6a-Epipretazettine, 192, 195
6a-Epipretazettine *O*-methyl ether, 192
3-Epitazettadiol, 195
3,4-Epoxyfuran, 207
2,3-Epoxysuccindialdehyde, 207, 209
3,6α-Epoxy-1α*H*,5α*H*-tropan-6β-ol, 204
6,7β-Epoxy-1α*H*,5α*H*-tropan-3-ol, 204
Ergolines, 86
Ergot alkaloids, 71, 85, 86
Ergot-like syndromes, in cattle, 75
Erinine, 109, 110

Erpinal, 110
Ervatamia heyneana, 128
Ervitsine, 112, 113
Erysodine, 82
Erysodinophorine, 82
Erysophorine, 82
Erysorine, 82
Erythrina arborescens, 82
Erythroxylaceae alkaloids, 200, 228—233
Erythroxylum australe, 231
Erythroxylum coca, 204, 228
Erythroxylum dekindtii, 230
Erythroxylum ellipticum, 232
Erythroxylum monogynum, 229
Erythroxylum truxillense, 204, 228
Erythroxylum vacciniifolium, 233
Eserine alkaloids, 84
4,5-Etheno-8,9-methylenedioxy-6-phenanthridone, 156
3-Ethoxybut-2-enoyl chloride, 221
N-Ethoxycarbonyl-L-prolinamide, 13
Ethylene dialdehyde tetraethylacetal, 208
2-(2-Ethylthiovinyl)heterocyclic quaternary salts, 294
N-Ethyltropinone, 213
Euphorbiaceae, 15
Euphorbiaceae alkaloids, 200
Evodia rutacaecarpa, 82
Exozoline, 77, 78

Ferrugine, 222
Ferruginine, 222
Filter dyes, 291
Five-membered heterocyclic compounds, indigo group, 253—266
Flindersia fournieri, 140
Flustra foliacea, 84
Flustramines A and B, 84
3-Formylcarbazole, 76
16-Formyl-16-epistrictamine, 109
N-Formylloline, 67, 68
N-Formylnorloline, 67, 68
N'-Formylnortropanyloxycarbonylnortropan-3α,6α-oxide, 217
N(4-Formylphenyl)aza[15]crown-5, 293
N-Formyl-L-proline, 24
Fourier transform proton nmr, 277
Fuchsisenecionine, 36, 38
Fukinotoxin, 54
Fulvine, 33
Fulvinic acid, 27, 61
Fumaric acid, 175
Fumaric dialdehyde tetramethylacetal, 209
Fumitremorgins, 92
Fungal metabolites, from tryptophan, 71
Furan, 209
2-Furoic acid, 230

Gabunia eglandulosa, 127
Galanthamine, 155, 156, 163, 164
—, synthesis, 179, 180
Galanthane bases, 151
Galanthine, 153, 155, 158
Gardneramine, 114
Gardneria multiflora, 114
Geissoschizoic acid, 139, 140
Gelatin, linked to cyanines, 283
Geranyl-geranyl pyrophosphate, 72
Gerrardine, 3
β-Glucosidase, 100, 102
L-Glutamic acid, 205
Glyceraldehyde, reaction with indole, 75
Glycoalkaloids, 97—99
Glycolozine, 76
Glycosides, monoterpenoid alkaloid, 71
Glycozoline, 76
Glyoxylic ester hemiacetal, 6
Goleptine, 158
Gonioma malagasy, 118, 145
Goniomine, 118, 119
Guettarda eximia, 99

Haemanthamine, 152, 155, 156
Haemanthidine, 155, 156
Haemanthus coccineus, 152
Half-chair conformation, 4
Half-wave potentials, cyanines, 273
Halichondria melanodocia, 75
Hallucinogens, 129
Halogenated ergot alkaloids, 85
Halogenated indoles, 71
Hamayne, 160
Hantzsch pyridine synthesis, 178
Haplophyta cimicidum, 148
Haplophytine, 148
Hastanecine, 22
Havanine, 155, 160
Heliotric acid, 45
Heliotridine, 23, 45

Heliotropium curassavicum, 35
Heliotropium eichwaldii, 40
Heliotropium europaeum, 39
Heliovicine, 35, 38
Heptamethine pyrilium dyes, 293
Heptazolidine, 76, 77
Heterocyclic quaternary salts, 270
Heterophylline, 58, 64
Heteroyohimbine, 97
Heteroyohimbine alkaloids, 99—106
Hexa-3,5-dienoic ester, 172
Hexamethylphosphoramide, 181
Heyneatine, 128
Hippacine, 155, 163
Hippadine, 155, 163
Hippafine, 155, 163
Hippagine, 155, 163
Hippeastidine, 157
Hippeastrine, 155, 156, 165
Hippeastrum ananuca, 157
Hippeastrum bulbispermum, 163
Hippeastrum vittatum, 155, 163
Histidine, 94
Hobartine, 134
Hodgkinsine, 135
Hofmann degradation, 3, 6
Holstiine, 116
Homolycorine, 155—157
Homoviridifloric acid, 26, 27, 34, 38
Hordenine, 155
Hörhammericine, 121
Hostiline, 116
Huang-Minlon reduction, 76
Hunteracine, 118
Hunteria congolana, 109
Hunteria eburnea, 118
Hunteria elliottii, 109, 141
Hydrocanthines, 148
10-Hydroxyakagerine, 115
3-Hydroxy-4-bromo-2,5-diethoxy-2,5-dihydrofuran, 207
3-Hydroxy-4-chloro-2,5-dimethoxytetrahydrofuran, 207
3-Hydroxycoronaridine, 127
6-Hydroxycrinamine, synthesis, 186
3-Hydroxy-2,5-diethoxytetrahydrofuran, 207
17-Hydroxydihydrocathenamine, 99, 100
1-Hydroxydihydropyrrolizinone, 63
2β-Hydroxy-1β-hydroxymethyl-8α-pyrrolizidine, 22
α-Hydroxyhyoscine, 239
6β-Hydroxyhyoscyamine *N*-oxide, 243
Hydroxyindolenines, 118
2-Hydroxyisovaleric acid, 45
10-Hydroxykribine, 115
6β-Hydroxy-7β-methoxyteloidine, 209
1-Hydroxymethylpyrrolizidine, 18
5-Hydroxy-*N*-methyltryptamine, 79
6β-Hydroxy-1α*H*,5α*H*-nortropan-3α-yl tiglate, 231
4-Hydroxy-L-proline, 18
4-Hydroxypyridinium ions, 232
Hydroxyrhazinilam, 124
2′-Hydroxyroxburghilin, 12
Hydroxysenkirkine, 52, 62
3-Hydroxystachydrine, 2
19-Hydroxytabersonine, 122
Hydroxytropanes, 206
Hydroxytropane heterodiesters, 239
6β-Hydroxy-1α*H*,5α*H*-tropan-3-one, 207, 229
6β-Hydroxy-1α*H*,5α*H*-tropan-3α-yl tiglate, 231
3-Hydroxy-2,3,4-trimethylglutaric acid, 27
11-Hydroxyvittatine, 152
Hyella caespitosa, 78
Hyellazole, 78
Hygrine, 1, 4
Hygroline, 244
ψ-Hygroline, 244
Hymenocallis arenicola, 155, 160
Hymenocallis caribea, 155
Hyoscine, 199, 202, 204, 207, 209, 212, 235—242, 246, 248, 249
Hyoscine *N*-oxide, 249
Hyoscine *N*-oxide hydrochloride, 241
Hyoscyamine, 199, 202, 204, 234—236, 238, 240—242, 246, 248
Hyoscyamine *N*-oxide hydrochloride, 241
Hyoscyamus sp., 199, 239
Hyoscyamus alkaloids, 242
Hyoscyamus niger, 243
Hypaphorine, 81, 82
Hystadine, 94

Iboga alkaloids, 127—137
Iboluteine, 130, 131
Ibophyllidine, 129, 130
Iboxyphilline, 129, 130
Inaequidenine, 56, 63
Incanine, 33, 49
Indicine *N*-oxide, 38

Indigo, 253
—, electronic absorption maxima, 260
—, polymeric, 258
—, reaction with hydrazine, 266
—, synthetic routes, 253—260
Indigo derivatives, *cis-trans* isomerism, 260
—, spectroscopy, 260
Indigo diimine, 266
Indigoid dyes, 253
Indigotin, 253
Indizoline, 76
Indoles, 147
—, halogenated, 71
—, substituted with isoprenoid units, 72
Indole, reaction with glyceraldehyde, 75
Indole alkaloids, 71—149
—, biosynthesis from loganin, 97, 98
Indole *N*-glycoside, 75
Indole metabolites, from pathogenic micro-organisms, 75
Indolenines, 132, 147
Indole polyols, 75
Indole sesquiterpenes, 72
Indolines, 132, 147
Indoline alkaloids, 94
Indoline chromophores, 107
2-(3-Indolyl)indoxyl, 265
Indoxyls, oxidative dimerisation, 255
Indoxyl alkaloids, 130
Integerrinecic acid, 32, 62
Iodoacetic ester, 203
1α-Iodo-2α-acetoxytropan-3-one, 218
Iodoarylalkylamines, 184
19-Iodotabersonine, 122, 125—127
Ion exchange resins, pyrrolizidine alkaloids, 16
Ionization energy, cyanines, 273, 275
Ipomeoea sp., 86
Iron ennacarbonyl, 210
Iron pentacarbonyl, 210
Isatin, 257
Isatinecic acid, 62
Isatoic anhydride, 253, 254
Isobellendine, 220
Isobutyric acid, 237
Isobutyryloxytropane, 249
Isocorymine, 109
Isocrocandine, 46, 61
Isocromaduric acid, 27, 34, 61
Isocromadurine, 27, 46, 61
Isodarlingianine, 5
Isoechinulin A, 90
Isoechinulins B and C, 90
Isoleucine, 205
cis-trans Isomerism, indigo derivatives, 260
Isometeloidine, 239
Isopentenyl groups, 73, 76, 89, 95, 140
Isopentenyl units, 77, 87, 91, 94, 96
6-Isopropenyl-5-methyl-1α*H*,5α*H*-tropan-3-ones, 215
2-Isopropylmalic acid, 64
Isopterophorine, 56, 63
Isoretronecanol, 18—21
Isosenaetnine, 55, 63
Isositsirikine, 103
Isosplendine, 117
Isovaleric acid, 221
Isovoacangines, 147, 148
Isoxazoles, 217
Isoxazolidines, 216, 217

Jaconecic acid, 30
Jequirity seeds, 81

Kava root, 10
Keto-lactams, 75
Ketopiperazines, 96
Kitraline, 122
Kitramine, 122
Kleinia kleinioides, 55
Knightalbinol, 226
Knightia alkaloids, 222
Knightia deplanchei, 222, 223, 225, 226
Knightia excelsa, 222
Knightia strobilina, 224, 226, 227
Knightinol, 224
Knightoline, 226
Kribine, 115
Kribine acetals, 115

Laburnine, 18
Lanciferine, 107, 111
Lasiocarpic acid, 45
Leguminosae, 3
Lencojum aestirum, 163
Lencojum vernum, 163
Leucine, 87, 205
Leurocolombine, 142
Leurosine, 142, 144
Ligularia dentata, 52
Ligularidenecic acid, 31, 34, 62
Ligularidine, 31, 52, 62
Lindelofia spectabilis, 39

Linoleic ester, 283
Lirio sanjuanera, 155
Lithium di-isopropylamide, 182
Littorine, 218, 234—236, 238—240, 242
Littorine alkaloids, 234
Lochnericine, 121
Loganin, biosynthesis to indole alkaloids, 97, 98
Lolidine, 67, 68
Loline, 17
Lolium cuneatum, 67
Lyaloside, 98, 99
Lycoramine, 156
—, synthesis, 182, 183
γ-Lycorane, 173—175
Lycorenine, 156
α-Δ-Lycoren-7-one, 175
Lycoricidine, synthesis, 188—189
Lycoricidinol, 156
Lycorine, 153—157, 159, 163, 164
—, synthesis, 168—173
—, X-ray diffraction analysis, 167
Lycorine alkaloids, synthesis, 168—176
Lycorine chlorohydrin, crystal structure, 166
Lycoris radiata, 156
Lycoris sanguinea, 155—157, 164

Macronine, synthesis, 186
Mahanimbidine, 77
Mahanimbine, 77
Maleic dialdehyde, 209
Malic dialdehyde, 207
Mancinella keineri, 263
Mandragora autumnalis, 241
Mandragora officinarum, 243
Mandragora vernalis, 241
Manganic trisacetylacetonate, 180
Mannich bases, 112
Marcfortine A, 96
Marcfortine B, 96
Marcfortine C, 96
Maritidine, 157
—, crystal structure, 166
—, synthesis, 179, 180
Mass spectrometry, cyanine dyes, 281
—, tropane alkaloids, 246—250
Meliaceae, 12
Melinacidins, 136, 137
Melobaline, 122
Melochia tomentosa, 74
Melodinus celastroides, 123
Melosatins A and B, 74
Mercaptotetra-azapentamethine cyanine dye bases, 289
Merocyanines, 268, 290, 291
Mesaconic acid, 243, 244
Mesaconic ester, 243, 244
Metabolites, mould, 87—96
Metalloprophyrins, 251
Meteloidine, 203, 208, 231, 234, 235, 237, 239, 240
Methine oxonols, 291, 292
Methuenine, 112
1-Methoxycarbazole, 76
1-Methoxy-3-carbomethoxycarbazole, 76
6-(*N*-Methoxycarbonylamino-5-methylenedioxyphenyl)cyclohex-3-ene-carboxylic ester, 177
2-Methoxycarbonyl-*t*-6-(3,4-methylenedioxyphenyl)cyclohex-4-ene-carboxylic acid, 177
1*R*,2*R*,3*S*-2β-Methoxycarbonyl-1α*H*,5α*H*-tropan-3β-yl benzoate, 205
15-Methoxy-14,15-dihydroallo-catharanthine, 125
10-Methoxyeglandine-*N*-oxide, 128
9-Methoxygirinimbine, 77
3-Methoxyindolin-3-one, 256
2-(3′-Methoxy-2′-methylpropanoyl)-1α*H*,5α*H*-tropan-3-one, 220
3-Methoxy-2-methylprop-2-enoyl chloride, 220
S-Methoxysuccinic acid, 203
Methoxytazettine, 155
6′-Methoxytrichostachine, 11
Methoxytropanes, 206
N-Methyl-*O*-acetylindoxyl, 255
Methylamine hydrochloride, 206
N-Methylanthranilic acid, 82
1*S*-8-Methyl-8-azabicyclo[3.2.1]-octan-3α,6β-diol, 203
8-Methyl-8-azabicyclo[3.2.1]octane, 200
8-Methyl-8-azabicyclo[3.2.1]-octan-3-one, 200
8-Methyl-8-azabicyclo[3.2.1]-oct-3α-yl acetate, 202
2-Methylbutanoic acid, 12
3-Methylcarbazole, 76
Methylcocaine, 228
Methylcycloheptan-3-one, 246
3-Methylcyclohex-2-enone, 215

4-Methyl-*N,N*-diethylaniline, 293
Methylecgonidine, 228, 230, 231
Methylene blue, 280
3,4-Methylenedioxy-6-methoxybenzaldehyde, 11
3,4-Methylenedioxyphenylacetone, 192
1-(3,4-Methylenedioxyphenyl)-but-3-en-1-ol, 175
3-(3,4-Methylenedioxyphenyl)-tetrahydrophthalic anhydride, 175
2-Methylene-1,3,3-triethylindoline, 290
2-Methylene-1,3,3-trimethylindoline, 290, 294
Methyl (*p*-hydroxybenzoyl)acetate, 220
N-Methylindigo, 255
N-Methylindoxyl, 255
N-Methylloline, 67, 68
O-Methylnorbelladine, 153
1-Methyloctahydroindigo, 256
N-Methyloctahydroindigo, 256
3-Methylpent-2-enoic acid, 26
N-Methylphenylalanine, 135
N-Methylpyridinum benzoate, 231
N-Methylpyrroles, 212, 233
N-Methylpyrrole-2-carboxylic acid, 233
N-Methylpyrrolidine, 200, 244
N-Methylpyrrolid-2-one-5-acetic acid, 205
2-Methyl-1-pyrroline-1-oxide, 6
N-Methylstrictosidine, 98
O-Methyltazettine, 162
1-Methyltetrahydroindigo, 256
N-Methyltetrahydroindigo, 256
5-Methyl-1αH,5αH-tropan-3-one, 215
Michael addition, 246
Minovincine, 126, 127
Mitrogyna speciosa, 98
Moffat-Pfitzner oxidation, 193
Monocrotalic acid, 28
Monocrotaline, 33
Monocrotalinine, 48, 61
Monoester alkaloids, 35—38
Monomethine cyanines, 278, 282
Monomethine oxonols, 291, 292
Monoperphthalic acid, 210
Monoterpene indole alkaloids, 67
Monoterpenoid alkaloid glycosides, 71
Montanine, 152
Morphine, 199
Mould metabolites, 87—96
Mukonidine, 76
Mukonine, 76
Mupamine, 77
Murex, 262
Muricidae, 262
Murraya exotica, 77
Murraya koenigii, 76
Mycosporine-2, 8
Mycotoxins, 92
Myrtopsis myrtoidea, 79

Narciclacine, 156
Narciprimine, 156
Narcissamine, 164
Narcissidine, 153
Narcissus pseudonarcissus, 153
Narcissus tazetta, 155
Narcotine, 199
Narwedine, 155
Naureline, 110
Necic acids, 26—32
—, C_6-acids, 26
—, C_7-acids, 26
—, C_8-acids, 26—28
—, C_9-acids, 28, 29
—, C_{10}-acids, 29—32
—, physical properties, 34
Necines, 17—25
—, dihydroxylated derivatives, 22—24
—, monohydroxylated derivatives, 18—21
—, physical properties, 25
—, trihydroxylated derivatives, 24
—, unhydroxylated derivatives, 17
Nef rearrangement, 254
Nemorensic acid, 29, 34, 64
Nemoresine, 57, 64
Nemoresine *N*-oxide, 64
Neoechinulins, 89, 94
Neoechinulins A and B, 89
Neoechinulin C, 89, 90
Neoechinulin D, 89
Neoechinulin E, 83, 90
Neopetasitenine, 54, 63
Neosenkirkine, 52, 62
Neosidomycin, 75
Nicandra sp., 200
Nicotine, 236, 246
Nitroaldehydes, 216
1-Nitrobut-3-ene, addition to acrolein, 216

Nitrones, cycloaddition to carbon-carbon double bonds, 216
—, 1,3-dipolar cycloaddition to alkenes, 23
2-Nitrophenylselenocyanate, 176
Nitropolyzonamine, 67
β-Nitrostyrene, 172
Noratropine, 238
Norbelladine, 152, 180
Norcimiciphytine, 148
Norecgonine, 204
Norecgonine methyl ester, 217
Norgalanthamine, crystal structure, 166
Norhyoscine, 235, 237—240, 242
Norhyoscyamine, 235, 239, 240, 245, 249
Nornicotine, 236
Norpluviine, 153
Norruspoline, 6
Norruspolinone, 6
Nortropan-3α,6β-diol, 232
1αH,5αH-Nortropan-3α,6β-diol 6-tiglate, 232
Nortropane, 248
1αH,5αH-Nortropan-3α-yl isovalerate, 231
1αH,5αH-Nortropan-3α-yl 2-methylbutyrate, 231
Nor-ψ-tropine, 216

Obscurinervine, 123
Octahydroindigo, 256
Octahydro-1*H*-naphtho[2,1-c]indoles, 161
Okolasine, 11
Olokuto, 230
Opium, 199
Ornamine, 159
Ornazamine, 160
Ornazidine, 160
Oscine, 204, 240
Otonecine, 62, 63
Otonecine alkaloids, 15
Otosenine, 54, 63
2,2′-Oxacyanines, 273, 274
Oxaline, 94—96
Oxidative coupling reactions, 151
N-Oxides, alkaloids, 241
—, tropanes, 241
Oxindoles, 81, 96, 139, 257
Oxindole alkaloids, 104, 105, 114, 139
5-Oxocoronaridine, 128
6-Oxocoronaridine, 128
14-Oxolyaloside, 99
Oxonols, 268, 291
2-Oxo-3-(3-oxo-1-isoindolinylidene)indoline, 257
3-Oxo-2-(3-oxo-1-isoindolinylidene) indoline, 257
5-Oxostrictosidine, 98
3-Oxotabersonine, 123
19-Oxotabersonine, 126

Paecilomyces varioti, 13
Paliclavine, 86
Panacratium maritimum, 155
Pancratine, 155
Pandaca sp., 127
Pandaca boiteau, 112
Pandaca mintiflora, 141
Pandoline, 129, 130
Paramuricea chamaeleon, 79
Parsonsia heterophylla, 58
Parsonsia spiralis, 58
Parsonsine, 58, 64
Paspaclavine, 86
Paspalicine, 73
Paspaline, 72
Paspalinine, 73
Paxilline, 73
Pauridiantha lyalli, 98
Pauridianthoside, 99
Peduncularine, 67, 134
Peepuloidine, 10
Penicillium commune, 96
Pencillium islandicum, 85
Penicillium italicum, 87
Penicillium ochraceum, 88
Penicillium oxalicum, 94, 96
Penicillium paxilli, 73
Penicillium roqueforti, 94—96
Penicillium verruculosum, 92
Pentadecamethine dyes, 270
Pentamethine dyes, 270
Pentamethine oxonols, 292
Pentyl nitrite, 282
Pertungstic acid, 26
Petasinacine, 22
Petasinecic acid, 30, 34, 63
Petasinecine, 38, 66
—, physical properties, 25
Petasinecine diacetate, 22
Petasinine, 22, 36, 38
Petasinoside, 22, 65, 66
Petasitenine, 30, 54, 63
Petasites japonicus, 36, 54, 65

Phenylalanine, 205
6β-Phenylcarbamoyloxy-1α*H*,5α*H*-tropan-3α-ol, 209
α-Phenylglyceric acid, 239
Phenyl-pentyl units, 74
Phenyl phenylthiosulphonate, 182
Phenyl tetra-azapentamethine cyanines, 288
1-Phenyltetralin-1,4-dicarboxylic discopine esters, 245
Phenylthiol, 286
Phenylthiotrimethine cyanines, 286
Phosphacyanines, 290
β-Phosphatrimethine cyanine, 290
Photochemical *cis-trans* isomerisation, 260
Photocrinamine, 168
Photo-oxidation, cyanines, 280
Photosensitizers, cyanines, 283
Phthalaurin, 257
Phthalocyanines, 252
—, coordination compounds, 252
Phthalorubin, 257
Physalis sp., 201
Physalis alkekengi, 243
Physalis peruviana, 246
Physochlaina alaica, 243
Physoperuvine, 246
Picraline, 108, 110
Picraline-type alkaloids, 107—110
Pigments, pyrrole, 251
—, tetrapyrrole, 251
Piper sp., 10
1-Piperethylpyrrolidine, 11
Piper guineense, 10, 11
Piperic acid, 10
Piperidines, 141, 200
Piper methysticum, 10
Piper nigrum, 10
Piperonyl nitrile, 184
Piperoyl chloride, 10
Piper peepuloides, 10
Piper trichostachyon, 9—11
Platynecine, 38, 64
Pleiocarpamine, 145
Pluviine, 153
Polonovski reaction, 143, 144
Polyalthenol, 72
Polyalthia oliveri, 72
Polyester fibres, 291
Polymeric indigos, 258
Polymethines, stabilized state, 267, 268
Polyzonium rosalbum, 67
Porphyrins, 251
—, coordination compounds, 252
Powelline, 163
Precondylocarpine, 118, 119
Precriwelline, 195
Presecamines, 120, 141
Pretazettine, 154—156, 162
—, from tazettine, 195
Procerine, 36, 38
1-Proline, 12, 13
L-Prolinol, 9
Propiolic ester, 18, 24
Prosopis nigra, 79
Proteaceae alkaloids, 200, 219—222
Proteinase casein hydrolysate, 87
Protonation, cyanines, 275
Protostrychnine, 117
Pseudo-*Aspidosperma* skeletons, 128, 129, 146
Pseudo-aspidospermidine, 147
Pseudolycorine, 155, 159
Pseudo-yohimbine alkaloids, 105
Psychotria beccarioides, 135
Psychotridine, 135
Pteridium aquilinium, 7
Pterolactam, 7
Pterophorenecic acid, 31, 32
Pterophorine, 31, 56, 63
Ptychodera flava layanica, 263
Pummerer rearrangement, 253
Purpeline, 111
Pycnanthine, 145
Pyracrimycin A, 6
Pyranotropanes, 220
δ-Pyranotropanes, 222
Pyranotropane alkaloids, 227
Pyridinium chlorochromate, 181, 182
Pyridinium phenyllactate, 232
Pyridinium tiglate, 232
Pyrilium dyes, 293
L-Pyroglutamic acid, 205
Pyrroles, 69, 212
Pyrrole-2-carboxylic acid, 233
Pyrrole-*N*-oxides, 124
Pyrrole pigments, 251
Pyrrolidines, 139, 172
Pyrrolidine, 9, 10, 12
Pyrrolidine alkaloids, 1—14
Pyrrolidine bases, 1—7
3-Pyrrolidinol, 176
4-Pyrrolidinopyridine, 193
Pyrrolidones, 7, 8
2-Pyrrolidone-*N*-acylation, 13

1-Pyrroline-1-oxide, 217
—, cycloaddition to dihydrofuran, 21
Pyrrolizidines *see also* Pyrrolizidine alkaloids
—, diastereomeric, 21, 23
—, spectroscopic properties, 15
—, *endo*-1-substituted, 17
—, ultraviolet spectra, 15
Pyrrolizidine alkaloids *see also* Pyrrolizidines, 15—69
—, aromatic esters, 65, 66
—, conversion to pyrroles in the liver, 69
—, detection, 16
—, extraction, 16
—, ion exchange resins, 16
—, mass spectra, 15
—, mass spectrometry, 16
—, miscellaneous types, 67, 68
—, pharmacology, 68, 69
—, separation, 16
—, silylated derivatives, 16
—, structure elucidation, 33
Pyrrolizidine esters, 20
—, infrared spectra, 15

Quadrigemines, 135
Quaternary lactone salts, 203
Quaternary salts, reaction with bis-aldehydes, 270
Quaternine, 108
Quaternoline, 108
Quaternoxine, 108
Quinoline, 278

Ranunculaceae, 15
Rauflexine, 111
Rauwolfia canescens, 98
Rauwolfia reflexa, 111
Reaction mechanisms, cyanines, 294—296
Redox potentials, cyanines, 273, 274
Reflexine, 111
Reserpine, 104
Retro-Diels Alder reaction, 161
Retroisosenine, 29, 57, 64
Retronecanol, 66
Retronecine, 16, 17, 23, 38, 45, 61, 62, 64, 66
—, macrocyclic diesters, 33
Retronecine-7,9-dibenzoate, 65, 66
Retrorsine, 33
Reverse Mannich reaction, 103
Rhazinaline, 108, 109
Rhazinilam, 119, 124
Rhazya stricta, 97, 108
Rhizophoraceae, 3
Rhizophoraceae alkaloids, 200
Rhodophiala bifida, 152
Rindline, 116
Robinson reaction, tropinone, 206
Robinson synthesis, 209
Roquefortine, 94—96
Roxburghilin, 12
Roxburghines, 139
Rugulovasines A and B, 85
Ruspolia hypercrateriformis, 6
Ruspoline, 6
Rutaceae, biosynthesis of carbazoles, 76

Salpichroa origanifolia, 1
Salpiglossideae, 237, 243
Sanguinine, 155, 164
Sarpagine-ajmaline alkaloids, 111—114
Sarpagine alkaloids, 145
Schizanthin A, 243
Schizanthin B, 244
laevo-Schizanthin, 243
Schizanthus, 243
Schizanthus hookeri, 244
Schizanthus pinnatus, 243
Scopine, 204, 209, 212
Scopinone, 207
Scopolamine, 199
Scopolia carniolica, 243
Scopolia lurida, 243
Scopolia tangutica, 1
Scrophulariaceae, 15
Secamines, 141
2,3-Secoakuammigine, 100, 101
Secodine, 120
Secologanin, 99, 100, 102
—, incorporation with tryptamine, 97
Seco-strychnine alkaloids, 116
Secotropane alkaloids, 246
Secoyohimbine alkaloids, 99—106
2,2′-Selenacyanines, 273, 274
Semi-conduction, 272
Sempre avanti daffodils, 153
Senaetnine, 55, 63
Senampeline A—B, 42
Senampeline C—G, 43
Senecic acid, 62
Senecio sp., 24
Senecio aetnensis, 55
Senecio aucheri, 55

Senecio auricola, 52
Senecio barbertonicus, 55
Senecio cissampelinus, 42, 43
Senecio doronicum, 57
Senecio fuchsii, 36
Senecio inaequidens, 56
Senecio kirkii, 52
Senecio mikanoides, 43
Senecio nemorensis, 29, 57
Senecio procerus, 36
Senecio pterophorus, 56
Senecio pulviniformis, 56
Senecio vernalia, 51
Senecioic acid, 38, 243, 244
Senecioic *N*-methylpyrrolidine ester, 244
Senecivernic acid, 31, 34, 61
Senecivernine, 31, 51, 61
Senkirkine, 52, 62
Serine, 72
Serotenine, 79
Serratoline, 132, 133
Sesquimeric bisindole alkaloids, 138
Sesquiterpene pyrophosphates, 72
Shihunine, 2
Sitsirikine, 103
Smenospongia aurea, 79, 83
Smenospongia echina, 79
Sodium azide, 288
Sodium cyanoborohydride, 21
Solanaceae, 221
Solanaceae alkaloids, 200, 202, 234
Solandra sp., 1, 238, 239
Solandra alkaloids, 238
Solanum carolinense, 1
Solvatochromism, merocyanines, 290
Sorelline, 134
Spectral desensitization, 273
Spectral sensitizers, acetylenic dyes, 286
—, cyanine dyes, 267
—, mechanism, 272
Spiracine, 58, 64
Spiraline, 58, 64
Spiranine, 58, 64
Stabilized polymethine state, 267, 268
Stachydrine, 1, 2
Stemmadenine, 118
Streptomyces sp., 6, 67
Streptomyces hygroscopicus, 75
Striatic acid, 29, 61
Strictalamine, 108, 109
Strictamine, 108, 109
Strictosidine, 97, 98, 101
—, tritium-labelled, 98
Strigosine, 26
Strobiline, 227
Strychnine, 116, 117
Strychnobaridine, 139
Strychnofendlerine, 116, 117
Strychnofoline, 139
Strychnopentamine, 139
Strychnos sp., 139
Strychnos alkaloids, 116—119
Strychnos dale, 115
Strychnos decussata, 115
Strychnos elaecarpa, 115
Strychnos fendleri, 116
Strychnos gossweileri, 98
Strychnos henningsii, 116
Strychnos holstii, 116
Strychnos icaja, 117, 149
Strychnos nux-vomica, 117
Strychnosplendine, 117
Strychnos splendens, 117
Strychnos tchibangensis, 139
Strychnos usambarensis, 138
Styrene, reaction with acetic anhydride, 254
Styryl dyes, 292, 293
Succindialdehyde, 201, 206
Sungucine, 149
Supinidine, 18, 23
Surugatoxin, 83
Symlandine, 41, 45
Symphytum uplandicum, 40, 41
Syneilesis palmata, 53
Syneilesine, 30, 53, 62
Syneilesinolides A—C, 30, 34, 62

Tabernaelegantines, 147
Tabernaelegantinines, 147
Tabernaemontainine, 112
Tabernaemontana accedens, 145
Tabernaemontana albiflora, 130
Tabernaemontana divaricata, 128
Tabernaemontana elegans, 147
Tabernaemontana iboga, 129
Tabernaemontana subsessilis, 129
Tabersonine, 124, 125
meso-Tartaraldehyde, 203, 208
Tartaric dialdehyde acetal, 209
Tazettadiol, 195
Tazettine, 154—156, 162
—, synthesis, 192—196
Tchibangsenine, 139, 140
Teloidine, 203, 208, 212, 229, 230

Teloidine 3,6-ditiglate 7-isovalerate, 239
Teloidine esters, 230
Teloidine heterodiesters, 234
Teloidine mixed esters, 239
Teloidine 3-tiglate 6,7-diisovalerate, 239
Teloidine 3-tiglate 6,7-di(2-methylbutyrate), 239
Teloidine 3-tiglate 6-(2-methylbutyrate), 239
Teloidinone, 208, 209
Tetrabromoacetone, 210, 211
Tetrabromo-6,6′-dimethoxyindigotin, 264
Tetra-*n*-butylammonium fluoride, 193
Tetrachloromethane, 286
Tetrahalogeno-*p*-benzoquinones, 281
Tetrahydroalstonine, 99—101
Tetrahydroclividine, 178
Tetrahydroclivonine, 178
Tetrahydrocarbazole, 139
Tetrahydrodarlingianine, 5
Tetrahydroindigo, 256
Tetrahydroindirubin, 257
Tetrahydroindoxyl, 256, 257
Tetrahydrometinoxocrinine, 184
Tetrahydropresecamine, 141
Tetramethylethylenediamine, 181
1,2,3,4-Tetramethyl-1-*H*-indolium salts, 294
Tetrapyrrole pigments, 251
Thaisidae, 262
Thermoactinomyces sp., 80
2,2′-Thiacyanines, 273, 274
—, electron affinity, 275
—, ionisation energies, 275
—, protonation, 275
Thiadiazoles, 81
Thiazolic acids, 80
Thiazolium compounds, conversion into cyanine dyes, 267
Thiochromanones, 259
Thiochromone, 259
Thioindigo, 259
Thioindigo derivatives, electronic absorption maxima, 262
—, *cis-trans* isomerism, 260
—, spectroscopy, 260
Thioindoxyl-2-carboxylic acid, 259
Thiophthalimide, 257
Tiglic acid, 117
Tigloidine, 201, 235, 238, 240, 242
Toluene-4-sulphonic acid, 188, 255
Toluene-*p*-sulphonic anhydride, 209
Toluene-*p*-sulphonyl chloride, 209
Tosic acid, 209
Trachelanthamidine, 21, 38
Trachelanthic acid, 38, 45, 64
Tribromo-6-methoxyindigotin, 264
Tributylphosphine, 176
Tri-*n*-butyltin hydride, 215
Trichodesma incanum, 49
Trichodesmic acid, 61
Trichodesmine, 33
Tricholeine, 11
Trichonine, 9
Trichostachine, 10
Tridecamethine dyes, 270
Tridecapentaenedial, 270
2,3,5-Triethoxytetrahydrofuran, 207
Triethyloxonium tetrafluoroborate, 168, 186
N-Trifluoroacetylnorbelladine, 179
Trifluoroperacetic acid, 210
2,3,5-Tri-isopropyltetrahydrofuran, 207
Trimethine cyanines, 277
Trimethine dyes, 284, 293
Trimethine oxonols, 291, 292
3,4,5-Trimethoxybenzoic acid, 230, 233
3,4,5-Trimethoxycinnamic acid, 111, 229
Trimethoxycinnamoyl chloride, 229
2,3,5-Trimethoxytetrahydrofuran, 207
Trimethyloxonium tetrafluoroborate, 290
2-[(Trimethylsilyl)oxy]-cyclohexa-1,3-diene, 214
Triphenylphosphine, 124
2,3,6-Trisubstituted tropanes, 225
2,3,7-Trisubstituted tropanes, 226
$1\alpha H,5\alpha H$-Tropan-$3\alpha,6\beta$-diol, 203, 204, 209, 227, 229, 230, 235—237, 240, 243
$1\alpha H,5\alpha H$-Tropan-$3\alpha,6\beta$-diol 3-acetate 6-isobutyrate, 221
$1\alpha H,5\alpha H$-Tropan-$3\alpha,6\beta$-diol 6-angeloyl ester, 244
$1\alpha H,5\alpha H$-Tropan-$3\alpha,6\beta$-diol ditiglate, 203, 240
$1\alpha H,5\alpha H$-Tropan-$3\alpha,6\beta$-diol 3,6-ditigloyl ester, 241
$1\alpha H,5\alpha H$-Tropan-$3\alpha,6\beta$-diol esters, 221, 233
$1\alpha H,5\alpha H$-Tropan-$3\alpha,6\beta$-diol heterodiesters, 234

1αH,5αH-Tropan-3α,6β-diol 3-isobutyrate, 237
1αH,5αH-Tropan-3α,6β-diol 3-isobutyrate 6-acetate, 221
1αH,5αH-Tropan-3α,6β-diol 3-isovalerate, 203
1αH,5αH-Tropan-3α,6β-diol 6-tiglate, 235
1αH,5αH-Tropan-3α,6β-diol 3-tiglate 6-acetate, 239
1αH,5αH-Tropan-3α,6β-diol 3-tiglate 6-(2-methylbutyrate), 239
1αH,5αH-Tropan-3α,6β-diol 3-tiglate 6-propionate, 239
1αH,5αH-Tropan-3α,6β-diol 3-senecioyl ester, 244
Tropanes, 212
—, 2,3-disubstituted, 223
—, 3,6-disubstituted, 227
—, N-oxides, 241
—, synthesis, 216
—, 2,3,6-trisubstituted, 225
—, 2,3,7-trisubstituted, 226
Tropane, 200, 202, 248
Tropane alkaloids, 199—250
—, from *Darlingia*, 222
—, in *Datura* roots, 240
—, mass spectroscopy, 246—250
—, synthesis, 206—219
1αH,5αH-Tropan-6-en-3-yl acetate, 209
1αH,5αH-Tropan-3α,6β-epoxide, 209
1αH,5αH-Tropan-2α-ol, 218
1αH,5αH-Tropan-3α-ol, 202
1αH,5αH-Tropan-6β-ol-3α-yl acetate, 209
1αH,5αH-Tropan-3-one, 215, 220, 221
1αH,5αH-Tropan-3α,6β,7β-triol, 203, 208
1αH,5αH-Tropan-3α,6β,7β-triol 3,6-ditiglate, 235, 240
1αH,5αH-Tropan-3α,6β,7β-triol 3,6-ditigloyl ester, 241
1αH,5αH-Tropan-3α,6β,7β-triol 6-tiglate, 239
1αH,5αH-Tropan-3α,6β,7β-triol 3-tiglate 6-isovalerate, 239, 240
1αH,5αH-Tropan-3α-yl acetate, 238, 240
1αH,5αH-Tropan-3α,6β-yl 3,6-acetate, 209
1αH,5αH-Tropan-3α-yl benzoate, 224
1αH,5αH-Tropan-3α-yl n-butyrate, 237
1αH,5αH-Tropan-3α-yl 2-fuorate, 230
1αH,5αH-Tropan-3α-yl 2-hydroxy-3-phenylpropionate, 234
3α-Tropanyl-3′-hydroxy-3′-phenylpropionate, 218
1αH,5αH-Tropan-3α-yl isobutyrate, 236, 245
1αH,5αH-Tropan-3α-yl isovalerate, 230
1αH,5αH-Tropan-3α-yl 2-methylbutyrate, 236
N(Tropan-3′α-yloxycarbonyl)-nortropan-3α-ol, 217
N(Tropan-3′α-yloxycarbonyl)-nortropan-3α,6α-oxide, 217
1αH,5αH-Tropan-3α-yl phenylacetate, 230
1αH,5αH-Tropan-3α-yl tiglate, 235, 236, 238, 240, 242
1αH,5αH-Tropan-3α-yl tiglate N-oxide, 243
1αH,5αH-Tropan-3α-yl 3,4,5-trimethoxycinnamate, 232
Trop-6-enes, 210, 212
Tropene oxide, 209
1αH,5αH-Trop-6-en-3α-ol, 209, 211
2,3-(2′,3′-Tropeno)-5-methyl-α-pyrone, 202
2,3-(2′,3′-Tropeno)-5-methyl-δ-pyrone, 220
1αH,5αH-Trop-6-en-3-one, 209, 217
Tropic acid, 202, 204, 239
Tropine, 201—203, 211, 215, 217, 224, 225, 230, 231, 234, 235, 238, 240, 242, 244, 248
ψ-Tropine, 201, 211, 215, 224, 230, 235, 238, 240, 242, 248
—, cyclo-adduct derivatives, 216
Tropine acetyl esters, 237, 239
Tropine benzoyl ester, 228
ψ-Tropine benzoyl ester, 228
Tropine cinnamoyl esters, 218
Tropine isobutyryl ester, 246
Tropine isovaleryl ester, 246
ψ-Tropine-like base, 226
Tropine phenylalanine ester, 218
Tropine tigloyl esters, 237
ψ-Tropine tigloyl ester, 201
Tropine-3,4,5-trimethoxybenzoate, 229
Tropine 3,4,5-trimethoxycinnamate, 229

Tropinone, 200, 213, 224, 248
—, synthesis, 201, 206
Tropinone methochlorides, 213
Truxillines, 228
Tryptamine, 79, 99, 100, 102, 140
—, incorporation with secologanin, 97
Tryptamine alkaloids, isolation, 79
Tryptamine pentamer, 135
Tryptamine units, derived alkaloids, 79—134
Tryptophan, 72, 87, 135
—, biosynthesis to ergot alkaloids, 85
—, derived fungal metabolites, 71
Tryptophan alkaloids, isolation, 79
Tryptophyl bromides, 104
Tryptoquivalines, 92, 93
Tsilanimboana, 116
Tsilanine, 116
Tungstic acid, 210
Turneforcidine, 24, 61
Tyrian purple, 262
Tyrindoxyl, 263
Tyrindoxyl sulphate, 263
Tyriverdin, 263
L-Tyrosine, 179

Uluganine, 41, 45
Ulugbekia tschimganica, 41
Uncaria elliptica, 139
Undecatetraenedial, 270
Ungernia sp., 164
Ungernia sewerzowi, 157
Ungernia spiralis, 155, 162
Ungernia vvedenskysi, 155, 162
Ungminoridine, 155
Ungminorine, 155
Ungspiroline, 155, 162
Ungvedine, 162
Uplandicine, 41, 45
Usambarensine, 138, 139
Usambaridines, 138, 139
Usambarine, 138
Uvaria elliatiana, 72

Varadine, 155, 160, 161
Valerine, 203, 212
Valeroidine, 203, 221, 231, 236, 237
Valine, 205
Valtropine, 236, 238
Variotin, 13
Velbanamines, 141, 142
Venantine, 104, 105
Verrucologen, 92
Verrucologen *O*-dimethylallyl ether, 92
Verticillin, 136
Vilsmeier formylation, cyclic ketones, 270
Vinamidine, 142
Vinblastine, 141, 142, 144
Vincadifformine, 129
Vincadioline, 142
Vincamajine, 111
Vincamidines, 142
Vincamines, 124
Vincamine-eburnamine alkaloids, 71
Vinca rosea, 141
Vincoline, 122
Vincoside, 97, 98, 101
Vindolines, 141—144
Vindolinine, 125, 145
Viridifloric acid, 26, 27, 34, 38, 45
Vitamin B_{12}, 251
Vittatine, 152, 155, 163
Voacangarine, 128
Vobasine, 145, 146
Vobasine *N*-oxide, 112, 113
Vobasinol, 145, 146

Wieland-Gumlich aldehyde type intermediates, 116
Wisanidine, 11
Wittig reaction, 9, 272
Wolff Kishner reduction, 218

Yamataimine, 51, 62
Yohimbine, 102
β-Yohimbine, 102
Yohimbine alkaloids, 99—106

Zaidine, 155, 160, 161
Zea mays, 80
Zephyrathes carinata, 155, 158
Zephyranthine, 170
Zwitterionic dyes, 278

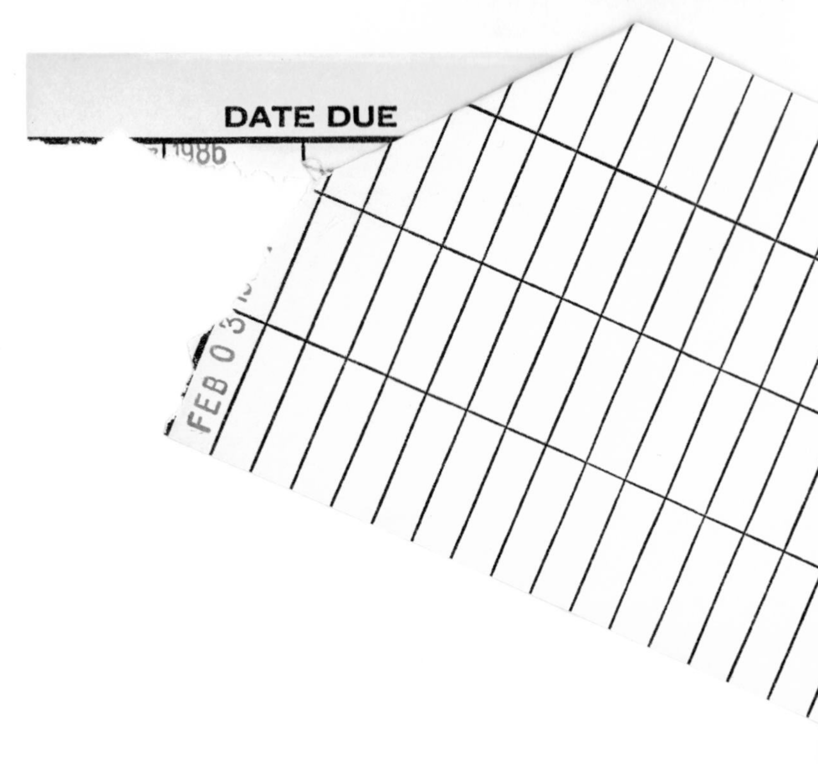
DATE DUE
1986
FEB 0 3